KNOCK ON WOOD

KNOCK ON WOOD

Nature as

Commodity in

Douglas-Fir Country

W. Scott Prudham

ROUTLEDGE
NEW YORK AND LONDON

Published in 2005 by
Routledge
270 Madison Avenue
New York, NY 10016
www.routledge-ny.com

Published in Great Britain by
Routledge
2 Park Square
Milton Park, Abingdon
Oxon OX14 4RN U.K.
www.routledge.co.uk

Library of Congress Cataloging-in-Publication Data
Prudham, W. Scott.
    Knock on wood : nature as commodity in Douglas-fir country / W. Scott Prudham.
        p. cm.
    Includes bibliographical references and index.
    ISBN 0-415-94401-5 (hardback : alk. paper) — ISBN 0-415-94402-3 (pbk. : alk. paper)
    1. Lumber trade—Environmental aspects—Northwest, Pacific. 2. Environmental policy—Northwest, Pacific. 3. Forests and forestry—Economic aspects—Northwest, Pacific. I. Title.
    HD9757.N95P78 2004
    333.75'09795--dc22
                                                                        2004011850

Contents

Acknowledgments

This book is the result of the contributions of many. I want first to acknowledge the generous financial support of the Social Sciences and Humanities Research Council of Canada, which furnished me with a doctoral fellowship and a postdoctoral fellowship. I also was fortunate to be supported under a grant from the John D. and Catherine T. MacArthur Foundation. The University of California at Berkeley provided me with a Muriel McKevitt Sonne Graduate Fellowship and Non-Resident Tuition Scholarship and the Chancellor's Dissertation Writing Fellowship. Berkeley is a wonderful place to be a graduate student, as is the city by the Bay, and I wish to thank numerous people whom I cannot name here who help make it so—professors, staff, students, residents. I loved it there, and I miss it.

I owe special thanks to the excellent staff of the Energy and Resources Group: Kate Blake, Donna Bridges, Sandra Dovali, Tony Folger-Brown, LaShaun Howard, and Lee Tajkakhsh; I would not have survived without your help. I also benefited immeasurably from graduate seminars and the miscellaneous contributions of graduate students, particularly the amazing students in the Energy and Resources Group. Courses offered by Peter Berck, Anthony Fisher, Louise Fortmann, Michael Watts, Rachel Schurman, Richard Walker, and Peter Evans were very helpful, and I owe my thanks to these people for helping me organize and situate my ideas. I also received valuable academic advice from Laura Enriquez, Louise Fortmann, Richard Howarth, Anne Kapuscinski, Laura Nader, Nancy Peluso, and Geoff Romm. Dick Norgaard was extremely patient with me and remained my advocate throughout my years in the Energy and Resources Group. Dick Walker gave me excellent and timely advice even though I was not a geography student, he respected me enough to tell me the truth (even when it hurt!), and he encouraged me even when the products of my efforts were less than good. Dick also has been instrumental in seeing this book published, and I owe him a deep debt of gratitude for continuing to help me after I left Berkeley.

Rachel Schurman was a stalwart, a constant enthusiast, a careful critic, and a good friend. I also had the benefit and privilege of working with an excellent study club at Berkeley in preparing my research prospectus. Thanks for the inspiration, the perspiration, and the fun along the way to the members of "What Is the Question?": William Boyd, Sidney Bob Dietz II, Navroz Dubash, Denny Kelso, James McCarthy, Dara O'Rourke, Paul Sabin, and Thomas Sikor. In some ways, you were every bit my supervisory committee.

I also want to acknowledge Steve Lonergan of the University of Victoria who stands tall among the weeds and who encouraged me to pursue my studies at Berkeley way back when. Trevor Barnes read my dissertation carefully and offered invaluable help in the revisions for this book. Juanita Sundberg was very supportive and helpful through the writing of much of this book. Thanks to the numerous people who helped me in my research, including those who consented to being interviewed and those who assisted me in other capacities. This includes the people of the Illinois Valley, and particularly those of Takilma, for allowing me into their community. I had the privilege of working and playing with Erik Jules, Marcus Kaufmann, and Matt Kaufmann during the summer of '96, and with Matt again in '97. Thank you TORCH; flame on.

Thanks also to a number of very helpful professors and researchers at Oregon State University and helpful staff members of the U.S. Forest Service, the Bureau of Land Management, and the International Woodworkers of America. I also owe thanks to Alex Murphy and Carolyn Cartier (formerly) of the University of Oregon's Geography Department and to Bill Robbins of Oregon State University for their support during fieldwork. Robin Jane Roff worked with me as a research assistant on this and helped me immeasurably. We have not heard the last of Robin I think.

I have been working with some great students at the University of Toronto these past few years, including Kim Beazley, Frank Donnelly, Sharlene Mollett, Angela Morris, and Nicki Simms, all of whom have helped me work through relevant ideas and literature; thanks. I also have drawn tremendous political and intellectual inspiration from my experiences with the United Automobile, Aerospace, and Agricultural Implement Workers and the Association of Graduate Student Employees/UAW. As Bruce Springsteen put it, we busted out of class (in more ways than one), and I am sure I learned more from Mary Ann Massenburg, Tanya Mahn, Ricardo Ochoa, Jill Hargis, Mike Miller, Dan Garcia, Jim Freeman, and other people with the union than I ever learned in school (or ever will).

Thanks to my editor David McBride who is incredibly solid, smart, patient, and kind. I wish to express my deep gratitude to my parents Wil and Lyn, my sister Heather, and my brother-in-law Rob for providing the kind of uncondi-tional love and support that I have needed and in some ways abused for the sake of projects like this one. Ditto for Brad, Di, Josh, Mikaela, and Kiera, my other family. Finally, thanks to sweet Celia for helping me see this through. I know I have been a pain in the ass about it. So it goes.

1
The Political Economy
of an Ecological Crisis

Of Owls and Options

Running west from Bend to Eugene, Oregon State Highway 20 crosses the crest of the Cascade Mountains at perhaps the most magnificent point in the range. To the south, three rugged volcanic peaks of the Sisters, the rounded summit of Mt. Bachelor, and the jagged crags of Broken Top thrust through snowy blankets toward summer skies of cobalt. To the northwest, Mt. Washington's classic volcanic form slips in and out of view. As the road snakes toward Santiam Pass, the oppressively hot and dry summer air of Oregon's high eastern desert relents to the cool breezes of the Pacific. Up the eastern slope, stands of lodgepole pine (*Pinus contorta*) grow straight in tight clusters, surrounded by magnificent, old-growth ponderosa pines (*Pinus ponderosa*), with their corpulent cones and furrowed orange-brown bark. As the road crests and descends toward the fertile Willamette Valley, the landscape quickly changes. Broken canopies and sunny forest floors of the east side give way to brooding, wet, and densely packed forests of the west side. This transition marks a traveler's entrée into the Douglas-fir region.

The Pacific Slope[1] of the Cascade Mountains and almost all of the Coast Range from Astoria to Ashland comprise Oregon's contribution to an extensive belt of forests stretching from southwestern British Columbia well into northern California; an area widely referred to in forest industry and policy circles as the Douglas-fir region (see Figure 1.1).[2] Though the region is home to a diversity of softwood and hardwood tree species, its name is that of a remarkable variety of conifer that is pervasive throughout and that grows in pure stands across vast areas. It is by far the most significant commercial forest tree in the region.[3]

Forests here are singular for many reasons. One reason is that they compose a significant part of the temperate rainforest zone of North America, where

1

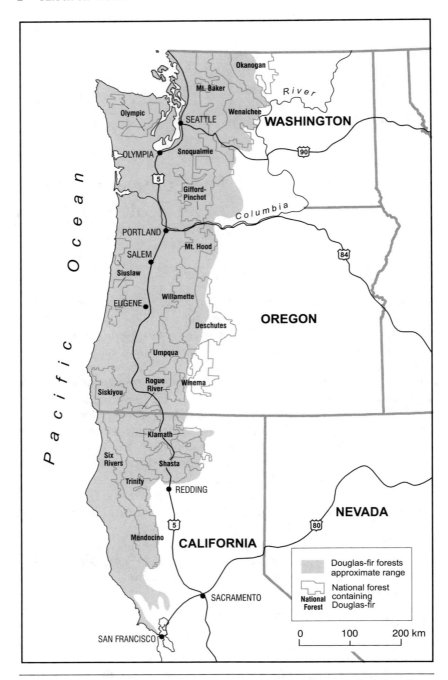

Fig. 1.1 The Douglas-fir Region. Source: Adapted from the U.S. Department Agriculture Forest Service and Bureau of Land Management 1994. Final Supplemental Environmental Impact Statement of Management of Habitat for Late Successional and Old-Growth Forest Related Species Within the Range of the Northern Spotted Owl, Volume 1. Portland, Oregon. Adapted by the Cartography Office, Department of Geography, University of Toronto.

warm, moist currents, prevailing onshore breezes, and rain-inducing topography sustain lush, diverse, and highly productive forests unlike those anywhere else in the world. Moreover, these forests occupy a singular position in the nation's political ecology. The Douglas-fir region is the most biologically productive commercial softwood region in the United States, and it also is home to the largest accumulated volumes of commercial sawtimber in the nation. According to the most recent Forest Service assessment of U.S. forest resources,[4] approximately 55 percent of Pacific Northwest Douglas-fir and Hemlock-Sitka spruce forestlands (the two dominant conifer forest types in the region by land area and commercial significance) are capable of annual growth in excess of 120 cubic feet per acre, the highest rated productivity class used in the inventory (see Figure 1.2). This incidence is higher than anywhere in the United States outside of California's north coast redwoods, and it is almost three times the rate of occurrence in U.S. timberlands overall. Even more remarkably, almost 70 percent of Douglas-fir forestland owned by the forest industry in the Pacific Northwest is rated in this highest productivity class.[5] And even after decades of intensive industrial harvesting targeting the largest trees for lumber, plywood, and pulp and paper production, the forests of the western slope of the Cascades retain on average the largest trees and the largest timber densities (i.e., volumes of timber per unit area) in the United States.[6] In short, this is timber country.

Beginning with the establishment of the Northwest's first sawmill in 1827,[7] conversion of the old-growth forests of the Douglas-fir region into commodities has sustained a diverse regional assemblage of wood-products manufacturing. This industry made the Douglas-fir region among the most important timber-producing and wood-products manufacturing areas in the United States—and indeed in the world—through much of the twentieth century.[8] A long boom

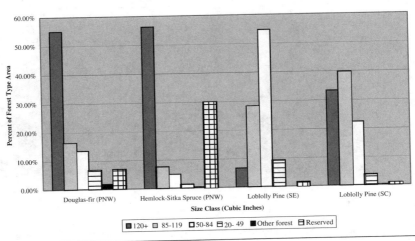

Fig. 1.2 Productivity Class Distribution of Forest Lands in the Pacific Northwest, Southeast, and South Central Regions, 2002. Source: United States Department of Agriculture Forest Service, 2003.

first sustained by Washington and California later propelled Oregon to promi-nence as the foremost state in the nation in the production of softwood timber and solid wood products, and from the 1930s onward, Oregon led all states in the aggregate production of solid wood products. Perhaps not surprisingly, the wood-products sector has been central to Oregon's economic development and cultural identity, particularly in rural, wood-products–based communities. Even as recently as the mid-1960s, more than 45 percent of the state's manufacturing workers were wood-products employees, representing 15 percent of the state's workforce.[9] And throughout much of the post–World War II period, the timber boom, particularly in Oregon, was sustained by the region's vast federal lands and their rich forests, governed by the principles of sustained-yield scientific forest management (see Figure 1.3).

Endangered Species, Endangered Spaces

But on June 23, 1990, the forest industry's unchallenged supremacy in the North-west woods ended. That day, the U.S. Fish and Wildlife Service confirmed the addition of the northern spotted owl (*Strix occidentalis caurina*) to its list of threatened species under the Endangered Species Act, bringing to a boil the most significant crisis of environmental conservation and preservation in the United States since Hetch-Hetchy eighty years earlier. Following the owl's listing, a convoluted and politically charged sequence of intensive lobbying, lawsuits, and scientific inquiries ensued, punctuated by dramatic court injunctions issued

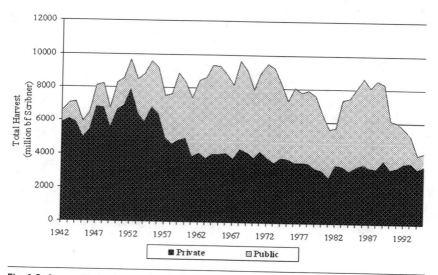

Fig. 1.3 Oregon Timber Harvest by Tenure, 1942–1995. Source: Oregon Department of Forestry (1997) along with other years of the same publication.

by Western Washington Federal District Court judge William Dwyer against the federal government, blocking logging and timber sales on federal lands in the region.[10]

With a persistent state of crisis prevailing over the forest economy of the Northwest, in April 1993 newly elected president Bill Clinton presided over the Northwest Forest Conference in Portland, following through on a campaign promise to deal with the old-growth issue. The summit produced not only a lot of conflict but also a deal aimed at protecting old-growth and so-called late-successional forests, along with the species dependent on them (including spotted owls).[11] The plan eventually adopted called for a reduction in annual probable timber sale quantities—the "PSQ"[12]—on federal lands within Oregon, Washington, and California by roughly 75 percent compared with levels characteristic of the 1980s. In Oregon, sales of public timber dropped from a high of 5.2 billion board feet[13] during the 1980s to just more than 790 million board feet in 1995 (see Figure 1.4).[14] The era of liquidating old-growth forests had essentially come to a close.[15]

Old-Growth and the New Politics of Nature

The implications of the owl crisis were and remain hotly contested.[16] Environmentalists championed reductions in public timber sales and pressed for still more. They relied on not only new ecological research, including spotted owl biological surveys, but also new studies of old-growth ecosystems to highlight

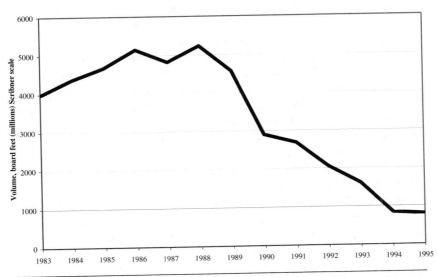

Fig. 1.4 Public Timber Sales in Oregon 1983–1995. Source: Oregon Department of Forestry (1997) along with other years of the same publication.

the diversity of life in older forests. Forestry scientists, many increasingly questioning the sustained-yield model that had underpinned industrial-scale clear-cutting in the region's federal forests for half a century, followed the lead of environmental social movements and embraced the "New Forestry" and "ecosystem management."[17] No more was old-growth decadent, a wasting asset, as it had been portrayed by conventional, industrial forestry.[18] Now it was a precious heritage, the "ancient forest," pristine and wonderful, with inherent value, and in scientific terms, rich in biomass and biodiversity, an important source of habitat for unique species.[19] These views reflected and reinforced the perspectives of an increasingly affluent and urban Northwest, as more people experienced forests as integral not to livelihoods but rather to leisure. Still, others felt disenfranchised, abandoned by government forest policy and offended (with some justification) by environmentalists seemingly indifferent to workers and their families. Workers and communities threatened by reductions in federal timber sale programs struck back with rallies, protests, and lawsuits. In Forks, Washington, they hunted spotted owls, displaying the corpses in protest.[20] Public discourse became consumed with the perception that jobs and environment were incompatible, a theme reinforced by conservative and industry groups seeking a "green backlash" to curb the muscular wilderness lobby.[21] And the stakes were indeed large; protecting one breeding pair of spotted owls, for example, would mean setting aside perhaps four hundred to eight hundred hectares of old-growth, which in the mid-1980s was worth on the order of $4 to $8 million in timber alone.[22] Not surprisingly, in Washington, D.C., pressure mounted to weaken the Endangered Species Act, drawing in liberals and conservatives alike.[23]

Because of the economic stakes involved, the drama of court cases and protests, and the intrigue of congressional hearings and forest summits, and because the conflict so clearly pit country versus city, worker versus environmentalist, preservation versus exploitation, and nature versus culture, the spotted owl became a touchstone for much broader debates about the politics of nature, science, and society in the late twentieth century. Some commentators, including noted environmental historian Richard White, observed the ways in which the spotted owl conflict highlighted limitations of an environmentalism that seemed increasingly callous, and even antihumanist.[24] Public discourse and scholarly commentary focused on the apparent trade-offs between jobs and environment as a signature conflict over social and cultural values where nature is concerned. Under this rubric, one focus became how different kinds of identity (e.g., environmentalist, scientist, logger, politician, woman, man, European, Aboriginal, etc.) shaped ("socially constructed") the meaning of "nature"; that is, which landscapes are desirable and, ultimately, which kinds of ecosystems will be privileged.[25] To some, the old-growth conflict was defined as a struggle over culture and identity as expressed through nature.[26]

This is an important perspective to be sure. People of different cultural backgrounds unquestionably understand and act differently in relation to nature based on their identities, however those identities are conceived and acquired.[27] Moreover, social groups often contest access to and control over nature in terms of these contrasting moral, political, and cultural values and subjectivities. But to accept jobs versus environment debates and discourses as such, to pit alternative views on the meaning of environmental change against one another and outside of their deeper historical context, risks idealism. It risks taking the controversy surrounding the spotted owl as an emergent sign of changing cultural values (as important as these may be) divorced from the historical, institutional, and material context in which these values came to be pitted against one another. In short, this perspective interrogates how different groups of people think about and want different things when it comes to nature, but it does not ask why they are forced to do so, nor why we cannot all have what we want when it comes to nature. It also does not fully emphasize who has actual power to shape and transform nature and who does not, or why. More fundamentally, focusing on the politics of difference does not direct our attention to who or what is causing the perceived threats to nature about which everyone seems so concerned. Yet if we do not understand the social origins of environmental change along with the proliferation of meanings associated with these changes, if we do not know how and why biophysical nature is being transformed, how can we hope to understand the various controversies surrounding these transformations?

The Misadventures of Capitalist Nature

I became interested in the spotted owl controversy when I moved to the Pacific Coast of North America in the summer of 1990, soon after the U.S. Fish and Wildlife Service listed the spotted owl. In my early twenties and commencing graduate studies, I wanted to understand what all the fuss was about and how it might be possible to comprehend the complex and intersecting political, economic, environmental, and cultural dimensions of the controversy. And as I read and talked with more people, and spent more time in the region, I increasingly believed that this conflict was not so much about nature per se as it was about capitalism.

That is, the spotted owl controversy pointed to a fundamental tension running through capitalist societies having to do with what might be called capitalist nature. Biophysical nature comprises a set of crucial inputs for capitalist production. Industry requires sources of raw materials, sinks to absorb wastes, and spaces to accommodate the production and circulation of commodities. Yet, at the same time, biophysical nature is that which almost by definition is not produced (at least not entirely) by capitalist production. No matter how far down the current commodification of life and the associated remaking of biological reality goes, from salmon crossed with tomatoes to the specter of the new eugenics, capitalism will always rely to some extent on nonproduced inputs. And that means despite its inherently expansionary tendencies, capitalism cannot sustain itself alone; that is, by simple means of individual firms competing and responding to various price signals.[28] As Karl Polanyi noted in his celebrated critique of liberal capitalism, a self-regulating capitalist market, able to call forth the nature it needs based on price signals alone, is a bourgeois pipe dream.[29] This in and of itself seems jarring—an economic system so pervasive as the basis of the contemporary globalizing world that we typically take it (i.e., capitalism) for granted, and yet it fundamentally relies on nonproduced inputs. It is a system so impressive in its capacity for growth, productivity increase, and innovation, yet in this respect it is curiously dependent and vulnerable, often without our acknowledgment. Capitalism needs nature.

Even more problematic is that we all need nature. There are various and competing social claims on the environment more generally—for water, for food, for shelter, for air, for social space; in short, for life. So although capitalism needs nature, capitalism does not really make nature, and, moreover, capitalism and the rest of society must compete to some extent politically for access to and control over nature. Might these also be important points of departure for understanding environmental politics? Might this point to a thoroughly problematic relationship between capitalist commodity production and biophysical nature?

The spotted owl controversy, or so it seems to me, points in these directions; namely, on the one hand, it points out that an industrial commodity production complex had developed over the course of more than 150 years based on the conversion into commodities of a highly distinct kind of regional forest, "resources" produced not by capital but by ecological processes (with some seldom-acknowledged help from Aboriginal land management practices). Moreover, and at the same time, the landscape-altering conversion of trees into commodities in the Pacific Northwest remade the region's forests, and not only in ways that increasingly undermined the commodity production system relying on these distinct forests but also in ways that scientists, environmentalists, and increasing numbers of the region's residents came to find unacceptable. Capitalist and social nature had come into conflict.

Anyone who has walked through a stand of old-growth Douglas-fir or Sitka spruce and compared it with younger growth, even those stands up to one hundred years old, has some intuitive sense of the significance of the transformation in Douglas-fir region forests wrought by industrial logging over the past 150 years. Interpretations of the significance and meaning of this transformation are inescapably subjective. But this cannot be allowed to obscure the very real material transformations involved. One need only look to the Forest Service's periodic timber inventories to see that the incidence of the largest trees (a crude measure but a measure nonetheless) in the region has been halved in the past half century.[30] One Forest Service estimate suggested that two-thirds of the approximately six million hectares of old-growth that existed in the nineteenth century in the Pacific Northwest was logged by 1985.[31] Another estimate put this figure closer to 90 percent.[32] Whether opponents of industrial logging specifically name "industry" or "capital" as the targets of their ire (as opposed, say, to loggers, corrupt bureaucrats, or yuppies), this fight points most certainly and centrally to the highly problematic (economic, political, cultural, and ecological) prospect of making commodities out of nature. The old-growth controversy and struggle is in this sense a highly significant and poignant example of conflict over the set of political and ecological problems or "crisis tendencies" that are not incidental to but rather inherent in the production of commodities from biophysical nature. My goal in this book is to examine the manufacture of wood commodities in Oregon's Douglas-fir region and the political ecology of this project as an example of the misadventures of capitalist nature.

Self-Regulating Markets and Crisis Tendencies in the (Re)Production of Nature

I do not pretend to be the first to recognize that securing sufficient natural resources and environmental inputs is a distinct and important problem for capitalist societies, as indeed it has been for many kinds of societies in human history. There is a rich literature spanning political and ideological perspectives. In traditional political economy, discussion of the problems associated with natural resource depletion and environmental degradation have been dominated by a now stale dichotomy. On the one hand, a laissez-faire view[33] holds that escalating prices in a market economy will induce the development of substitutes for scarce resources (i.e., technological change will offset resource depletion), including development of techniques for augmenting resource stocks (e.g., the development of aquaculture to offset wild fisheries' decline or of various wellhead injection systems for enhancing rates of oil recovery). This theory provides the foundation for a rather smug faith among neoliberals who dismiss the notion of ecological limits, even when confronted with apparent tautologies and contradictions in the use of market prices as indicators of scarcity.[34] On the other hand is the neo-Malthusian or Cassandra school—epitomized by the Limits to Growth studies—which tend toward a rather all-encompassing pessimism.[35] Neither school of thought is particularly well-known for careful attention to historical or geographical context, and neither offers particularly penetrating analyses of social causes and consequences or the political and politicized responses to environmental change. Moreover, neither school of thinking is known for engaging the rich variation of biophysical environments and how these environments shape and constrain human geographies.[36]

Seeking a way out of this stale debate, many have sought to locate the social origins of environmental change; the political, economic, cultural, and bio-physical implications; and the various human responses under the rubric of what has come to be known as political ecology. Over the course of the past two decades or so, political ecologists have built an impressive canon of work looking at the intersection of specific ecological conditions and social relations and institutions prevailing over the use of natural resources and environmental changes, with most of the focus on peasant communities in the global South.[37] In their seminal volume on the subject, Piers Blaikie and Harold Brookfield defined what they called "a regional political ecology"[38] approach as one aimed at understanding the "constantly shifting dialectic between society and land-based resources, and also within classes and groups within society itself." This defining approach to political ecology emphasizes particular societal articulations with nature, drawing on a lineage of Marxist and neo-Marxist studies exploring local social formations being integrated into an expanding global capitalism, and adding to this explicit and careful attention to ecological outcomes and processes.[39]

The approach I take follows Blaikie and Brookfield's interest in regionally specific processes of economic and environmental change within a broad political economy framework.[40] However, responding in part to the need to attend to the capitalization of nature per se in political ecology,[41] I focus on commodity production in the forest sector as the locus of social and environmental transformations, and associated political, economic, and ecological contradictions. Motivated by contemporary debates about the environment, and the relationship between society and nature, I proceed to address such issues in terms of not only the representations of what nature is or is not but rather how and why nature has been and continues to be produced and transformed and with what political, economic, and ecological implications.[42]

Nature as Fictitious Commodity

I begin with the basic insight that production of commodities from natural inputs under capitalism is an inherently problematic or contradictory undertaking because of the fictitious or false character of nature as a commodity. The idea of nature as a category of fictitious commodity comes from Karl Polanyi's *Great Transformation*.[43] According to Polanyi, liberal capitalism is defined by the free circulation of commodities (i.e., things that are produced exclusively for sale) within a self-regulating market (i.e., one not subject to nonmarket, social controls over the allocation of goods and services).[44] Yet this self-regulating market relies centrally on three forms of what Polanyi called "fictitious commodities"; that is, things that circulate in the market as though they are commodities (produced exclusively for sale) when they are not. These fictitious commodities are money, labor, and land. The self-regulating market, according to Polanyi, is predicated on an artificial separation between the economic and social spheres and, more specifically, on the privileging of market forces over social forces more generally.

For Polanyi, the development of modern capitalist economies is thus fraught with tensions that arise from contradictory social forces, what he called the "dual movement," which he applied specifically to land as a fictitious commodity. He posited that on the one hand is the pull of a self-regulating market to govern the allocation of land (i.e., nature), contested on the other hand by competing social claims (e.g., community ties restricting the mobility of labor, social claims on clean air and water, etc.). Because a self-regulating market cannot truly create or fully control the production of natural inputs, it cannot resolve these competing claims with price signals, giving rise to fundamental or structural tensions or crises around the politics of nature. A similar logic applies to labor as a fictitious commodity.[45] In each case, there is a dual movement toward and against market coordination.

Here, then, is perhaps the earliest structural theory of environmental social movements and politics in capitalist society.[46] Polanyi's insight that biophysical nature is not (wholly) produced by market coordinated processes is deceptively simple, yet remains remarkably salient as an observation on the problem of environmental change and its politics in an age of an ever-more global and often self-regulating capitalism. The fictitious commodity argument provides a theoretical entrée for examining how and why commodity production from natural inputs, as well as the commodification of nature per se,[47] is an inherently problematic undertaking. Moreover, Polanyi's point of departure—that nature is not wholly produced by human hands—may be extended into the realm of production. In this light, not only is the circulation of nature as a commodity problematic because of the dual movement issue but also the social "production of nature," increasingly evident in capitalist society,[48] is never completely social. There is at the most basic level a necessary discontinuity between capitalist production and biophysical nature.

This is the essential basis of James O'Connor's extension of Polanyi's work; what O'Connor called the theory of capitalism's "second contradiction" and the underproduction of nature.[49] O'Connor's theory of ecological crisis closely echoes Polanyi's; nature under capitalism (specifically "capitalist nature") is treated as a commodity when it is not. However, taking a more production-centered line of argument (in contrast to Polanyi's emphasis on circulation or exchange), O'Connor argued, "The point of departure for an 'ecological Marxist' theory . . . is the contradiction between capitalist production relations (and productive forces) and the conditions of capitalist production."[50] Among these conditions of capitalist production is the provision of inputs (raw materials) and the assimilation of outputs (waste and pollution). The problem for

O'Connor is that capitalism fails to reproduce nature as an external condition, resulting in the underproduction of the environment.[51]

Again, it bears noting that this problem has two distinct dimensions, one leading to a social or political crisis, the other to a more narrowly economic or production-centered one arising from capital's inability to fully capitalize material nature. Together, O'Connor and Polanyi therefore offer points of departure for understanding the problematic capitalization of nature—including commodity production from nature and the necessarily (incomplete) commodification of nature—and the wider political and social dimensions of ecological crisis tendencies arising from conflicts between capitalist and social nature. What remains is to situate these broad insights and attend to the historically and geographically contingent ways in which ecological crisis tendencies work themselves out. This is one of the chief aims of this book.

Production and Nature

If the spotted owl crisis points to a fundamental problem for capital in coordinating the allocation and reproduction of material nature as a set of inputs (including the consumption of waste assimilating capacity), how can this problem be conceptualized in more concrete terms? What are the various ways in which this problem is manifest, and what tendencies exist in its resolution, however partial? After all, if the disjunction between social and ecological production is a necessary problem for capitalism, and particularly nature-centered or resource-based activities, it is not a static problem. As O'Connor wrote, "Capital limits itself by impairing its own . . . environmental conditions, hence increasing the costs and expenses of capital, and thereby threatening capital's ability to produce profits, that is, by threatening to bring on economic crisis."[52] This is a dynamic process, subject to all manner of responses by capital and conditioned by the seemingly infinite variability that exists in material nature.

Despite his predilection for metanarratives of capitalist transformation (begging the question as to how nature's difference can actually matter at all), David Harvey argued that it reflects the essence of historical materialism to embrace the intertwining of nature and culture through historical analyses of particular ecological transformations, paying attention to the difference that material nature makes.[53] Enrique Leff expressed a similar sentiment:

> The availability of nonbiotic resources and the conditions for biological reproduction of different ecosystems affect the form and appropriation of natural resources. These factors also establish potentials and set certain limits to the expansion, reproduction, and sustainability of capital. These then, are the reasons to insist on thinking about *how ecological processes are inscribed in the dynamics of capital.*[54] (emphasis added)

A productive avenue for pursuing such inscriptions is with the development of a theory of ecological crisis, whereby the destruction of particular natures under capitalism results not in absolute or final "limits" but instead in ongoing political struggle over the social (re)production of new natures, both for the purposes of renewed accumulation and for broader societal goals. How might some order be imposed on such a messy collision? One way to conceptualize this more concretely is to identify three aspects of industry's confrontation with and reliance on ecological production processes: nature as land, nature as time, and nature as form.

Land

In *The Urban Experience* Harvey stated, "In order to overcome spatial barriers and to annihilate space with time, spatial structures are created that themselves act as barriers to further accumulation."[55] Because production takes place in physical space, a certain amount of capital must be removed from circulation to secure suitable locations. However, once locations are secured, the stickiness or inelasticity of spatial configurations acts as a constraint on the circulation of capital. This form of constraint, politically reinforced by excludable private property rights, is the genesis of rents in land and of obstacles to reconfiguring the geography of production. Though Harvey's insights are intended for the urban context, with all its specific institutional dimensions, the idea of physical space as constraint may also be applied to nature-centered industries.

In agriculture, for example, capturing economies of scale in production (i.e., increasing efficiency in production costs associated with expanded levels of output) faces the natural constraint of land as a physical space whose useful services are diffuse and difficult to concentrate. As Karl Kautsky observed,

> In industry any expansion of the enterprise also presents an increasing concentration of productive forces, with all the advantages which this brings—savings in time, costs, materials, easier supervision and so on. By contrast, in agriculture, other things being equal, any expansion of the enterprise *means the same methods of cultivation being applied over a larger area.*[56] (emphasis added)

Not only does agricultural production tend toward dispersal but it also is easily fragmented by accumulated historical patterns of ownership. Thus "the unique nature of land under private ownership is a major obstacle to the development of large agricultural enterprises in every country with small-scale land-ownership, irrespective of how superior large farms may be—an obstacle which industry never has to face."[57] These challenges profoundly affect the efficiency with which labor and machinery can be deployed in industrial agriculture.

Forestry and agriculture are directly analogous in this respect.[58] Just as Kautsky's farm faces the problem of coordinating extensive production from a central location (i.e., the farmhouse), modern wood-products mills are large and immobile and must draw timber from an extensive surrounding hinterland.[59] The fixity of land as space blocks consolidation of certain production activities, most notably logging itself (see chapter 3). For example, the average Oregon sawmill, if integrated backward into timberland ownership, requires access to about one hundred thousand acres of commercial timberland to remain in operation over the long term. Moreover, any increase in economies of scale at the mill translate into a greater appetite for raw materials and a consequent increase in the area of timberlands necessary to feed the mill. Because the costs of collecting and transporting raw materials generally increases with distance, not to mention the aforementioned problems of coordinating production over longer distances, land constraints may act as a direct impediment on economies of scale in forest commodity manufacture in exactly the way Kautsky identified in agriculture.

At the same time, in forest commodity production, securing access to raw materials is a key issue. The forest industry has, on the whole, exhibited more of the character of extraction than cultivation, as firms have opted to move from place to place in search of standing stocks of timber.[60] In this context, firms must also choose how they will secure access to land. They have two choices. If they integrate backward to own forestlands, then the long growth cycle of forest trees confronts them with a highly specific manifestation of the problem of land as fixed capital. Not only does ownership lock up capital and place it at risk of being devalued by changes in economic conditions (e.g., declining fiber prices, increasing interest rates, political opposition to management practices) but also capital as standing timber faces the peculiar risk of being "devalued" by extreme events such as high winds, fire, or disease outbreak. But if firms rely on market transactions in either private or public timber (i.e., choice two), they relinquish the security of long-term supply. Which strategy is chosen is a product of locally and historically specific issues; for example, forest growth rates under prevailing cultivation regimes (biology and technology), financial returns on growing timber (economics), local landownership patterns (institutions), and the degree to which the state is willing to socialize the costs of forest management (politics).

Time and Rates of Natural Production

Most commercial forest trees grow slowly by capital market standards. Recognizing this as a significant political economic challenge for capital, Marx wrote in the second volume of *Capital: A Critique of Political Economy,*

The long production time (which comprises a relatively small period of working time) and the great length of the periods of turnover entailed make forestry an industry of little attraction to private and therefore capitalist enterprise.[61]

In this and related ways, biological time may be seen to act, like land, as a problem for capital in and of itself. Thus, Barbara Adam discussed time as a key problem facing the industrialization and modernization of nature, noting that social theory has given too little notice to the formation of distinctive "timescapes" through the collision of social and natural time under capitalist modernity. She wrote specifically in this regard, "A timescape analysis is not concerned to establish what time is but rather what we do with it and how time enters our system of values."[62]

Agrarian political economists also have written extensively about the time problem, including the way seasonal rhythms of animal and crop growth confront industrial capital with a challenge to the continuous deployment of capital and labor.[63] And these insights may be adapted to all nature-based industries, because they all rely on ecological processes with their own rhythms, albeit to varying degrees.[64] This includes the Douglas-fir industry and its social regulation. The seasonality of crop cycles, for example, has a parallel in reforestation. In tree planting, there is a distinct season determined by the period during which seedlings are dormant and can be safely transplanted. This in turn propels a pronounced unevenness in seasonal labor demands that have been met in a variety of ways. Biological time also presents other challenges. As the Douglas-fir industry moves out of an extractive mode and into one of cultivation, attempts to intervene in and rationalize forest growth to suit the interests of more efficient commodity production are met with numerous obstacles having to do with biology as time. Douglas-fir in western Oregon, as noted, are slow growing, taking on average between sixty and eighty years to reach commercial harvest age. This implies long lags between any investments in tree breeding and a return on these investments through sales of wood products. Such delays are a major obstacle for firms pursuing intensive tree-breeding programs, though, as I discuss, this has not prevented firms from taking on industrial tree breeding and forest cultivation. It has merely helped shape the institutional strategies they have chosen to do this.

Form

Nature-centered industries must also confront raw material inputs whose physical form is shaped by natural production processes. Of course, material differences in production across sectors is one of the sources of diversity in patterns of industrial organization. As Andrew Sayer noted, "Steel, clothing, insurance policies, or foreign holidays . . . can each be produced within a variety of forms of organization. However the economic effects of doing so will differ, precisely because they are materially different activities, which is largely why their industrial organization tends to differ."[65] For example, as economic geographers have long understood, a commodity's ratio of value to weight is an important consideration in the spatial configuration of production and distribution networks.

Yet a crucial distinction of natural resource industries, including agriculture, is the logical separation or confrontation between natural and social production. Because a certain amount of production by definition takes place prior to human intervention or design, technologies and the deployment of labor to some extent revolve around natural properties.[66] Characteristics that may be important include material and energy density, volatility, phase (i.e., gas, liquid, or solid), mobility, solubility, sheer strength, and so on. It is extremely hard to circumscribe what issues are important or how they matter.[67] This is exactly the point; difference matters.[68]

In the forest sector, the physical properties of logs and fiber are crucial to the relative quality of manufactured products. But these properties are also critical to the development and deployment of production technologies and processes and to the capture of economies of scale and scope in the industry. In fact, the issue of log heterogeneity may make the wood-products industry unique in the variability of its raw material, and this has significant implications for industrial geographies and ecologies.[69]

The Capitalization and Commodification of Nature

Although nature—as land, time, and form—acts as a powerful constraint in industries that are centered in the transformation of biophysical inputs, this does not mean the confrontation between social and natural production should be understood in terms of rigid obstacles, limits, or determinants of social organization and technological development. This could amount to "a damaging material pessimism," ignoring the ways capital takes hold of biophysical nature, to integrate and transform it as an accumulation strategy.[70] It is hardly by accident that Enrique Leff (quoted previously) pointed to the ways that ecological processes are inscribed in the dynamics of capital.[71] In fact, O'Connor's thesis of capitalism's second, specifically ecological, crisis involves a discussion about how capital responds to the problem of underproducing nature. Significantly, he cited conversion from old-growth to plantation forestry as an example of how nature is first degraded as a condition of production, then capitalized by "the increased penetration of capital into the conditions of production (e.g., trees produced on plantations)."[72] As O'Connor noted, examples of these tendencies abound in contemporary society, whether it be the depletion of fisheries around the world, degradation of soils under intensive agriculture, depletion of groundwater, and of course the transformation of forested landscapes. In each case, the implications of these tendencies may be witnessed in the production of a more thoroughly, though never completely, capitalized or commodified nature. Contemporary developments in biotechnology, including the ways in which industrial strategies and markets form around the capitalization of nature, are an obvious example.[73]

Others too have noted the significance of nature's increasingly socially produced character under capitalism as a counter to the stale dichotomy of nature as rigid limits to growth and development on the one hand and complete disregard for biophysical nature on the other.[74] For David Goodman, Bernardo Sorj, and John Wilkinson, the nature-society interface in capitalist agriculture is constantly under assault, subject to transformation by two related firm strategies they called appropriation and substitution. They used the idea of appropriation to refer to processes by which discrete aspects of farm production are carved off and become the basis of industrial production processes. These industrially manufactured products are then reintroduced into farm production as "inputs or produced means of production." The key is that this involves piece-by-piece augmentation or outright displacement of discrete aspects of farm production, while leaving the farm-based character of production essentially intact. A classic example is industrial fertilizer manufactured from hydrocarbon fuels then introduced to farms as a replacement for animal dung and crop waste. By contrast, substitution represents a more radical intervention by industry, involving replacement of rural processes and products with industrial ones. Thus, although appropriation inserts industrial inputs into rural production, substitution involves the "elimination of the rural production process, either by utilizing non-agricultural raw materials or by creating industrial substitutes for food and fibres." A classic example of substitution is the production of synthetic fibers such as rayon.[75]

Jack Kloppenburg Jr. focused on the more specific dynamics of capitalist development and biological reproduction in agricultural crops. He argued that plant reproduction has acted historically as an obstacle to and a vehicle for the extension of capital accumulation into the realm of crop-variety production. The basic problem of crop-variety production and reproduction from an industry standpoint is that crops act in a dual capacity as means of production and product. However advantageous in other respects (!), this makes plants poor candidates as the objects of excludable property rights, if such rights cannot be enforced. This problem is evidenced historically by the practice of farmers' saving seed from one year to the next. Through seed saving and sharing among farmers, advanced varieties rapidly became de facto common property under the traditional social and biological organization of crop breeding that prevailed until relatively recently in American agriculture. In this realm, public science played the critical role of generating varieties for use by farmers.[76] Yet Kloppenburg shows how private seed companies have overcome these obstacles, helping to bring us to the age of plant biotechnology.[77] Biotech in this respect represents the culmination of capital's extension into the realm of biological reproduction accomplished by way of a set of technological innovations and bolstered by the reinforcement of property rights over life-forms. Echoing Goodman, Sorj, and Wilkinson, Kloppenburg thus locates the origins of negative social and environmental consequences of agroindustrialization in an increasingly capitalist nature.

I attempt to build on these basic insights into the ways that biophysical nature comprises a set of obstacles, opportunities, and surprises to capital arising from the disjunction between social and ecological production, using as a kind of extended case study commodity production from nature and gestures toward the commodification of nature in the Douglas-fir region's forest industry.[78]

The Broader Politics of Nature and Its Regulation

However, what of the second dimension of the Polanyi-O'Connor thesis dealing with political conflicts over capitalist and social nature? Specifically, in what directions do Polanyi's dual movement and O'Connor's theory of ecological crisis point for understanding environmental politics and environmental regulation? What is the relationship between the material transformations of biophysical nature on the one hand, the meaning conferred on these transformations, and the development of social control—state centered or otherwise—over nature's transformation at the hands of capital and the market on the other hand? Polanyi, though offering a theory for the specific origins of environmental social movements in market capitalist societies, says little more. There is in his theory a strong suggestion that resistance will directly and explicitly oppose self-regulating or laissez-faire market coordination of environmental change and the allocation of environmental inputs and services. Surely, however, the politics of capitalist nature are more contingent than Polanyi allows.

Accounts of ecological crisis that use particular states of nature as inherently desirable, bypassing the subjectivity of these claims, are common in environmental narratives. Consider any campaign that starts with "save the." Yet recent and important scholarship makes a compelling case that nature cannot be understood apart from its discursive representation—indeed this also is a major preoccupation of recent work in political ecology.[79] Ideas of nature are constitutive of, and therefore help to construct, what is understood as nature, and, in turn, are constitutive of environmental problems and their purported solutions.[80] This is not to say that nature does not exist or that all representations of nature are equally valid. Rather, it is to say that all representations are just that, representations mediated by culturally and politically specific perspectives resulting in the inexorable intertwining of the subjective and objective. On the one hand, it is plain that O'Connor understood this well; he specifies that ecological crises always arise from the material transformations of nature and the inescapably subjective political processes of interpreting, defining, and politicizing (i.e., "naming") these transformations. A similar idea also runs through Neil Smith's production of nature thesis. He contends that bourgeois ideologies are a key influence on prevailing meanings (social constructions) conferred on nature and environmental change, including, for example, wilderness. Such subject-object distinctions make the interpretation of nature and, in particular, the comparison of alternate states of nature, inescapably politically and discursively constituted.[81] After all, as Raymond Williams famously noted, " 'Nature' is perhaps the most complex word in the [English] language."

Despite the abstract character of these observations, there are some quite practical implications for environmental politics. As biologist Richard Lewontin noted, the environmental movement often has invoked the idea of a stable, harmonious nature in order to bypass explicitly political—and therefore subjective—statements about right and wrong in relation to the natural world. In his words,

> Any rational environmental movement must abandon the romantic and totally unfounded ideological commitment to a harmonious and balanced world in which the environment is preserved and turn its attention to the real question, which is, how do people want to live and how are they to arrange that they live that way.[82]

More generally, each environmental problem or ecological crisis needs political subjects to construct and confer meaning.[83] And the response to such ecological crises has as much to do with these political subjects and the dynamics of their struggles over meaning—within an institutional, regulatory, economic, and cultural milieu—as it does with the underlying or structural character of the problem itself. Again, this is not to suggest that the material character of environmental struggles is unimportant. I essentially agree with Proctor's critical realist stance; that is, empirical evidence does not speak for itself, but it can and should matter all the same.[84] However, as the conflict over the northern spotted owl indicates, environmental problems often become very active struggles over meaning, and mediating the veracity of different meanings is fraught with problems.[85]

In this discussion, my focus is less on struggles over meaning per se and more on their implications for understanding the politics of ecological crisis tendencies; in particular, how do the struggles over the meaning of nature (i.e., what nature is, which natures are preferable) shape and reshape environmental regulation? Considering the crisis of the northern spotted owl and logging in old-growth forests, new and powerful meanings were clearly conferred on old-growth forests by environmental activists and social movements on the one hand and changing ideas in scientific forestry on the other. These new meanings were contested in the courts, in the newspapers, in town hall meetings, and also in academic analyses. But they also helped to remake forest policy in the region during the 1990s.[86] In this respect, it is important to see that the struggle for meaning arose as a facet of a crisis of legitimacy for the regulatory framework that in part enabled resource extraction and nature's commodification up to that point. Moreover, the spotted owl crisis was essential in putting in place a new model of forest regulation. How might these aspects of the crisis be approached?

I turn to the Regulation School for an approach to understanding the integrated relationship between particular tendencies in capital accumulation (and specifically nature-based capital accumulation) and the social regulation of environmental change. A key strength of the Regulation approach, at the broadest level, is its purchase on the various and contingent ways in which crisis tendencies in capitalist economies and societies are ameliorated through particular, highly specific combinations of the trajectories of capital accumulation with political, social, and cultural tendencies more generally. This is obviously quite broad. I emphasize a Regulation approach to analysis, seeking explanation in the ways crisis tendencies emerge and are—at least temporarily—addressed in the relations between regimes of (capital) accumulation and modes of social regulation and in ways that are temporally and spatially contingent.[87]

According to Regulation Theory, these combinations, once established, may persist for years and even decades, giving rise to relatively coherent periods in historical political economy.[88] The goal, in the simplest possible terms, is to understand the specific dynamics of capitalist crises that arise from the social relations and institutions of capitalist commodity production and circulation in particular places and at particular times, and the ways in which states, citizens, social movements, and various economic and extraeconomic institutions respond to and address these tendencies, again, in specific places and at specific times.[89] Critically, however, if the combinations of regimes of accumulation and modes of social regulation are contingent, they are also nevertheless functionally related. That is, although the broad tendencies of capital accumulation and the particular dynamics of crisis do not determine or call forth the ways in which they will be socially regulated, nevertheless, certain combinations of social regulation just seem to work and thus endure for a time at least.[90]

Although Regulation Theory did not emerge from the analysis of ecological crisis tendencies and their politics, there is nothing preventing Regulationist concepts from being adapted to this purpose; indeed, there have been various attempts to do this.[91] The basic point of departure is that if there are ecological crises or contradictory tendencies of capitalism in the production of commodities from and the commodification of nature, then it should be possible to examine how such tendencies are addressed, with varying degrees of success, within particular regulatory formations, and reflecting the ways in which environmental change is contested in the social realm. This includes looking at the key arenas of state policies and practices in natural resource and environmental management; at the formation and tactics of social movements, their discursive and institutional strategies and efficacies; and at the role of science and scientific constructs and discourses.

Returning to the Polanyi-O'Connor thesis, this suggests a way of examining the idea of the dual movement in relation to nature (and also labor)[92] as commodities. In particular, how do the politics of promoting and resisting the commodification of nature work themselves out in the realm of environmental regulation at particular junctures? How do capital, the state, science, and civil society act to shape the regulation of nature's commodification? In what manner does the construction of new meaning surrounding environmental change by political subjects shape the development and adoption of new approaches to regulation and the social control of nature, or what has been called ecosocialization?[93] Critically, in relation to O'Connor's thesis of a second contradiction of capitalism, does the struggle over meaning directly identify and challenge capitalist commodity production per se, or is this one of the contingencies of environmental politics to be worked out in the social realm?[94] I suggest the latter. Specifically, returning to the political ecology of the Douglas-fir region, and in particular to the history of regulating forest commodity production in the region through the spotted owl crisis, it seems apparent that the politics of nature and their influence on social regulation must be considered carefully and in case-specific fashion. As O'Connor argued, and I paraphrase, there is no crisis without someone to call it one.[95] And how it is named will influence how it is addressed.

Overview

In *The Limits to Capital* Harvey discussed the way in which Marx developed his analysis in *Capital* from a variety of standpoints or "windows"—the commodity, surplus value, circulation, the labor process, and so forth.[96] I have, in all humility, drawn inspiration for the organization of this book from these comments. I do not mean to suggest that this book will provide the missing link as it were in reconciling neo-Marxist political economy, economic geography, and the politics of nature and environmental change, as much as this link needs to be forged.[97] However, in addressing the politics, economics, and ecology of forest commodity production and its social regulation in the Douglas-fir region, I do aim to provide an extended case study of the various problems inherent in capitalist nature. Each chapter comes at the issue differently, looking at (1) the labor process, (2) the changing regional geography of commodity production, (3) the dynamics of scale and scope in commodity manufacture, (4) the industrialization of forest growth through tree improvement, and (5) the historical political economy of forest regulation. If, as I suspect, some readers find the narrative slightly episodic, this is why.

In the next chapter, I examine the problem of nature as it is manifest in relations between workers and their employers in the logging and reforestation sectors. Specifically, I consider the ways in which the "ecoregulation" of social labor plays out in these two arenas of the Douglas-fir industry.[98] These activities share in common that they require that workers toil directly in a forest environment, where nature as time and as a source of unpredictable variations in production conditions helps to dissuade "conventional" capitalist wage relations. Although the translation of this problem into specific labor relations has multiple possible outcomes, it is nevertheless important to see the influence on production relations that working in the forest has in these sectors. I discuss not only the organization of labor but also who bears the brunt of the risks and uncertainties in production and how these are distributed in production relations.

In chapters 3 and 4, I consider how raw material characteristics unique to the Douglas-fir region (and to the Douglas-fir specifically) have influenced the historical geography of solid wood-products manufacture. In chapter 3 I deal more with issues of industrial ecology in Douglas-fir region wood-products manufacture; that is, how the highly specific qualities of Douglas-fir region forests are reflected in particular patterns of commodity specialization in the region and how the gradual exhaustion of old-growth timber has propelled product and process innovations in solid wood-products manufacture, with significant environmental implications. In chapter 4, I explore the relationship of generally low economies of scale and scope in the production of lumber and other solid wood products, with the problem of raw material dependence. Here, the constraining influence of an extensive geography is critical, as is the heterogeneity of logs as a raw material input.

In chapter 5, I discuss the industrialization of forest growth through tree improvement (i.e., commercial forest tree breeding) in the Douglas-fir region. A historic pattern of cooperative and state-sponsored institutions has helped deal with specific biological problems or challenges confronting tree improvement. Yet possibilities opened up by tree improvement research and development, and particularly by recent forays into the genetic engineering of forest trees, suggest that a more complete, capitalist commodification of improved tree varieties may not be far off. However, this will occur only if regulatory obstacles and political opposition based on environmental impacts and the corporate control of life-forms can be addressed.

In chapter 6, I deal specifically with questions surrounding the social regulation of forest access and control and the contradictions of commodifying nature during the boom years from the late 1930s to the late 1980s. Specifically, I examine the emergence of sustained-yield forest management as an approach to addressing the fictions of labor and nature as commodities, drawing on a case study of southwestern Oregon's Illinois Valley. Sustained-yield doctrine, intended to regulate the flow of timber and thus the rate of old-growth liquidation, was aimed at consolidating and perpetuating the forest industry as a basis for social and economic development in rural communities of the Pacific Slope, thereby addressing the fictions of nature and labor as commodities. However, as evidenced by the experience of many communities, including the Illinois Valley, the apparent failures of this approach to social regulation are rooted in fundamental ecological and political economic contradictions that run through the sustained-yield approach.

In the epilogue, I return to the crisis of the spotted owl, and examine it, albeit more briefly, as a crisis of capitalist environmental regulation brought on by the contradictions of sustained-yield forestry. During and after this crisis, forests (particularly old-growth ones) were discursively reinvented as ecosystems, and were regulated as such. This reinvention and reregulation was built on two parallel avenues of ecosocialization. The first avenue was the emergence of a new way of seeing old-growth forests within the forestry science community, under the guise of what has been called the New Forestry. The second avenue was the increasing involvement of environmental nongovernmental organizations in the social regulation of forests, particularly public forests, consolidated institutionally with lawsuits and court injunctions against the federal government. During the spotted owl crisis, these twin avenues converged to provide the basis of reenlisting forests to industrial commodity production and to reregulating forestry in the guise of ecosystem management.

2
Working the Land
Production Relations in Logging and Reforestation

Nature and Flexible Production

Logging and reforestation share the distinction of being the only segments of the forest industry in which production takes place in the woods. In the Douglas-fir region, it is also common for logging and reforestation to be conducted as separate operations, with arms-length, contract relations of production connecting the mills to work crews in the woods. This is true even of some of the industry's largest and otherwise most vertically and horizontally integrated firms, including Weyerhaeuser and Willamette Industries (now combined). In this chapter, focusing specifically on the state of Oregon, I explore why logging and reforestation are so often pursued by means of relatively arms-length industrial labor relations. Drawing on agrarian political economy, and specifically Ted Benton's notion of the "ecoregulation" of labor processes in nature-centered activities,[1] I examine alternatives to fixed hourly wage relations as strategies for achieving production flexibilities inspired by the "in situ" character of production in logging and reforestation. Specifically, I argue that variability, risk, and uncertainty associated with the intimate engagement of social and ecological production in logging and reforestation create problems in predicting and rationalizing rates of production in these sectors, prompting capital to seek ways to displace risks and uncertainties while inducing rationalized production.

In this chapter, in keeping with a main theme of the book, I explore the intersection of and disjunction between social and ecological production, specifically through the lens of labor processes and production relations in logging and reforestation. Yet I stress that although the confrontation between social and ecological production explains why many firms prefer to offset certain risks and uncertainties in production, it does not sufficiently explain how particular social relations, including contracting out and piece wages, develop. This is the realm of politics, contingency, and history. Thus, in my account of social relations in logging and reforestation, I examine structural issues—specifically the confrontation between social and ecological production in logging and reforestation—and how human agency and contingency—specifically the politics and particulars of logging and reforestation labor markets—operate in dialectical relation with one another.[2] If I give short shrift to the agency of workers to contest relations of production here, it is because for the most part resistance to date has not held back the spread of contracting in logging, and contracting, piece wages, and the exploitation of immigrant workers in reforestation. I do not intend to suggest, however, that such effective resistance cannot or will not develop; quite the contrary. It is my hope and faith that it will.

Confronting Nature in Logging and Reforestation

There is a considerable literature on contract production networks under late capitalism. Such networks—specifically their apparent resurgence—underpin notions of post-Fordist production systems and are critical to debates over the dynamics of social power in contemporary industrial economies. Many explanations of contracting have been offered. For example, contracting has been linked to the development of "leaner" production systems; that is, stripped-down operations that capitalize on a firm's core strengths.[3] Transactions cost theory has been used widely to explain which activities are contracted out and which are not, based on neoclassical ideas of cost minimization and, ultimately, on Ronald Coase's well-known contrast of vertical integration (make) and market transactions (buy) as alternative firm strategies.[4] And in particular, transactions costs have been invoked to address internal and external economies of scope.[5] At the same time, the proliferation of contracting has been tied to possibilities opened by new information technologies, allowing better just-in-time coordination of supply chains over increasingly far-flung production networks.[6]

In this context, the efficiency of specific regional contracting systems has been theorized in terms of capturing knowledge economies within dense networks of localized competing contracting firms and their contract suppliers. Repeat contracting and negotiated prices among parent firms and contract producers who gain familiarity with one another is a noted aspect of the geography of numerous manufacturing districts. Michael Storper explained these regional networks of input-output relations based on so-called "untraded interdependencies."[7] Specialized contractors familiar with the needs of contracting firms compete in markets that occupy a space between pure, open competition and vertical integration or formal partnerships. One of the main reasons for some degree of continuity and loyalty in these regional systems of contracting stems from the development of knowledge economies and trust. Knowledge economies vis-à-vis specific production processes develop through repeat contracting between firms, and this knowledge can improve quality while also simplifying negotiations and reducing transactions costs. At the same time, repeat contracting builds trust, which is central to efficient markets.[8]

The flexibility literature thus offers important insights about the comparative efficiencies of contracting, repeat contractor networks, and various and more flexible forms of integrated wage relations. However, the literature generally does not provide an adequate explanation as to why particular activities would be subject to pressures for restructuring and contract production. For example, Alan Scott noted, "Scope effects (i.e., internal and external economies of scope) are ultimately defined by transactions costs, which in turn have both institutional and technical foundations."[9] Yet what are these foundations, and how is it that particular activities come to be isolated for potential outsourcing? How is it that some labor processes are designated lean and others "lard"? Addressing this issue requires careful consideration of specific production processes, consistent with a problem-oriented approach to industrial organization.[10] In logging and reforestation, this requires addressing what I refer to as nature-centered production.

Although very little of the flexibility literature deals with natural resource or nature-centered activities, some research has addressed industrial restructuring and the pursuit of various forms of flexible production in the forest products sector, largely with a Canadian focus.[11] Yet little has been written about logging or reforestation that deals with the question of nature and production relations per se. Because there is a general (if misguided) perception that natural resource industries are the hallmark of economic backwardness, it should come as no surprise that research on the new social economy would generally overlook such sectors.[12]

But there is a strong basis for the argument that nature-based production activities involve certain variabilities, risks, and uncertainties that encourage firms to seek flexibility in production relations. Susan Mann, for instance, specifically argued that agriculture's nature-centered character presents obstacles to the penetration of agricultural capital. Drawing on Marx and Kautsky, Mann called for an agrarian political economy that emphasizes "how various natural features of certain branches of agriculture can serve as impediments to the efficient use of advanced technology and wage labor."[13] One example of such mechanisms is provided by seasonal fluctuations and delays in animal and crop maturation, and crop cycles that extend production time, creating unevenness in the demand for wage labor. These are examples of "timescapes"[14] that place the continuous deployment of capital and labor and the social transformation of production processes in opposition to biophysical nature. They thus provide incentives for agrarian capital to achieve flexibility in production relations by exploiting contract and household labor, passing on the risks of production variabilities and the impetus to achieve production rationalization with contracts and piece wages. In short, timescapes and other challenges stemming from agriculture's nature-centered character provide a rationale for production relations with what Mann referred to as "a much different complexion than that found in industry proper."[15]

Using a very similar logic, Ted Benton argued that agricultural labor processes are examples of what he referred to more generally as "ecoregulatory" activities. He identified four central features of such activities: (1) labor is applied to optimizing conditions of transformation, that is, organic processes, which are relatively impervious to intentional modification; (2) labor is primarily aimed at sustaining, regulating, and reproducing rather than transforming those conditions and processes; (3) the organic character of production shapes the temporal and spatial distribution of social production; and (4) certain natural conditions of production (e.g., water, sunlight) act as the conditions of production and the subjects of labor.[16] Although the development of these kinds of ideas is strong in agrarian political economy,[17] similar thinking may be applied to a range of nature-based industries, particularly with respect to the ways that social relations coalesce.

My argument here is that the pursuit of more flexible production relations in Oregon logging and reforestation comprises specific examples of feature 3 of Benton's typology of ecoregulation. In particular, the difference that nature makes lies in breaking down the regularities typical of factory production regimes by means of a number of decisive disruptions. These disruptions act not to block or prevent capital accumulation per se but to shape and constrain production relations in important ways. Natural obstacles become the vehicles or organizing principles around which firm strategies and organizations coalesce.[18] Specifically contracts and piece-wage relations may be understood as key instruments by which capital imposes discipline and subordinates labor in the context of highly specific "biological and geographic peculiarities,"[19] achieving in the process a form of production flexibility.

Contract Production in the Oregon Logging Sector

The degree of vertical (dis)integration in Oregon logging is an important issue. Despite dwindling supplies of old-growth timber and significant reductions in federal timber sale programs in the region, as of 1997 Oregon led all states in logging employment and value added (see Figure 2.1).[20] Although logging accounts for a relatively small proportion of the state's overall employment, it remains important to the state's hinterland, where the forest sector is prominent in many small, remote communities.

In this context, it matters a great deal to many workers, their families, and their communities that there are substantial differences in wages, benefits, and

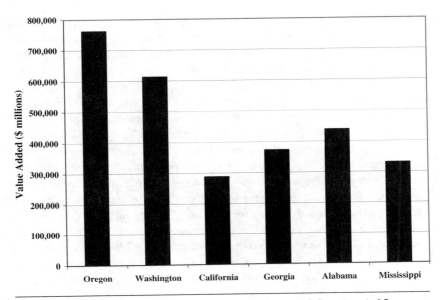

Fig. 2.1 Value Added in Logging in Leading States, 1997. Source: U.S. Department of Commerce Bureau of the Census, 1999.

working conditions when one compares company logging "sides" (crews) with those of the so-called gyppo, independent, or contract outfits.[21] Although wages are lower on gyppo sides, there are also extreme differences in job security and seniority protections. The character of logging being what it is (i.e., difficult, arduous, and dangerous), there is a premium on youth and skill, and workers deemed too old to work tend to be discarded quickly by contractors. These wage and job security differences in turn are achieved in part because contracting in the Oregon logging sector is virtually synonymous with the use of nonunion labor in the woods; unionized loggers are company loggers, and gyppo loggers are nonunion. Thus if you are an Oregon logger, what kind of a job you have is in large measure determined by whether you work directly for a mill or for a gyppo.

Not only is contracting central to the social distribution of surplus in Oregon logging but at the same time contracting is important because it represents a strategy for firms to manage risks, uncertainties, and rigidities in production. In particular, contracts allow wood-products firms to displace the costs of dealing with these problems onto gyppo contractors and their employees. In this sense, contracts are fundamentally not a meeting of equals in exchange. Rather they are an instrument of power used to achieve flexibility through shorter term commitments to gyppo loggers and, at the same time, to pursue integration and control through specific contract terms, various asymmetries in regional market competition, and differential control of assets.[22]

Yet, despite the salience of contract logging, information directly indicating the number of gyppo logging sides versus company sides in Oregon is not available. This is due in large measure to establishment-based reporting in the census of manufactures that does not differentiate between the two types of firm organization. However, data compiled by the Associated Oregon Loggers (AOL), a group representing gyppos, in conjunction with census data and interviews with key industry observers and participants indicate that company crews now account for no more than 10 to 15 percent of the industry, or between eight hundred and twelve hundred employees statewide.[23] The remaining seven thousand to eight thousand workers labor with gyppo outfits. This is not to say that vertical integration between wood commodity production and logging does not exist; it does. Some large firms with substantial regional landholdings, notably Weyerhaeuser and Willamette Industries (before they merged), operate some of their own company crews or "logging sides," as I discuss. Yet the predominant form of organization is vertical disintegration. Although Weyerhaeuser has recently decided to buck the trend in logging and revert to company logging sides entirely, it is significant to the argument I make here that the company has done so on the basis of restructured wage relations that make logging costs more transparent and pass significant production risks along to workers.

The dynamics of contracting become all the more intriguing when one considers that contracting, at least in the Northwest, was not always so widespread. Well into the twentieth century, a large proportion of logging crews worked predominantly in camps operated by lumber companies.[24] For example, according to a 1918 U.S. Department of Agriculture report on logging in the Douglas-fir region, the industry was then divided about evenly between independent and company sides.[25] However, most of the independents at that time bought land or cutting rights and sold logs to mills. By contrast, contemporary gyppo outfits usually own neither land nor cutting rights and work on a pure contract basis for mills who control the fiber.[26] Although Vernon Jensen in his landmark study of labor in the lumber industry noted the existence of gyppos in the Northwest during the mid-1940s, he too stated that the dominant pattern was for logging operations to be integrated with milling.[27]

Several historical factors may have played an enabling role in the expansion of gyppo logging. First, sawmills increased in average size following World War II (see chapter 4). The resulting centralization of lumber manufacture increased spatial separation between mills and the woods and likely decreased the advantages of integrated operations. Moreover, the onset of gasoline- and diesel-powered trucks to transport logs, workers, and equipment allowed for greater geographic reach and flexibility for logging crews as compared to the more capital intensive and less flexible logging railroads. And road construction and diffusion of the automobile made daily access to the woods from nearby towns much more feasible.[28] The chain saw also became increasingly common for falling and bucking during the late 1940s and early 1950s, decreasing the labor intensity of logging while allowing the fragmentation of large crews,[29] which made smaller logging operations economically viable and increased the geographic reach and flexibility of logging.[30] Thus, as Michael Williams noted, a whole set of changes in the technology of logging during the 1930s and 1940s "meant that the gyppos could handle all operations from stump to mill nearly single-handed."[31]

Social and Natural Production in Logging

Although these factors describe some of the circumstances under which contracting became more prevalent, they say nothing about why firms would want to use independent loggers in the first place. My argument is that the nature-centered character of logging generates a number of risks and uncertainties by impeding the continuous deployment of machinery and labor, the rationalization of the labor process, and the prediction and regulation of production costs and returns. Specifically, logging presents three basic kinds of ecoregulation: (1) extensive geographies, (2) frequent relocations and landscape heterogeneity, and (3) variable weather. All of these factors confront the continuous deployment of labor, undermine labor supervision, and impede the predictability and rationalization of production. They also contribute specifically to extraordinarily high accident rates in the industry, which, in addition to their human costs, act as obstacles to more rationalized production.

Most logging systems in use in the Pacific Northwest involve an extensive geography of labor deployment that makes coordination difficult.[32] In the woods, workers are typically deployed over a wide area and are constantly on the move. In a heterogeneous forest environment, particularly in the uneven topography of the Douglas-fir region, this makes supervision and the imposition of labor discipline extremely difficult. The most extreme example of this is provided by the workers who actually fell the trees ("cutters"). Because of logistical issues and the dangers involved, cutters typically work separately from the rest of the logging crew, often cutting as much as a month ahead. This creates obvious coordination and supervision problems, and many gyppos actually subcontract cutting to independent cutters.

Yet even at the landing site, coordination and control are hard to achieve. Workers must gather or "yard" downed logs to a central area to prepare them and load them for shipping. Once the logs are gathered, they need to be cleaned up and cut ("bucked") into specific lengths, depending on the requirements of different mills. The logs also need to be sorted and graded according to log quality in relation to the specifications of different kinds of mills (e.g., sawmills versus veneer mills) and the prevailing market demand for different kinds of logs (respectively, sawlogs and "peelers"). Finally, the sorted logs need to be loaded onto trucks for transport out of the woods. These tasks all require a high degree of coordination and the combination of speed and skill on the part of workers. Yet, at the same time, the different jobs on-site must be executed in a maze of downed logs, scrub brush, uneven terrain, and noisy equipment.

Coordinating these various tasks is difficult to achieve from processing facilities because logging sites are typically far removed from the mills. In fact, as wood-processing efficiencies have improved and as scale economies in mills have escalated, transportation costs have declined, making it possible for mills to source timber from greater and greater distances. For example, one mill in Eugene has hauled logs from as far away as Sacramento, California, a distance of more than 450 miles. Although such distances are not typical, the average spatial separation between logging sites and mills has certainly increased over time (see chapter 4).[33] For mills employing their own logging sides, this creates logistical challenges to monitoring and rationalizing production in logging operations.

This problem of extensive geographies is only compounded by inconstant geographies. Because commercial tree species in the Northwest have extremely long rotation or maturation ages, the distinction between biological production time and labor time dwarfs the seasonal problem in agriculture, with significant geographic implications.[34] That is, although logging parallels agriculture's land-based deployment of labor with attendant problems for monitoring and control, logging operations also confront the challenge of relocating on a regular and frequent basis to renew timber supplies.

Not only does frequent relocation exacerbate the labor control problem for capital, it also leads to wide variations in production conditions from one logging site to another. In the uneven terrain of the Pacific Northwest, these variations are significant indeed. Local timber stands vary widely, for instance, in the density of tree growth (measured in either stems per acre or total volume per acre). Species composition also varies considerably, as does the quality of the timber. Even fluctuations in the average size of the trees (diameter and height) matters, because the costs of using machinery to handle logs is generally insensitive to log size (within the range of the machine's capabilities), yet large trees yield a higher relative and absolute volume of merchantable wood from each log (see Figure 3.3). Moreover, each logging site has its own particular logistical challenges (e.g., steep slopes, unstable ground), which can be greatly compounded by increasingly stringent environmental standards.[35] All of these variations undermine what James Scott referred to as the "legibility" of nature, and they comprise unpredictable sources of ecoregulation that affect the rate of production in ways that can be quite difficult to assess *ex ante*.[36]

Although extensive geographies, frequent relocations, and local site variability all present certain challenges to labor monitoring and control and to the predictability and rationalization of production, variable weather creates further problems. In the Pacific Northwest, the weather is at its worst and most unpredictable in winter, with wet, cold, wind and snow making logging very difficult. In contemporary logging, increasing mechanization has facilitated and also motivated more consistent production. Yet poor and unpredictable weather, particularly at high elevations, continues to depress and render unpredictable the amount of actual production time in the sector and plays havoc with daily rates of output. As one logger noted,

> One of the things you are concerned about is the wind. If there is timber around, whether in a partial cut or if there is timber where you are logging, and the wind is coming, that is probably one of the most unsafe areas you can be in. If there is a lot of wind, don't go. You go out there, look around, and crank up the crummy and go home.[37]

In summer, hot dry spells in the otherwise humid Northwest also can cause delays and interruptions in logging as the result of risks that dry forests will be ignited by sparks from the equipment.

The implications of weather delays include daily and seasonal fluctuations in production and reduced total labor time in the sector. Loggers work fewer days and fewer hours on an annual basis than do their counterparts in other industries, including other wood-products sectors. On a national basis (see Figure 2.2) logging employees worked an average of about 190 fewer hours per year per worker between 1977 and 1992 than sawmill employees, and just more than 300 fewer hours per year than softwood-plywood mill employees. This translates into just less than five fewer weeks and just more than seven fewer

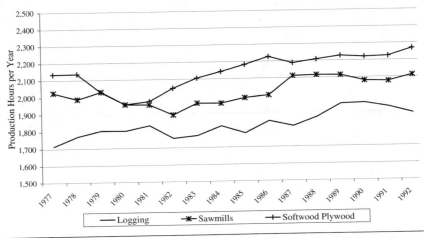

Fig. 2.2 Annual Hours of Production for Workers in Logging, Sawmills, and Softwood Plywood Mills, 1977–1992. Source: United States Department of Commerce, 1995.

weeks, respectively (based on a forty-hour workweek). In Oregon in 1992, loggers worked an average of 1,878 hours per employee, compared with 2,116 hours for sawmill workers, and 2,181 hours for softwood-plywood mill workers. The difference translates into the equivalent of between five and six weeks of work at forty hours per week.[38]

Fluctuating rates of productivity from site to site and depressed labor time in logging indicate variabilities and uncertainties related to logging's nature-centered character. But no single measure more clearly (and gruesomely) reflects variability and uncertainty in logging than accident rates. High mortality and morbidity rates have long been associated with the logging sector.[39] And logging remains arguably the most dangerous of all industries. According to 1995 occupational health and safety data (see Table 2.1), loggers rank behind only commercial fishers and sailors in the frequency with which they are killed on the job. Other occupations with a reputation for hazard—including structural metal work, law enforcement, and farming—do not compare in terms of the risk of death at work. In fact, besides seafaring occupations, only commercial airplane pilots and miners can be considered to be in the same range of mortality risk. In Oregon in 1996, the logging sector accounted for 15 percent of the state's occupational fatalities (nine out of fifty-four) despite accounting for less than 1.5 percent of the state's total employment.

Logging's human toll cannot be attributed to any particular technology or logging system.[40] Instead, the danger of logging is primarily due to the inherent variability and unpredictability of the work itself; in fact, evidence from around the United States and from other countries indicates that in logging, the main source of danger is actually the timber.[41] Fully 70 percent of the fatal accidents in logging nationwide from 1992 to 1997 resulted from workers' being struck by trees or logs. Moreover, an estimated one-quarter of all logging injuries take place during the actual felling of trees.[42]

Table 2.1 U.S. Occupational Death Rates by Selected Occupations, 1995

Occupation	Fatalities per 100,000 Employed	Relative Standard Error (percentage)
Sailors and deckhands	115	20.6
Fishers	104	15.6
Timber cutting and logging	101	10.6
Airplane pilots and navigators	97	9.8
Mining machine operators	78	17.5
Structural metalworkers	64	13.6
Farmworkers, including supervisors	30	3.6
Police and detectives, including supervisors	17	3.2

Source: U.S. Department of Labor, Bureau of Labor Statistics, 1997.

Though horrific in their own right, such accidents carry economic costs in terms of production slowdowns, lost labor, wages, health care, and even lawsuits stemming from injury-causing accidents.[43] In short, because accidents impact the variability and predictability of production, they comprise one source of risks and costs that are socially distributed in part through production relations, including contracts.

Production Contracts

As a response to logging's nature-centered character, contracts help wood-products firms achieve three things. First, contracts displace onto gyppos the risks of uncertain and discontinuous production generated by extensive and variable geographies. Second, contracts place the onus on contractors (through the explicit or implicit threat of market bidding) to achieve more continuous throughput and enhanced productivity—that is, to rationalize production—in the face of such variabilities and uncertainties. Third, contracts encourage the development of expert, experiential knowledge among contractors about how costs fluctuate across different sites and the incorporation of such knowledge into contract bidding. Although specialized knowledge can and does develop among company wage employees, firms have largely been unable to develop or control this knowledge under conventional wage relations of production. This is because the price they pay for labor is insensitive to variations from site to site in advance of production (i.e., the firms find out that logging a particular site is costly only after the fact), and direct monitoring for the purposes of learning to predict rates of production in advance (as a function of local conditions and to rationalize labor processes in ways that reflect local conditions) is difficult and expensive in the woods, far from mills. By contrast, competitive contract bidding can act as a way of capturing expert *ex ante* estimates of variations in production costs from gyppos.

There is, however, a catch. At least some of the expert knowledge loggers develop is firm specific, driven by geographic and technological factors in the contracting firm's commodity manufacturing facilities and by the specific characteristics of the firm's lands. This can give rise to firm-specific information economies of scale in logging. Thus, two distinct patterns of subcontracting have emerged in Oregon. The first strategy is relatively open bidding, involving more arms-length relations between contracting firms and gyppo logging outfits. This strategy has the advantage of using market competition to rationalize production and to act as a surrogate for expert knowledge by forcing contractors to bid against one another, but it does not allow economies of familiarity to develop. The second strategy, echoing Storper's notion of regional networks linked through untraded interdependencies, involves repeat contracting between mills and gyppos and features much greater familiarity and continuity than the more open-market strategy. This has the advantage of allowing gyppos to build up specialized expertise that is specific to the contracting firm's production and land portfolios. Not surprisingly, firms opting for the second strategy tend to be larger, horizontally and vertically integrated firms. For these firms, the complexity of their operations generates greater demand for highly specialized expertise.

Risk, Rationalization, and Information

The typical contract arrangement is one in which a gyppo is retained to log timber that is owned by the mill or has been purchased through a separate transaction. Gyppos in turn employ wage laborers who work together on one or more logging sides operated by the gyppo. Workers are predominantly paid on an hourly basis. The contracts stipulate production levels at given locations for fixed prices per volume and grade and are usually (but not always) negotiated between gyppos and the contracting firm.

What these contracts give to the mill is insulation from risk and an associated flexibility in their relationship to logging operations, as well as the capacity to use market signals to generate and track information about logging costs. Management of risk is one of the most commonly identified reasons in the flexibility literature for explaining why firms pursue contracting in many industries, because (1) dense contracting networks can pool and reduce aggregate risk[44] and (2) large firms can displace risk onto smaller and more vulnerable subcontractors. Typically, the source of these risks is cited as rapidly changing or fragmented market conditions.[45] However, in this case, the risks involved are not market risks but endemic production risks related to the unpredictability of production costs in logging.

In a regime where the rate of output is so highly uncertain and uneven, production contracts make sense for firms because they allow them to peg their costs only to the timber they receive, displacing the risks (i.e., costs) of uncertain production schedules onto the contractors. By setting the costs prior to logging a site, the firm places a ceiling on production costs and is therefore able to insulate itself from the risks of production slowdowns and cost overruns. With employees paid at fixed hourly rates, these risks are displaced onto the gyppo. In this risk-price regime, contracts also place the burden squarely onto the contractor to impose labor discipline and rationalize production under conditions that impede employee monitoring and make the assessment of production logistics difficult and unpredictable.

Not only do contracts insulate firms from the risk of cost overruns due to site variation but they are also asymmetric with respect to the risks (i.e., costs) of accidents. Contracts allow firms to displace accident costs and restructure risks in several ways. The first of these pertains to insurance costs for workers.

Workers' compensation legislation has generally been structured to socialize accident costs through standard payroll deductions. Because these deductions are in principle indifferent to contract versus vertically integrated social relations, as long as contractors pass their costs to firms, the distribution of accident costs should be indifferent to company and gyppo logging. Yet industry compliance with the requirement that all employees be covered by insurance is not uniform. Despite the availability of pooled insurance among gyppos, insurance costs are still very high. In response, a considerable number of contractors opt for cash wages paid under the table for some or all of their employees.[46] And this in turn may act to reinforce contracting if gyppo loggers use savings from this practice to suppress contract bid rates.

Second, accidents drive up insurance rates. By contracting for logging, firms displace the risk of escalating insurance premium hikes by linking the higher rates to the contractor rather than the mill. When a contractor's rates increase because of accidents, wood-products firms can avoid having to pay for the higher premiums by shifting to a different contractor with lower rates. In this way, mills use contracts to lower their exposure to the potential costs of accidents.

Third, the use of production contracts preserves an incentive among loggers to "get out the cut" in the face of strong incentives to lower the rate of accidents. That is, insurance premiums tied to accident rates combined with fixed wages or salaries create an incentive in company crews to increase safety even at the expense of production. By contrast, use of contractors allows firms to have it both ways. Gyppos face a cost price squeeze, because they pay insurance rates that rise in proportion to the number of accidents, but they pay wages on a strict hourly basis and receive contract prices that are fixed to volume alone. If the gyppos push the workers too hard, they will save on wages and receive higher profits from volume-based contracts but exacerbate the risk of accidents and thus production delays and higher insurance premiums on subsequent jobs.

Contracting out for logging also takes advantage of information economies generated by market competition among gyppo contractors. That is, production contracts have the advantage over conventional, integrated wage relations of allowing firms to capture expert knowledge among gyppos through contract bidding. Variation in the rate of production owing to topographic variation and changes in forest types (age, species, density, etc.) can be translated into more or less accurate *ex ante* assessments by knowledgeable, experienced loggers. This type of knowledge and information (i.e., conception) is very difficult for the vertically integrated firm to capture, even with careful accounting for the rates of production and costs within each division of the firm. This is because conventional fixed wage and salary employees do not bid for jobs based on their specialized knowledge, nor is there any indication when the job is done of how actual costs and potential costs compare because of wide variation in local conditions. By contrast, competitive contract prices reflect expert assessments of the costs of logging a site and the value of the timber. The point here most emphatically is not that mills are unable to measure the costs of their company logging operations. They clearly do so. They also know the value of the timber coming from each site and can therefore perform detailed *ex post* calculations of profit rates for each site and, for that matter, for each logging side. But, translating this knowledge into *ex ante* assessments of logging costs is impeded by natural variability from site to site. In turn, this impedes prediction of how the costs of logging each site will vary, which in turn impedes systematic rationalization of on-site production.

Of course, contractors do not report actual costs to the mills either. However, they do bid for contracts in markets that place downward pressure on logging prices. Mills may therefore rely on market competition as a check on inflated logging prices and as an indirect way of capturing an expert logger's ability to "read nature" as a form of information or knowledge. The combination of this expertise with the incentive to apply expert knowledge to improving on-site efficiencies propels a drive for contract relations in Oregon logging.

The Practice of Contracting: Nontraded Interdependencies and Repeat Contracting

If production contracts help mills offset various risks and uncertainties, induce rationalization, and capture expertise through contract bid markets, the expert knowledge contractors develop can also to some degree be highly specialized to the mix of lands and manufacturing facilities held by particular integrated forest products companies. This leads to information economies of scale between contractors and larger companies. Precisely because of these information economies, many contract relations between wood commodity firms and gyppo contractors do not conform to an ideal type of arms-length transaction; that is, open bidding with the contract awarded strictly according to price. Instead, numerous firms practice repeat contracting.

A significant number of Oregon firms do indeed simply announce jobs and accept the lowest credible bid (e.g., Seneca Forest Products in Eugene). Yet numerous prominent firms do not. Instead, some firms work with a set group of contractors to whom they offer jobs and accept the lowest bid (e.g., Roseboro Forest Products in Eugene, and Roseburg Forest Products of Roseburg).[47] This strategy maintains competitive relations among a subset of gyppos but also encourages familiarity between the parent firm and the contractors, encouraging the development of information economies of scale in their interactions. Other firms retain gyppo loggers on an annual basis and maintain an implicit understanding with contractors that relationships will be renewed (given sufficient work) in the absence of any truly competitive bidding. In these cases, it is very common for job prices to be negotiated between the wood products firm and the gyppo(s). Although such practices deviate from an ideal typical open and blind bidding process, they do so in ways that further reflect attempts to manage information economies in the logging industry that have much to do with its nature-centered character.

In logging, knowledge of a wood-commodity firm's particular needs combined with the degree of trust that accumulates through past performance creates a niche for gyppos with experience working with particular firms. One logger summed up why Willamette Industries—one of the firms with the largest land base in the state—typically works with a small group of contractors with the following anecdote:

> [Willamette has] a loyal commitment. It is good for them. Like this year, they were complaining that their cutting costs were too much. So they got some cutting contractors and they got a cheap rate, so that they could use the leverage on the few cutters they already contracted with. You know "we can get it done for this rate." Well it kind of backfired on them. They ended up with the best job that was available this year and got a cheap rate. Well, they've got special ways that they want their job done. And it turns out that the contractor didn't have enough people to do it, so he hired people and he got lower quality people, people that weren't familiar with the lengths and what they wanted [in terms of] diameters and grades, and they'd send somebody for a couple of days and then somebody else for a few days. So it is a major headache. And [Willamette's] quality control guy has got to keep going back and he's not getting what they want, and here's the best job and the best timber and it's getting butchered out there. You know, it wasn't worth [it].[48]

A specific aspect of specialized knowledge stems from the interaction between variation in the forest and the ways that each wood-products firm chooses to deal with such variation. The issue here is that logs vary in quality not only from site to site but also at a given site. Yet specific knowledge can be accumulated about a particular firm's lands (e.g., tree age, species composition, spacing, location, etc.) and about how the firm wants the lands managed. Compounding geographic variation in the woods, each wood-products company has its own particular way of sorting and processing logs according to its mix of production facilities in the area and the relative strength of the parent firm's margins in its nearby mills. That is, not only does the geographic distribution of each firm's plants vary but some firms profit more from plywood than lumber, others the reverse, and so on. Different kinds of logs (i.e., species, size, wood density, prevalence of knots and other flaws) are generally suitable to different kinds of processing, yet such decisions are to some extent a function of a particular company's mills, specialization, and market position. The specialized knowledge required to sort logs to suit the company develops among employees through extended employment with a firm and gyppos who contract repeatedly with the same company. But manufacturing firms cannot expect the same of gyppos if they are retained through truly arms-length market transactions.

By practicing repeat contracting with known gyppos, in some cases over periods in excess of fifteen to twenty years, firms can help to create and capture specialized knowledge among contractors while maintaining the added flexibility of contracting over wage employment. The merits of this approach to contracting are particularly apparent to large, horizontally and vertically integrated firms seeking to capture economies of scope in log processing. These firms require the most complex and firm-specific information to be assessed by their logging crews. Thus it is telling that the region's two largest and most diversified firms—Weyerhaeuser and Willamette Industries—have been particularly prone to repeat contracting. These firms have the largest landholdings among private forest companies in Oregon, and they also operate a diverse set of wood-processing facilities in the state. For them, repeat contracting in their gyppo logging operations makes sense because the information economies of scale in doing so are much greater than for smaller firms, with little or no land of their own and with only one or two mills to supply.[49]

To Contract or Not to Contract: Weyerhaeuser's Competitive Logging Program

Contract logging is a strategy for achieving flexible production relations motivated in response to logging's nature-centered character, a structural problem for firms to grapple with. Yet contract logging is not a unique solution to the flexibility dilemma; rather, the translation of structural problems arising from ecoregulation into production relations is ultimately contingent, negotiated, and contested within the dynamics of particular contract and labor markets. Indeed, perhaps the best demonstration of the influence of ecoregulation on production relations in logging may be seen in the history of Weyerhaeuser's competitive logging program. Under this program, the company opted to experiment with contracting and restructured wage relations as a strategy for rationalizing its logging operations. The results not only provide a direct indication of how much trouble firms have controlling production costs in logging under fixed hourly wage relations but also substantiate that there is more than one institutional solution in the pursuit of production flexibilities.

One of the problems with fixed hourly wage relations and contracting in the logging sector is that both present obstacles to firms seeking to directly monitor and rationalize production costs in logging. By using production contracts, firms can induce market competition as a surrogate for rationalizing production, but they cannot directly monitor how the prices they are paying for contracts correspond to production costs. Instead, firms depend on the competitiveness of the contracting market to keep logging costs under control. Yet if contractors make substantial profits on a particular contract, the firm has no way of knowing unless the contractor makes the books available to the firm (which is unlikely). If the local market for contractors is not competitive, the firm faces the risk of inflated contract prices. Firms that pursue more open, competitive bidding processes need not worry as much. But firms that prefer repeat contracting with a small set of gyppos exacerbate the risk of inflated contract prices. Because of this, some of the larger firms consciously manipulate contractor markets to make sure that, as one logger put it, there is "always one more local gyppo than they need to do all the logging."[50] However, this requires the exercise of considerable market power and manipulation. On the other hand, fixed hourly wage relations do not allow for direct monitoring and rationalization, for the reasons already discussed. Firms can certainly account for their production costs and returns from a given site after it is logged, but they do not know how closely their costs conform to how cheaply the job could have been done, nor does this necessarily help them predict the costs of future jobs. Thus, wood-processing firms seem caught between intersecting dilemmas.

Weyerhaeuser recognized and directly confronted these trade-offs by instituting a system of competitive bidding among contractors and between contractors and company logging sides. In so doing, the firm restructured its wage relations in logging and thereby challenged an industry-wide perception that contracting is always cheaper than company logging.

Weyerhaeuser's attempt to confront and directly monitor the costs of logging dates to the Northwest wood-products industry's labor dispute of 1986.[51] Using weak commodity markets and the collapse of an industry-wide agreement on wages and benefits to press for concessions in collective bargaining, Weyerhaeuser successfully negotiated a $4 per hour wage concession with its employees and initiated the competitive logging program. Under this program, the company began to set target prices for logging in each of its areas. These prices were offered to gyppos and to company sides based on Weyerhaeuser's estimate of what it should cost to log a site. Company crews were challenged to beat gyppo rates in the same area and to demonstrate that they could compete with the gyppo sides. Existing union contracts fixed the proportion of cutting done by company sides, but the company threatened to replace company sides with more gyppos in the longer term if the gyppos consistently outperformed the company loggers.

The key incentive in the system lay in restructured compensation packages to company employees. In addition to receiving their hourly wages, company loggers were offered performance incentives that guaranteed them any savings generated if they beat the targets. This induced rationalization undertaken by the company loggers, and at the same time allowed the firm to compare its internal bids or estimates with actual performance. What Weyerhaeuser found was that the company sides routinely beat the targets, often by substantial amounts.

Although the company was not saving money on individual contracts, Weyerhaeuser put in place a system that allowed it to better monitor, predict, and ultimately depress its production costs. Lowering production costs took place with expert-based rationalization among company sides. Although employees received the immediate benefits of beating targets, in the long run the company used the information generated to lower costs by improving management's capacity to estimate production costs *ex ante*. At the same time, the firm created more competitive pressures on its network of contractors; critical because the initially inflated targets were a direct indication of gyppo profit margins. That is, bonuses pointed to contractors whose profits were at least equal to, and likely higher, than the bonuses.[52] Weyerhaeuser then used information generated by the competitive logging program to improve its estimates and to ratchet down contract prices set in subsequent years.

As the competitive logging program progressed, company crews began to generate smaller and smaller bonuses, while Weyerhaeuser earned a reputation among Oregon gyppo loggers as a company whose contracts offer little hope of generating a profit. Said one Weyerhaeuser logging employee with knowledge of the program,

> I don't believe when [Weyerhaeuser] went into this that they realized that their [target] prices were that bad. This is how they found out. The first clue was when the crews were making these fantastic bonuses. . . . I am sure that they figured it out that if they were paying the crews these big bonuses, they were paying the gyppos too. . . . I think the company is getting what they want out of this. I am sure that they realize that they have gotten their logging costs down on both company and gyppo sides.[53]

Indeed, the program made Weyerhaeuser sufficiently comfortable with its ability to suppress logging costs that the company signed an agreement in 1997 with the International Association of Machinists and Aerospace Workers (Woodworkers Department) to phase out gyppo logging altogether.[54] Given the prevalence of contracting in the Oregon logging sector, and the fact that Weyerhaeuser is the dominant firm in western North America,[55] this was an extremely significant development. Critically, the agreement was predicated on the company's continued use of performance incentives to induce productivity improvements. Although the competitive logging program indicates that there is more than one way for capital to address the challenges of nature-centered production in the arena of social relations, it also indicates that there is considerable room for workers and their representatives to contest and shape the translation of structural obstacles into production relations, in this case with restructured wage relations as an alternative to independent contracting.

Oregon's Reforestation Sector

> The ten of us plant about 7,000 seedling trees every day, or about 700 "binos" apiece, enough to cover a little over an acre of logged-off mountainside each. . . . Two or three sideways hacking strokes [with the hoedag] scalp a foot-square patch of ground, three or four stabs with the tip and the blade is buried up to the haft. (Six blows 700 times amounts to 4,200 per day. At five pounds each, that comes to 21,000 pounds of lifting per diem, and many planters put in 900 to 1,200 trees per day). . . . The next tree goes in eight feet away from the last one and eight feet from the tree planted by the next man in line. Two steps, you're there. It's a sort of rigorous dance, all day long—scalp, stab, stuff, stomp, and split; scalp, stab, stuff, stomp, and split—every 41 seconds or less, 700 or more times a day. . . . The ground itself is never really clear, even on the most carefully charred reforestation unit. Stumps, old logs, boulders, and brush have to be gone over or through or around with almost every slash hampered step. Two watertight tree bags, about the size of brown paper grocery bags, hang on your hips, rubbing them raw under the weight of thirty to forty pounds of muddy seedlings stuffed inside.[56]

Ecological and Social Production in Tree Planting

The emergence of the reforestation industry in the state of Oregon is tied to the gradual depletion of old-growth forests and the politics of sustained-yield regulation. Oregon was the first state in the nation to require restocking of cutover forestlands, and when initial efforts by capital to comply using aerial seed bombardment failed (the squirrels were thrilled!), manual planting of seedlings took hold as the predominant methods of industrial reforestation (see chapter 5). And from the start, the distinct logistics of tree planting have affected its social organization. Much like logging, these distinct logistics are rooted in highly variable site conditions and an extremely discontinuous labor deployment schedule, all of which generate similar obstacles to the *ex ante* prediction of reforestation costs at each site and to the systematic imposition of labor discipline and rationalized coordination.

The first problem with tree planting is that deployment of labor is discontinuous and distinctly seasonal, much more so than in logging. Seasonality in Oregon tree planting is driven by the fact that most conifers, including Douglas-fir, are dormant during the coldest months of the year. When dormant, seedlings are best able to withstand the shock of being transplanted from nurseries to clear-cuts. Thus, peak reforestation labor demands in Oregon occur from

January to March. Reflecting this unevenness, workers in forestry services (which includes reforestation workers) in Oregon averaged a total of just 12.2 weeks of employment in 1993, with more than half of the industry's 7,600 workers employed for less than 8 weeks in total.[57]

The second problem in tree planting is the problem of space, as manifest in remote and rotating job locations and in the dispersed deployment of workers. Tree planting parallels logging in that production takes place in the forest, at locations that change frequently and are often quite remote. This creates coordination problems between mills and reforestation sites in direct parallel with those encountered in logging. A further parallel is that deployment of labor takes place on rugged terrain under constantly changing climatic and topographic conditions. Compounding this, workers tend to be widely dispersed at the site. Although there is little need for coordination among reforestation workers (unlike in logging), the inconstant geography of planting makes rates of production difficult to predict, whereas dispersal of workers makes productivity difficult to monitor and rationalize. Because trees are typically dispensed from a central location, it is easy to measure how quickly planters are working, yet the quality of planting—critical to reforestation success—is difficult to monitor; in fact, it is difficult to verify that seedlings are planted at all, a problem of considerable significance on the evolution of production relations in the sector.

There are also parallels between logging and reforestation in the ways that these nature-centered production conditions have been translated into production relations. In particular, a preponderance of vertical disintegration involving contract relations between landowners and tree-planning firms has been evident throughout the industry's history. As in logging, so too in reforestation, contracting out allows landowning firms and public forest management agencies to (1) avoid lasting commitments to tree-planting contractors and their workers, (2) pass along to the contractors the risks and uncertainties derived from constantly changing site conditions, and (3) place the onus on contractors to rationalize the labor process.

Yet, in other ways, production relations in reforestation have shown historical patterns quite distinct from those of logging. Repeat contracting has generally been less common in reforestation than in logging. Moreover, reforestation contractors have tended to rely on highly marginalized workers, with one observer characterizing the industry's workforce as "winos . . . and other poor quality workers."[58] This contrasts with the relatively more stable employment conditions in logging and, until recently at least, a level of social esteem afforded to loggers unknown to reforestation workers. Highly exploitive and miserable working conditions largely characterize reforestation, with low pay the norm and with extensive abuses, including overwork, denial of benefits, and, most recently, outright refusals by contractors to pay their workers (see below). Short-term contracts and pronounced seasonal fluctuations in labor demands also have created very fluid conditions, with worker turnover so high that firms have been reported to employ six workers during a season to keep a single planting job filled.[59] The industry is also somewhat infamous for the elaborate schemes devised by contractors to cheat private companies and state agencies. A central factor in all of this is the underlying remoteness and inconstancy of work, where abuses can take place without detection, combined with the generally low priority placed on reforestation, particularly acute during the earliest years of the industry.

Reforestation is also distinct from logging in the preponderance of piece-wage relations that have been the norm in the industry throughout its history. Piece wages reflect the unique circumstances of work in the reforestation sector. In particular, paying workers by the tree induces quick planting in an environment where more direct coercion is difficult, and also allows contractors to pass some of the risks of uncertain rates of production on to the planters. In this respect, piece wages are like microcontracts at the level of individual workers.

Even with these broad tendencies, however, the regime of contract and piece-wage relations typical of reforestation throughout its history in the Douglas-fir region is not without its own set of contradictions that have helped shape some historical shifts in the industry's organization. Specifically, piece wages are meant to induce worker self-discipline, but in reforestation they also lead to quality control problems. Under piece-wage regimes, employees have an incentive to cut corners on planting technique if this speeds up production and allows them to plant more trees. At the most extreme, workers have an incentive to ditch trees. Here again, the problem of monitoring workers is apparent. Piece-rate production is typically verified at the central point of seedling dispersal. Yet geographically extensive deployment of workers makes monitoring difficult for the contractors, whereas remote locations and rugged terrain afford ample opportunities for clandestine disposal of trees. Bunches of seedlings can be quickly discarded into rivers, streams, and lakes; over cliffs; down rocky crevasses; or under logs and stumps. But these practices obviously affect the success of reforestation efforts, or the lack thereof. In short, piece wages set up a trade-off between the pace and costs of tree planting on the one hand and the success of reforestation on the other. The resolution of this contradiction between quality and speed has shaped the social history of the Northwest's reforestation sector.

The Rise of Tree-Planting Cooperatives

Virtually from the outset, ditching trees became prevalent in the industry, so much so that ditching seedlings without being caught became celebrated among workers as a skill in and of itself. Initially, landowners were motivated primarily by regulatory and political pressures to reforest, not by direct incentives to grow trees (with some notable exceptions). Thus, they cared only about the number of acres successfully reforested or, more accurately perhaps, the number of acres that appeared to be reforested under the regulatory gaze. Reflecting this, most early reforestation contracts were actually based on the number of trees planted, not on the number of acres successfully reforested; under such a regime, contractors faced an incentive to inflate their planting totals and to turn a blind eye to ditching, and landowning firms did not intervene as long as they were appearing to meet regulatory obligations.[60] So long as the success of reforestation efforts was largely a chimera constructed from the number of trees apparently planted, contractors and their piece-wage workers were not in conflict with one another. However, with growing evidence of reforestation failures throughout the Northwest during the late 1960s and early 1970s, and mounting regulatory and political pressure on firms and public agencies to improve their results,[61] contractors were compelled to more carefully control their planters. And out of the emerging conflict between contractors and their piece-wage workers, a unique chapter in the region's reforestation sector, and in American labor history more broadly, emerged in the form of the Northwest's tree-planting cooperatives.

The cooperatively owned tree-planting companies were, as their name suggests, collectives formed by tree-planting workers. Specific models differed between cooperatives, and not all workers owned or controlled equal amounts of the operations or shared equally in revenues; still, the tree-planting cooperatives marked a distinct and colorful departure in a wood-products industry generally organized along more traditionally capitalist lines. Although the first of the cooperatives appeared in the late 1960s, they grew to occupy a significant profile in the Northwest during the 1970s (see Table 2.2), and at their peak accounted for as much as 25 percent of the Oregon tree-planting labor force.[62] The biggest and longest lasting of the cooperatives were southwestern Oregon's Green Side Up and the Hoedads out of Eugene.

These cooperatives offered several advantages over their more conventional competitors, advantages rooted in the peculiarities of tree planting. First, by giving every worker a stake in the company on more egalitarian terms, these firms could more easily retain experienced workers from season to season, a problem given the industry's high turnover and increasingly important with more and more emphasis on reforestation success, and thus worker skill and experience. Second, even while retaining piece-rate wages to encourage and reward individual production, the cooperative firms discouraged shirking—ditching or cutting corners—by giving every employee-owner a greater stake in the overall success of the company. Because every worker doubled as a manager, enhanced quality control and monitoring could also take place in the field among workers who might otherwise ignore and even exult in each others'

Table 2.2 Pacific Northwest Tree-Planting Cooperative Membership, 1972 to 1989

Year	Hoedads	All Cooperatives
1972	30	30
1973	125	125
1974	200	200
1975	250	300
1976	300	400
1977	350	500
1978	325	575
1979	300	650
1980	250	675
1981	200	650
1982	150	600
1983	100	500
1984	30	350
1985	25	225
1986	15	70
1987	30	85
1988	30	85
1989	30	85

Source: Mackie, 1990.

cheating. In these ways, tree-planting cooperatives combined individual productivity incentives with dispersed monitoring costs, making each worker responsible to and for each other. As Rick Herson, former president of the Hoedads cooperative stated,

> By being self-employed our labor is much more motivated to [do] quality work than any other segment of the reforestation labor pool. Each person shares in the responsibility of the work. The quality is directly related to the pay and consequently each worker is directly rewarded for top quality work.[63]

These aspects of cooperative organization played well against the backdrop of increasing emphasis on quality—not just quantity—in the reforestation industry. As standard practice among landowners began to switch in the late 1960s and early 1970s from per tree contracts to per acre contracts, independent contractors were squeezed between growing pressure to ensure restocking success and piece-rate employees whose incentive was hardly to ensure quality and whose ditching of trees had become common. Yet any attempt by contractors to police behavior among workers or switch to hourly wages was confronted with the peculiar nature of tree planting.[64] Widespread dissatisfaction with low wages and abusive labor practices in the industry also created strong incentives among the most experienced and skilled tree planters to work cooperatively, drawing some of the industry's top workers to the cooperatives. Thus, despite efforts by conventional contractors to drive the cooperatives out of business, they flourished. Mackie reported, for example, that Hoedads had a reputation for providing the highest quality work in the industry and at the same time could pay members on the order of $25 per hour and up to $25,000 dollars per year in 1980, well above industry standards, and it was on the way to converting tree planting into a form of living employment.[65] Rick Herson of Hoedads, in testimony before a 1977 congressional committee investigating reforestation failures, presciently noted the significance of the tree-planting cooperatives in precisely these terms; that is, as a way to generate living wages in reforestation and to create from tree planting a form of sustainable employment, while also noting,

> Co-ops and partnerships are stabilizing an industry that does not have a union to protect its workers from exploitation. This greater stability leads to a better understanding of silvicultural techniques and just what it takes to make a tree grow. A planter armed with his knowledge becomes a skilled technician, not just a stoop laborer. I have planted 250,000 trees in 5 years of tree planting.[66]

Synergies in the Agricultural and Reforestation Labor Markets

Unfortunately, the promise of living wages in the Northwest tree-planting sector introduced by the ability of the cooperatives to reconcile the conflicting imperatives of speed and quality was undone by the emergence of a new phase. Under this new labor market regime, price-based competition in reforestation contracts was reasserted by means of the exploitation of highly vulnerable and marginalized farmworkers. At the height of the cooperative period in 1979, approximately 90 percent of tree-planting laborers were U.S. citizens.[67] Yet, by this time, evidence of a new labor supply system in the region's reforestation industry was already becoming apparent, as contractors increasingly turned to migrant, predominantly Hispanic, farmworkers. Many of these workers were already in Oregon and other parts of the region to work on farms, including, for example, in the fruit orchards of southwest Oregon and in the rich lands of the Willamette Valley. Others were available for import courtesy of "coyotes" whose existing labor supply networks stretched throughout the region and extended back to Mexico and central America. Capitalizing on the offsetting seasonal peaks of labor demand in Pacific Slope agriculture and tree planting, private, noncooperative reforestation contractors were able to reestablish control of the industry by drawing on a highly marginalized labor force to compete with the cooperatives using cheap labor all on the backs of vulnerable workers.

The exact origins of the use of farmworkers in Northwest tree planting are not easy to identify. Details of this history are not readily available, because of the outright illegality of many of the practices characterizing this labor supply and management system and the extremely coercive conditions under which many of the workers toil; getting anyone to talk about it is difficult. However, in 1974 a Medford, Oregon, reforestation contractor named Robert Felix Gonzalez became the first person in the state to be convicted of employing nondocumented Mexican workers.[68] Moreover, Immigration and Naturalization Service (INS) records show an increase in arrests of undocumented workers in the Oregon reforestation industry from 112 in 1978 to 249 in 1979, without any increase in staffing resources devoted to the sector. By 1984 the proportion of the Northwest reforestation workforce who were permanent U.S. residents had, according to one knowledgeable observer, dropped to an estimated 30 percent.[69] And by 1993 an estimated one-half of the industry's workers were undocumented and in the country illegally.[70] The increasing use of undocumented workers in reforestation during the late 1970s is further indicated (albeit indirectly) by a suspicious drop in prevailing contract costs. Estimates of reforestation contract costs in Oregon by the Bureau of Land Management during 1979 to 1980 showed a systematic tendency to exceed actual costs; agency estimates exceeded actual low bids that year by an average of 23 percent (see Table 2.3). With its high labor intensity, about 80 percent of total reforestation costs are for labor; consequently, the tendency to overestimate costs is very likely a product of plunging wage rates. And these plunging wage rates were tied to increasing use of migrant farmworkers by reforestation contractors.

Some industry observers feel that the exploitation of undocumented Hispanic farmworkers in the Oregon reforestation sector was aided, albeit unwittingly, when the Carter administration ordered the INS to stop pursuing such workers in order to obtain a more accurate census count.[71] However, it is clear that the supply of these workers to reforestation contractors, and indeed the prevailing pattern of labor relations and abuses, is tied closely to the political geography of agricultural labor markets in Oregon, and the U.S. West more generally. Two principal threads connect what might superficially appear to be separate labor markets. First, the heavy reliance of the tree-planting industry on Hispanic workers, documented and undocumented, relies on established populations of Hispanic workers drawn to the state to work in agriculture. In particular, this labor supply relies on established networks of Hispanic labor recruitment delivering documented and undocumented workers for tree planting capitalizing on the offsetting seasonal peak labor demands in the respective industries. Second, the systematic abuses and coercive labor relations that came to pervade Oregon's tree-planting industry drew directly on practices in the state's agriculture sector, also relying on established networks and practices of labor supply and control.

Table 2.3 Estimated and Actual Reforestation Costs on Bureau of Land Management Land for Fiscal Year 1980

Contract #	Low Bid	BLM Estimate	Difference	Percent Error
43	$70,701	$109,236	$38,535	35.28
44	101,090	132,810	31,720	23.88
45	125,919	161,842	35,923	22.20
46	27,510	33,110	5,600	16.91
47	101,554	132,250	30,696	23.21
48	64,000	69,600	5,600	8.05
49	62,460	64,800	2,340	3.61
50	10,428	19,175	8,747	45.62
52	10,416	20,253	9,837	48.57
53	37,888	39,907	2,019	5.06
55	67,712	78,752	11,040	14.02
56	3,526	5,117	1,591	31.09
57	19,800	34,200	14,400	42.11
58	4,836	14,119	9,283	65.75
59	17,100	19,746	2,646	13.40
62	35,000	41,660	6,660	15.99
63	19,575	28,120	8,545	30.39
64	25,200	29,162	3,962	13.59
65	30,400	33,328	2,928	8.79
69	32,000	34,360	2,360	6.87
71	21,731	40,300	18,569	46.08
72	31,360	36,052	4,692	13.01
73	25,700	30,590	4,890	15.99
75	112,970	117,554	4,584	3.90
76	24,800	34,200	9,400	27.49
79	35,866	42,336	6,470	15.28
81	64,986	71,319	6,333	8.88
83	84,730	89,250	4,520	5.06
84	7,830	9,370	1,540	16.44
86	19,740	44,616	24,876	55.44
90	9,600	13,817	4,217	30.52
91	54,280	57,000	2,720	4.77
94	9,585	17,946	8,361	47.59
115	20,625	26,749	6,124	22.89
117	9,150	10,559	1,409	13.34
Totals	$1,400,068	$1,743,205	$343,137	
Average Error				22.87

Source: U.S. Congress House Committee on Agriculture, 1981.

Oregon's reliance on and supply of migrant agricultural labor is ultimately tied to U.S. programs designed to deliver a low wage, flexible (i.e., nonunion, vulnerable) workforce to American agribusiness. It was the *bracero* programs developed jointly by the Mexican and U.S. governments in 1942—in the context of wartime labor shortages—that created an officially sanctioned and orchestrated flow of Mexican agricultural workers northward. Under the bracero program, Mexican workers were officially licensed and allowed to work at specific locations under contract to growers in various industries. When the contracts expired, bracero workers were required to return to Mexico. Although the program gave an official sanction to the importation of Mexican workers, it also very directly influenced the flow of illegal or nonsanctioned workers into the United States, as agroindustry recognized the advantages of drawing on such a vulnerable workforce. Thus, although nearly four million braceros were recorded entering U.S. agriculture between 1942 and 1960, they were accompanied by millions more undocumented Mexican workers.[72] And although the bracero program officially ended in 1964, networks in a new geography of labor supply were by this time well established and able to draw workers in without the official assistance of the U.S. government. Thus, for example, Miriam Wells estimated that noncitizen immigrants comprised 80 percent of the California agricultural labor supply from the mid-1960s through the late 1980s, with about half the total workforce undocumented.[73]

The bracero program initially served California growers,[74] and the most popular and scholarly focus on the issue of migrant farmworkers has been directed at that state. Yet migrant Hispanic farmworkers have long been important in Oregon's agriculture sector. Numerous state studies conducted during the late 1950s and in the 1960s point to the important role of a generally migratory labor pool for the state's seasonal agricultural labor demands.[75] Out-of-state migrants comprised an estimated 20 to 30 percent of total seasonal labor in Oregon agriculture between 1958 and 1968,[76] and Hispanic workers formed a sizable contingent of the interstate workforce. According to the Oregon Bureau of Labor, roughly 25 percent of the total migrant workforce in Oregon in the late 1950s was Hispanic, and about 30 percent of this group was composed of Mexican citizens.[77]

The supply of migrant, and predominantly Mexican, Hispanic farmworkers in Oregon, as in many of the states of the U.S. South and West, was tied closely to networks of contractors and subcontractors who acted as intermediaries between growers and the workers. The contractors and subcontractors managed the supply networks with personal connections and also were able to exploit the vulnerability of foreign workers—many of whom spoke little English—to keep wages low while profiting from supply contracts signed with growers. In their turn, growers were largely pleased to have cheap help, and they turned a blind eye to abusive and coercive practices. At a broader level, the exploitation of farmworkers was considerably enabled by their exemption from protections (including the right to organize) under the National Labor Relations Act.[78] As a result of this structure of exploitation, inadequate conditions typically prevailed where the migrants worked, including poor quality housing, the lack of potable water, and the lack of sanitary facilities.[79]

There have been efforts to curb the worst of farmworker abuses in Oregon and in other agricultural states.[80] Moreover, large year-round populations of Hispanic farmworkers are now established in Oregon, including for instance, in the town of Woodbridge, south of Portland. More permanent settlements and worker and migrant organizing for assistance and social service provisioning have improved conditions considerably. Yet it is also true that although Hispanic farmworkers continue to form a crucial contingent of the state's seasonal farm labor market, they also remain among Oregon's most exploited workers. According to the region's farmworker union (Pineros y Campesinos Unidos del Noroeste), annual average income for Oregon farmworkers stands at $6,500, unchanged in more than a decade. Moreover, violations of minimum wage standards are still routine in prevailing piece-wage systems. Health care benefits are rare, housing for many workers and their families is poor, nutrition is often inadequate, and many workers are exposed to high levels of harmful agricultural chemicals. These problems are only compounded by the fact that the rights of Oregon farmworkers to organize remain unprotected.[81]

The salience of the Oregon farmworker story to the state's reforestation sector is that parallel patterns of labor supply contracting have developed in reforestation, capitalizing on the offsetting seasonal peak demands for labor in agriculture and reforestation. Moreover, some of the same tendencies toward coercive labor practices are apparent in reforestation work. Labor contractors, at least some of whom seem to have gained experience contracting for the supply of migrant farmworkers, have used and extended existing networks of labor supply to deliver Hispanic workers to reforestation jobs. Some of these contractors have participated directly in securing reforestation contracts on either federal or private forestland. But, more typically, separate labor contractors and subcontractors seem to work in tandem with tree-planting contractors. Although the latter secures the contracts with landowners (including the Forest Service and the Bureau of Land Management), the former takes responsibility for labor recruitment and supply. Undocumented workers, preferred by contractors because of their extreme vulnerability, are often brought directly from Mexico or are secured through populations of Hispanic farmworkers already in the region.

Because so little analysis of labor relations in the tree-planting sector has been done, there exists little documentation of the abuses. Yet anecdotal evidence points to exploitation at least as widespread and severe as what has been reported in connection with agriculture.[82] Because of the lower profile of reforestation (and thus less attention to the problem of abusive labor practices) combined with the remoteness of work sites and the winter conditions that often prevail, conditions in reforestation might actually be worse. Contracts between workers and their employers are almost uniformly verbal, with wages set on a piece-rate basis. Although federal and state minimum wage requirements stipulate that piece rates must result in the hourly average wage meeting or exceeding the minimum, this often is violated. Some contractors have reportedly required workers to put in ten- to fourteen-hour days, at piece rates that fall short of minimum wage levels and with no extra pay for overtime. Moreover, from such meager wages, the contractor frequently deducts the costs of transportation and food. Although it is illegal for contractors to profit from such deductions, it is evident that they do so by means of charging exorbitant rates. Moreover, workers often are housed and even forcibly detained at remote planting sites for the duration of the contract, living in what can be substandard shelters (sometimes even tents) without proper heat during winter months in the mountains. Potable water and proper sanitation are typically lacking. Despite these conditions, housing costs also may be deducted. At the end of the contract, or perhaps at the end of a two-week pay period, contractors have been reported to systematically violate the terms of their verbal contracts by underpaying workers and by going so far as to report undocumented workers to the INS. The INS then deports the workers, and the contractor can thus avoid payment altogether.

During an investigation into the use of undocumented workers in Oregon's tree-planting sector in 1980, Lane County commissioner Jerry Rust characterized the exploitation of undocumented workers as follows:

> They are often arrested and deported before they are paid, to the benefit of the contractor. They are charged exorbitantly for their travel, for their room and board. Oftentimes they find themselves owing their souls to the company store. They have no health or accident insurance and no unemployment benefits. It is even questionable whether the taxes they pay are being turned over to the State and Federal taxing authorities. Since no social security numbers exist for undocumented workers, this abuse is likely to be widespread. To whom could they tell their story? . . . I believe there to be an efficient, well-organized system for bringing undocumented laborers into the Northwest woods to perform contract work. Many undocumented workers have been arrested and deported time and time again.[83]

In the same session, David Audet, of the Oregon Legal Services Corporation (formed to handle legal issues for seasonal and migrant farmworkers) stated the following:

They are paid by piecework. Oftentimes, though, my clients have just received an absolutely unrelated sum at the end of the job. This is because of what we call improper deductions. The contractor will promise them, let's say $800. Well, oftentimes the deductions which we are talking about are the gas and oil for the chain-saws, money for rides, money for food, oftentimes money for transportation from Mexico. One of the worst abuses is when the contractors use the person known as the "'coyote,'" who is the person that actually transports the people. This person will bring them up from Mexico or from California and hold them there until they have paid off what they owe them.[84]

Owners of private forestland, including large companies such as Weyerhaeuser and Georgia Pacific, have unquestionably made extensive use of reforestation contractors who in turn employ and exploit undocumented workers. But perhaps what is most striking and appalling about the degree of coercion in the industry is the complicity of federal agencies, including the Bureau of Land Management and the Forest Service. At least since the late 1970s, there have been indications that Forest Service and Bureau of Land Management reforestation contracts were being granted to contractors who employed undocumented workers. Both the Bureau of Land Management and the Forest Service were required until very recently to award contracts to the lowest bidder in blind bidding processes. The reforestation contracts offered by these agencies were therefore susceptible to unscrupulous contractors who could win competitive bids by exploiting undocumented workers and thus suppressing their labor costs. But according to numerous observers, neither agency has been particularly concerned with stopping the exploitation of undocumented reforestation workers. Instead, the agencies tended to ignore labor relations in reforestation as long as the trees got planted. Commenting on the tacit approval given to extensive abuse of workers in the industry, one official noted, "It appears to me that the Federal Government itself through its contracts is quite possibly the largest single employer of undocumented workers in the Northwest."[85]

By all accounts, the INS has become more diligent and even-handed in pursuing this issue, tracking down not just undocumented workers but also contractors who thrive on exploiting and even cheating these workers.[86] However, a system of labor supply has been established in reforestation, one that relies on poorly paid, vulnerable, and highly exploited workers whose work in the woods is facilitated by at least sixty years of administrative ambivalence to the plight of migrant farmworkers. Until more concerted efforts to organize these workers are facilitated by more exacting labor standards applied to reforestation contractors and the public agencies and private forest products companies who retain them, and until systematic changes put a final end to the hypocrisy that allows immigration law to be used as a tool to exploit workers, these practices are likely to remain a staple in reforestation work. Moreover, the dream of living wages and sustainable employment for all reforestation workers, hinted at by the tree-planting cooperatives, will remain just that: a dream.

Conclusion

In this chapter, I argued that the intimate confrontation between natural and social production in logging and reforestation constrains the continuous deployment of labor and capital and the rationalization of production in these sectors. The resulting pressure on firms to displace the costs of discontinuous and unpredictable output propels development of social relations with a "different complexion."[87] Thus, in logging and reforestation, more flexible production relations have developed as responses to ecoregulation; these include restructured wage relations, quasi–arms-length contracting, and piece-rate wages. Production contracts allow firms to pass on risks, and restructure incentives in a production environment where ensuring continuous throughput and labor process rationalization are difficult. At the same time, in logging, repeat contracting between larger, more complex firms and a limited set of gyppos is a relatively common practice, allowing firms to capture expertise gained through familiarity among experienced gyppos and to capitalize on trust that develops over time through repeat contracting. In this manner, ecoregulation not only propels flexibility in logging but also underlies the particular manifestation of Storper's untraded interdependencies, linking firms and gyppos in repeat contracting networks. These same untraded interdependencies underlie Weyerhaeuser's competitive logging program, a different institutional solution involving restructured, incentive-laden wage relations between the firm and company logging sides.

Reforestation is characterized not only by the prevalence of contracting but also by the absence of repeat contracting and the use of piece-rate wages. Although the ecoregulatory rationale for pervasive contract relations of production in the reforestation sector is broadly parallel to logging—including inconstant geographies, unpredictable rates of production, challenges to labor control and rationalization—reforestation exhibits its own particular forms of flexibility, with generally more price-based competition for contracting and less repeat contracting. Moreover, piece wages, common in reforestation but not in Oregon logging, may be understood as microproduction contracts, passing on the risks and uncertainties associated with variable production conditions to reforestation workers. As in logging, so too in reforestation, these broad structural generalities linking contract relations of production to the problems of ecoregulation does not mean that there is only one unique solution to the ecoregulatory dilemma. Instead, at different points in the history of the sector, reforestation cooperatives and highly exploitative employment of migrant farmworkers represent distinct institutional strategies, with clear normative differences between them.

Thus, although I emphasize the salience of ecoregulation and the problematic conjuncture of social and ecological production in the labor process as an influence on production relations in both sectors, there is at the same time no unique or scripted solution to the problems posed by ecoregulation. Social relations cannot simply be read off of the structural constraints posed by nature-based production. This further underscores that any notion of ecological crisis tendencies in the production of commodities from nature and the commodification of nature must embrace the structural dimensions of these tendencies and also the politically, culturally, historically, and geographically contingent ways in which they are addressed.

3
Industrial Ecologies
and Regional Geographies

> In the forest products industries, new technologies have offset changes in
> species availability and growing stock, averting severe dislocations and
> significant price increases. Technological progress has thus helped keep
> the "Malthusian" limit on forest resources at bay.[1]

Converting raw material inputs into commodities is constrained by the character
of resource industries. Because these activities rely centrally on biophysical processes for their inputs, their industrial geographies and technologies are influenced by the timing, location, and quality of raw materials. These constraints
are neither invariant over time nor completely resistant to concerted efforts to
overcome them. But they are very real influences, despite postindustrial narratives suggesting the conquest of nature's limits.[2] This is most apparent in extractive industries such as mining in which capital must constantly confront an
exogenous geography of raw material availability.[3] Yet even in renewable resource
sectors, issues such as extensive land requirements and regional differences in
natural rates of productivity exert significant influences on the organization of
capitalist production. Institutional issues (e.g., property rights, taxation), environmental politics, technological change, and proximity to markets are also
important factors shaping industry geographies, yet natural resource processing
and conversion must still confront the basic realities of raw material dependence.[4] Converting timber into wood commodities is no exception.

In the following two chapters, intended as complements to one another, I
explore an "industrial ecology" of solid wood commodity manufacture in the
Douglas-fir region. By industrial ecology, I mean the relationships between
industry geography and organization on the one hand and biophysical nature—
including not only raw material characteristics but also environmental impacts

of various kinds—on the other.[5] In the next chapter I deal specifically with links between raw material dependence and economies of scale and scope, particularly in the lumber sector; in this chapter I examine more broadly the shifting relationships between regional specialization, timber supply, and commodity systems, linking geography, technology, firm strategies, and environmental impacts with the changing availability (in time and space) of raw materials.

Because of its distinct, and initially old-growth inventory of forest resources, for much of its history the Douglas-fir forest industry has occupied a distinct position within the nation's wood-products sector, characterized by specialization in solid wood-products manufacture. However, with the industrial liquidation of old-growth forests and their replacement by young-growth, this regional specialization has shifted. Continued regional specialization in solid wood commodities has drawn on a suite of product and process innovations whose general orientation is toward greater and greater degrees of log reduction and wood reassembly or reconstitution in the manufacture of new composite panel and lumber products. These products, the first of which was plywood, make use of a wider range of species and smaller logs and feature greater degrees of wood reduction—right down to individual fibers—in production processes.

These new reconstituted products and processes are a diverse group, yet together they may be seen as a response by capital to ecological crisis tendencies manifest as the limited availability in time and space of high-quality, old-growth timber once typical of the Douglas-fir region. Providing renewed opportunities for capital accumulation through the manufacture of new commodities, requiring significant energy inputs (to fuel high temperature and pressure processes) and industrially manufactured adhesives to bind wood fractions together, the new reconstituted products invoke the idea of the industrial appropriation of nature, proposed by David Goodman, Bernardo Sorj, and John Wilkinson as a way of explaining broad tendencies in industrial agriculture. There are limits to the parallels that may be drawn between forestry and agriculture in so much as Goodman et al. focus considerably on the displacement or reconfiguration not only of ecological processes and problems but also of farm labor and petty commodity producers by means of new avenues of industrial manufacture.[6] At the same time, their recognition that the dynamics of industrial agriculture revolve around a struggle by capital—never complete—to offset and supplant farm-based production (particularly specific ecological processes) using a variety of new products and processes resonates strongly with the ways in which industrial solid wood-products production has shifted over time in response to the changing availability of wood fiber. I adopt the appropriation idea here as a way to capture the dynamic relationship between capital and biophysical nature in the manufacture of specific commodity lines, transformed over time by ways in which capital has sought to reduce the constraints of a limited and heterogeneous timber supply through manufacture of increasingly reconstituted wood products. Although appropriation refers to this dynamic "industrial ecology" linking raw material characteristics with firm strategies and industrial geographies, it is important to stress that by no means is any final victory over nature implied (if such a notion makes any sense to begin with). Indeed, reconstituted wood products still rely on the availability of wood fiber and they are also accompanied by a set of environmental and human health risks tied largely to the use of synthetic organic adhesives for binding wood fractions together. I close this chapter with a brief survey of these concerns.

Regional Specialization

The Douglas-fir region's position within the historical geography of the American forest industry, and within the nation's lumber sector in particular, may be understood in some measure as a last frontier for capital in a westward march in search of new raw materials. The now familiar tale of an initially extractive industry moving from region to region and leaving denuded forests in its wake is one I examine in greater detail in the next chapter in reflecting on lumbering's historically extractive and migratory character and its significance to the development of economies of scale.[7] This migratory, extractive dynamic is also crucial context for the politics of sustained-yield regulation, a topic of the sixth chapter. Yet what sometimes garners less attention in analyses of the historical geography of the U.S. forest industry is the degree to which regional specialization in particular commodity lines revolves around spatial differentiation in forest types. This regional specialization has been and remains central to the Douglas-fir industry, where capital has long focused on the production of solid wood products and, in particular, structural wood commodities.

To understand the Douglas-fir region's unique place in the U.S. forest sector more generally, one must appreciate first how distinct the region's forests are. Current-day controversy over the extent (historical and contemporary) and character of old-growth forests[8] should not obscure the fact that at the turn of the century, the Pacific Northwest contained massive quantities of standing timber in amounts the likes of which no one had seen before, featuring larger trees and more densely packed stands than had been previously witnessed. The 1913 U.S. Bureau of Corporations report on the lumber industry, among the first official assessments of the state of the industry and its resource base, captured the qualitatively different character of Douglas-fir region forests as follows:

> On first going from the Lake States to the Pacific coast, cruisers made estimates far below the truth, because *stands per acre* were so enormous that men accustomed to eastern stands could not grasp or accept them.[9] (emphasis added)

The report's inventory substantiated these impressions, indicating that Pacific Northwest[10] forests averaged roughly thirty-two thousand board feet per acre. Along the 120- to 150-mile stretch of the Douglas-fir region, stands were even more densely packed, averaging on the order of seventy-five thousand board feet of timber per acre. By comparison, the average stand density of forests in the nation was estimated at about ten thousand board feet per acre, while no single state outside the Pacific Northwest averaged more than eleven thousand board feet per acre. Quite simply, the Douglas-fir forests were a timber bonanza for capital.[11]

To be sure, this contrast was in some measure a product of the old-growth character of Douglas-fir region forests and the cutover state of other forested regions. The accumulated product of centuries of growth, Northwest forests stood in marked contrast with those of other regions, exhausted by industrial extraction; in fact, the Pacific Northwest likely contained more than half of the nation's total standing timber at the turn of the twentieth century.[12] Yet it is important not to overstate the significance of an old-growth effect. It bears noting that other regions too had old-growth before the lumber industry got to them; thus, cruisers would presumably have seen large trees before. Yet, as the lumber industry report noted, experienced timber cruisers from other regions simply could not believe their eyes.

Moreover, despite a project of industrial liquidation now more than a century old in the Northwest, the difference of the region's forests remains quite evident.[13] Some sense of the enduring significance of Pacific Northwest forests as a distinct kind of timber resource in comparison to other forested regions of the United States (as well as the scale of forest transformation entailed by industrial extraction in the Northwest) is shown in Figures 3.1 and 3.2. The first

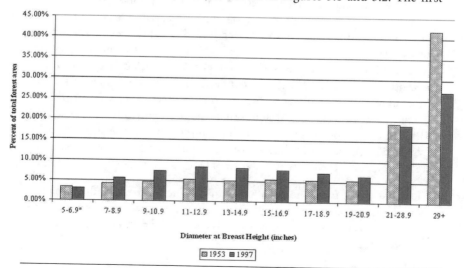

Fig. 3.1 Pacific Northwest Timber Inventory by Size Class, 1953 and 1997. Source: Simth et al., 2001.

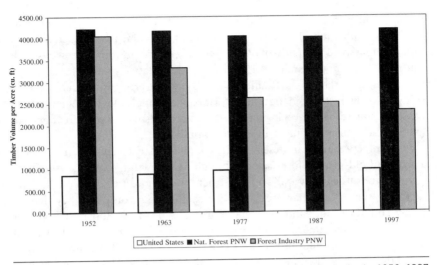

Fig. 3.2 Density of Softwood Timber on U.S. and Pacific Northwest Timberlands, 1952–1997. Source: Smith et al., 2001.

figure depicts the distribution of Pacific Northwest timber across various size (diameter) classes, tracking data from 1953 to 1997 using the Forest Service's periodic inventory of American forest resources. As the figure shows, the incidence of timber in the largest size classification—with a diameter at breast height (dbh) of twenty-nine inches or more—was reduced in the Northwest from 42 percent of all trees to 27 percent over this period as a result of conversion from old-growth to young-growth. Yet the latter figure compares with negligible amounts of timber in this size category elsewhere in the nation, including slightly more than 1 percent of the timber inventory in the Northeast and even less in the American South. In Douglas-fir stands, the contrast is even more stark, with approximately one-third of all trees in this largest rated size category as of the most recent inventory.

Figure 3.2 compares average softwood timber densities on U.S. timberlands— as measured by the volume per unit area—with comparable figures on private, industrial timberlands and National Forest timberlands in the Pacific Northwest for the years 1952, 1963, 1977, 1987, and 1997 (years in which the data are available from Forest Service inventories). Here too it is evident first that forests of the region have been dramatically remade during the sustained-yield era subsequent to World War II by industrial extraction and (particularly on private timberlands) industrial management practices with rotation periods far too short to regenerate old-growth. In short, these data are indicative of a rather dramatic example of the production of materially "new natures" by capital.[14] Yet what is also evident from the data is how densely stocked the region's forests remain in relation to U.S. timberlands; Pacific Northwest National Forests averaged about 4,200 cubic feet per acre in 1997, whereas private timberlands in the region averaged about 2,300 cubic feet per acre. By comparison, the overall average for U.S. timberlands in 1997 was about 960 cubic feet per acre.[15] Thus, despite a massive conversion of old-growth inventories to young-growth stands during the last approximately 150 years, and a relatively small total area of industrial timberlands, the Pacific Coast region holds roughly 45 percent of all softwood growing stock in the nation.[16] Moreover, Douglas-fir, the region's staple tree and source of the region's namesake, accounts on its own for approximately one-fifth of the nation's softwood timber.[17]

It bears repeating that these measures of regional distinction are not merely a legacy of old-growth forests, even though old-growth has been and remains significant. Rather, they indicate a distinct type of forest resource, with commercially significant characteristics. For example, not only are they densely stocked but the timberlands of western Oregon and Washington are also among the most productive softwood timberlands in the country (see Table 3.1). As noted in the introduction, Pacific Northwest (Oregon and Washington) timberlands are generally much more productive in terms of their annual growth rates than are the timberlands of any other region with the exception of the tiny belt of redwood forests along the northern California coast.[18]

At the same time, the size of trees being cultivated in the Douglas-fir region, though considerably reduced from those typical of old-growth stands, is quite large by contemporary plantation standards. Even on the region's private industrial lands, where old-growth timber was largely exhausted by the 1970s, management practices are reproducing relatively large trees by growing Douglas-fir and other important conifers on sixty- to eighty-year rotations on sites with exceptionally high rates of productivity. This is an extremely long rotation age, contrasting sharply, for example, with the twenty-five- to thirty-year rotations

Table 3.1 Forest Productivity Classification by Region, United States, 1997

Region	Productivity Class (cubic feet per acre per year)						
	Total	120+	85–119	50–84	20–49	0–19	Reserve Land
United States	746,958	75,778	105,074	172,853	149,959	191,410	51,883
Northeast	85484	3546	9893	23105	42379	1598	4963
North Central	84,842	4,211	16,186	31,895	28,218	1,470	2,862
Southeast	88,662	3,291	20,270	50,194	11,049	1,160	2,699
South Central	125,438	32,038	35,893	33,138	15,126	8,040	1,202
Great Plains	4,798	87	458	1,109	2,664	409	71
Intermountain	138,447	3,494	7,941	20,332	34,934	53,637	10,108
Alaska	127,380	2,109	1,202	761	8,323	105,148	9,836
Pacific Northwest	51,612	20,680	8,003	7,709	4,775	4,469	5,977
Pacific Southwest	40,296	6,323	5,227	4,611	2,490	15,480	6,164

Source: W. B. Smith et al., (2001).

used in the most productive loblolly pine plantations of the Southeast. These management practices comprise the social (re)production of a distinct timber resource, but they reflect also the distinctive growth dynamics of Douglas-fir, a species that adds volume comparatively late in life in relation to other conifers. Because of this distinct growth pattern, Douglas-fir reaches what is called its "culmination age"—the point at which the tree's average annual growth rate is maximized—later than other species. To obtain the maximum wood volume from plantations, companies tend to harvest it at this later culmination age.[19]

Its distinctive growth pattern is one of the characteristics of the Douglas-fir that make it a tree particularly if not uniquely well suited to producing sawtimber and solid wood products, including products intended for structural applications.[20] The fact that Douglas-fir logs tend to be larger on average than other logs, in part not only because of the old-growth effect but also because of the species' late culmination age, is important. "Size matters" in the production of solid wood products because bigger trees tend to yield more merchantable volume of structural wood per log, in absolute and relative terms. Indeed, this is to some degree endemic to any species grown on a longer rotation, because older trees add more dense wood as they grow than do young trees; this suits the production of commodities that require dense wood, primarily solid wood products.[21] Also, larger diameter logs allow for the production of larger dimension, single pieces (beams, etc.) than smaller logs, and there is a market premium for such pieces because of the rarity of the largest logs. Moreover, there is also a volume bonus to using larger logs for solid wood-products manufacture because as log diameter increases over the range from five to approximately thirty inches (dbh), the proportion of merchantable volume for structural wood increases; that is, the ratio of structural wood volume to the total log volume goes up (see Figure 3.3).[22] This is essentially a biologically driven economy of scale in solid wood-products manufacture. Because solid wood products usually offer a higher return per volume of wood than do pulp and paper products, sawtimber stumpage is typically priced at two to four times the price of pulpwood stumpage.[23] Again, although these relationships apply to any species of large-diameter timber suitable for solid wood-products manufacture, Douglas-fir has a particular association with solid wood products (1) because of the old-growth effect and (2) because of its distinct growth dynamics and late culmination age in plantation settings.

Yet log size—whether from old-growth or plantation settings—is not the only reason why the Douglas-fir makes good structural wood. Douglas-fir also grows in relatively dense stands in which trees tend to self-prune along their lower trunks as they age, reducing the number of branches in the lower reaches of the tree. As a result, larger and older Douglas-fir—particularly old-growth but also to a lesser extent mature, plantation-grown young-growth—produces more clear, straight-grained wood than other types of trees, with fewer of the

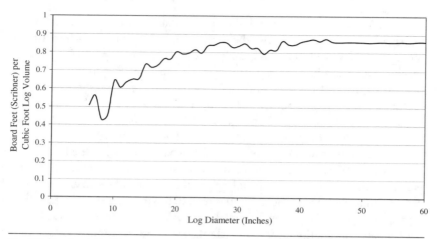

Fig. 3.3 Merchantable Volume versus Log Diameter, 20 Foot Log Standard. Source: Data taken from the U.S. Forest Service National Forest Log Scaling Handbook, FSH 2409.11, published September 1973, Washington, DC.

knots that result from branches. This kind of wood is prized for engineering and aesthetic reasons and produces higher grades of solid wood products.[24] Finally, with a high specific gravity and a relatively long fiber, wood from Douglas-fir is exceptionally strong relative to other conifers—particularly across the grain—yet is also highly flexible; this combination of strength and flexibility is rare and critically important, allowing the wood to flex and not break when bearing loads.[25]

The significance of these distinct resource characteristics is critical in understanding the Douglas-fir region's pattern of specialization in the manufacture of solid wood products—past and present. Consider Oregon first. As one of the most richly forested states in the nation, Oregon has long been a giant in the nation's wood-products sector and has consistently ranked among the leaders in softwood timber harvest. Despite reductions in the wake of the spotted owl crisis, Oregon still ranked third in softwood timber removals in 1997.[26] Moreover, Douglas-fir is by far the most important commercial softwood species harvested, accounting for about 70 percent of the state's timber harvest by volume. In western Oregon in particular, where the incidence of the species is concentrated, Douglas-fir accounts for closer to three-quarters of the total harvest by volume (see Figure 3.4).[27] Given the significance of the state's forest resources and timber harvest, it should not be surprising that Oregon would be a leading wood commodity producer. What is notable, however, is the degree to which the state's forest industry specializes in solid wood-products manufacture in particular. Consider, for instance, that Oregon has consistently been a leading producer of lumber during the post–World War II period (see Figure

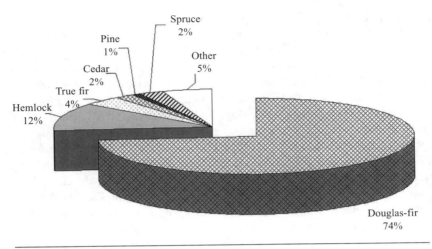

Fig. 3.4 Timber Harvest in Western Oregon by Species, 1995. Source: Oregon Department of Forestry, 1996.

3.5). Moreover, in 1994 sawmills and plywood mills in Oregon consumed a combined 91 percent of all of the state's roundwood or whole logs, whereas the pulp, paper, and paperboard industry accounted for only 2 percent of log consumption.[28] Although Oregon does have a substantial paper sector, with eighteen paper mills and twenty-two pulp mills in the state in 1995,[29] these massive, capital-intensive facilities are ultimately considered secondary to solid wood-products mills in their demands for fiber. That is, instead of using whole logs,

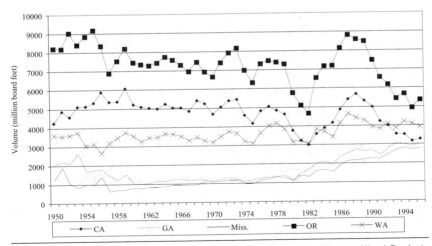

Fig. 3.5 Lumber Production in Leading States, 1950–1996. Source: Western Wood Products Association, 1998.

Oregon pulp and paper mills rely primarily on residual chips produced as a by-product of solid wood-products manufacture, obtaining on the order of 95 percent of their fiber supply in this manner.[30]

This bias toward solid wood-products manufacture is a regionwide phenomenon. The Pacific Northwest, for example, has consistently been a leading producer of lumber in the United States in disproportion to its timber harvest. Although the region's share of national lumber production has declined from a peak of more than 60 percent in the early 1950s, it remains higher than 40 percent.[31] Moreover, as shown in Figure 3.6, whole-log flow in the Pacific Northwest goes primarily to solid wood-products manufacture. Using data from 2001, more than 70 percent of all softwood logs in the Pacific Northwest in 2001 went to sawmills for lumber manufacture. Together, sawlogs and veneer logs (used for manufacturing plywood) accounted for 88 percent of all whole logs used. By contrast, sawlogs and veneer logs comprised about 57 percent of the total in the American South and 64 percent in the nation overall. In the Pacific Northwest only 2.2 percent of all logs went directly for pulpwood, whereas slightly less than 40 percent of softwood log flow in the South and 27 percent of the nation's overall softwood log flow went to pulpwood.

An additional indication of regional specialization may be seen by examining the ratio of employment in solid wood-products manufacture to employment in pulp and paper manufacture by state. The top states are rank ordered according to this measure ("index of solid wood products specialization") in Table 3.2. Note that Oregon is almost three times as specialized in solid wood-

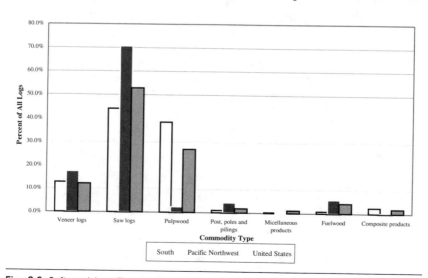

Fig. 3.6 Softwood Log Flow by Commodity Type and Region, 2001. Source: United States Department of Agriculture Forest Service, 2003.

Table 3.2 Rank Order of Top Ten States by Index of Specialization in Solid Wood-Products Manufacture

State	Index of Solid Wood-Products Specialization[a]
Oregon	6.02
Mississippi	2.40
Washington	2.01
North Carolina	1.71
Virginia	1.48
Indiana	1.46
California	1.41
Alabama	1.38
Florida	1.34
Arkansas	1.26

[a]The index is calculated by simply taking total employment in what used to be classified as the solid wood-products industry, standard industrial classification 24 (Lumber and Wood Products), and dividing this by total employment in what used to be classified as the pulp and paper industry, standard industrial classification 26.
Source: U.S. Department of Commerce, 1996. Only those twenty-nine states with total employment in wood-products production exceeding ten thousand employees are included in the analysis.

products manufacture as is Mississippi, the second-ranked state, whereas Oregon and Washington rank first and third, respectively, with a substantial gap separating Washington from the next state, North Carolina. The contrast between Oregon and Washington is also intriguing, and perhaps somewhat counterintuitive, given their proximity and common inclusion in the Douglas-fir industry. However, this gap is consistent with the argument I make here, that Washington generally retains less old-growth and smaller diameter timber but also relies less heavily on Douglas-fir per se, with hemlock and spruce being more prevalent, particularly in northwestern Washington.

Manufacturing Regional Advantages:
The Industrial Geography of Plywood Manufacture

Yet if the Douglas-fir region's forest industry is one that has specialized and continues to specialize in solid wood-products manufacture because of its distinct forest resources, it would be a mistake to view this as a static relationship determined by God-given natural endowments. Instead, firm strategies and technologies evolve in dynamic relation with biophysical nature. In particular, a dramatic transformation in the region's forests has taken place over the past 150 years, accelerated during the past 80 or so, as predominantly old-growth has been replaced by a young-growth timber inventory. If the history of regional specialization revolves around distinct raw material characteristics to significant degree, as I have tried to demonstrate, then it follows that a dramatic change in resource characteristics of the sort entailed by liquidation of old-growth should be reflected in new industrial geographies. This too is my argument. Specifically, over time, as old-growth has become more scarce, solid wood-products specialization in the Douglas-fir industry has been transformed. In particular, processes akin to industrial appropriation have to some extent offset the loss of old-growth by increasingly breaking wood down into smaller fractions and recombining or reconstituting them, aided by the use of chemical adhesives. The significance of highly distinct forest resources has not been erased but rather reconfigured by this broad process of industrial appropriation, as new combinations of wood fibers and industrial processes give rise to new commodity lines, each with their own industrial ecology. In significant measure, this suite of changes, and the industrial geographies and ecologies they entail, is foreshadowed by examining, albeit briefly, the history of plywood.

Plywood is a panel product made by gluing together three or more layers, or plies, of wood. The face is a continuous veneer strip peeled from logs with a lathe, while the inner and back plies can be veneer, lumber, particleboard, or hardboard. Modern, standardized plywood for structural applications in construction is composed exclusively of veneer layers. These layers are oriented so that the grain in each ply runs orthogonally to its neighbors, increasing strength and reducing the risk of splitting. The general principle employed in plywood manufacture is very old; there is evidence, for instance, that the Greeks and Egyptians made similar panels by layering strips of peeled wood together. However, modern veneer originated in the late nineteenth century, when the first patents were issued for plywoodlike composites and when the use of glued veneer layers in piano construction provided impetus to modern veneer manufacture.[32] Although the nation's first plywood mill was likely constructed in Oshkosh, Wisconsin, in 1904, large-diameter old-growth Douglas-fir and modern plywood were first linked through an exhibit at the 1905 Lewis and Clark Exposition in Portland, Oregon, celebrating the centenary of the Lewis and Clark Expedition.[33]

From this beginning, commercial plywood began to gain in popularity, initially as a material for constructing doors, with Washington far and away the nation's leading plywood state. Washington maintained this position throughout the first half of the century, relying on its extensive stocks of large-diameter old-growth Douglas-fir to sustain the industry. In fact, softwood plywood during this period was exclusively tied to large-diameter old-growth Douglas-fir; for example, the McCleary plywood plant in McCleary, Washington, in 1912 could make use only of so-called peeler logs[34] in excess of sixty-six inches in diameter—an incredibly large diameter by today's standards.[35]

By 1940, with national output in excess of 1.2 million square feet of plywood,[36] approximately 75 percent of all production still came from Washington, and most of the rest from Oregon.[37] Subsequent to World War II, a housing boom provided larger markets, and plywood came to be more widely accepted as a substitute for structural lumber in floors, walls, and roofs. With these rapidly expanding market outlets, production leaped by more than ten times between 1935 and 1955.[38] Yet, during this period, Washington State's private, old-growth timber began to show signs of scarcity, and production shifted increasingly to Oregon; by 1963, Oregon accounted for 60 percent of all U.S. plywood production and 56 percent of all plywood mills. Despite these intraregional shifts (foreshadowing what was to come as old-growth became more scarce), the industry continued to rely almost exclusively on large-diameter Douglas-fir (see Figure 3.7), with Oregon, Washington, and California together producing 97 percent of all softwood plywood in 1963. All of the region's output came from Douglas-fir.

Several factors explain the dominance of Douglas-fir in sustaining U.S. plywood output for so long. First, layers of veneer are produced by large lathes that strip off wood in sheets from logs rotated between two blocks. Efficiencies in this process naturally come from using bigger logs, because larger sheets can be stripped from bigger logs and with lower handling costs per unit area of stripped wood.[39] But, in addition, larger diameter logs also yield greater raw material recovery rates, because the lathe leaves behind a core of wood of standard size[40] independent of the diameter of the log. Thus, with larger logs, this

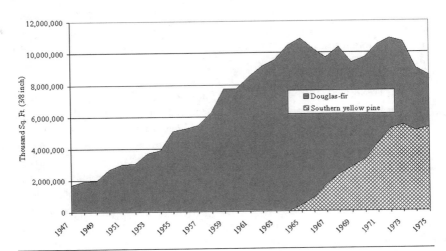

Fig. 3.7 U.S. Softwood Plywood Production by Species, 1947–1975. Source: United States Bureau of the Census, 1957; Ruderman, 1974–85.

core, a by-product of veneer stripping, comprises a lower percentage of the original log volume—meaning less waste as a proportion of usable veneer volume. In addition, however, the strength of Douglas-fir fiber confers engineering advantages to the final product; no other source of fiber could match the combination of strength and flexibility of Douglas-fir during the first phase of plywood's history. It was also the case that plywood was initially developed in the Northwest with Douglas-fir. With a set of processes developed for gluing layers of Douglas-fir veneer together and for drying the resulting plywood, path dependence meant that production was tied for some time to the species, and thus the region. Indeed, as Richard Baldwin noted in his oddly fascinating overview of plywood production practices, based on the experience of Douglas-fir region producers, "it was assumed that veneer logs *had* to be large and free of virtually all defects" (emphasis added).[41] Old-growth Douglas-fir fit the bill.

One of the firms most directly responsible for consolidating this regional specialization, and arguably its primary beneficiary, was the Georgia-Pacific Corporation. Georgia-Pacific began in 1927 as a small southern lumber wholesaling firm, the Georgia Hardwood Lumber Company.[42] In 1947 the company's founder, Owen Cheatham, bought the Bellingham Plywood Corporation of Washington, and in 1948 he moved the company's headquarters to the Douglas-fir region. The firm's name was changed to the Georgia-Pacific Lumber and Plywood Company that same year. Subsequently, Georgia-Pacific rose rapidly through the ranks of established forest products firms during the 1950s, based largely on the conversion of old-growth Douglas-fir into plywood. The company consolidated its position in Oregon by acquiring timberlands and mills in the Springfield, Toledo, and Coos Bay areas, and by 1963 Georgia-Pacific had become the nation's largest producer of softwood plywood, roughly doubling the output of its largest competitor. Georgia-Pacific also diversified quickly into lumber manufacture, becoming the fourth-largest manufacturer of Douglas-fir lumber by the early 1960s.[43]

Yet if Georgia-Pacific was integral to consolidating the Douglas-fir region's monopoly in softwood plywood production well into the 1960s, the firm was also integral to eroding this regional monopoly. As the plywood industry continued to expand, and as lumber production continued apace, Georgia-Pacific and other firms in the region confronted escalating Douglas-fir stumpage prices and increasingly tight supplies of private commercial timberlands stocked with old-growth. Although the limits of old-growth in the region were long foreseen, and were used to place considerable political and regulatory pressure on capital to reforest lands long before serious supply limits were actually reached (see chapters 5 and 6), the economic signs of increasingly scarce old-growth began to appear during the late 1950s, becoming more widely apparent in the 1960s and 1970s. Buoyant demand for Douglas-fir during the region's long boom, and particularly high pressure on old-growth large-diameter timber, propelled stumpage prices upward. Indeed, increases of as much as 100 percent between 1965 and 1973 have been suggested.[44] Another study observed upward tendencies in the price of Douglas-fir stumpage relative to the producer price index from 1950 through 1980, with particularly large increases during the mid-1970s.[45] And data compiled by Deborah Warren of the U.S. Forest Service's Pacific Northwest Research Station indicated that average stumpage prices for Douglas-fir sawtimber sold from National Forest timberlands in the Douglas-fir region increased more than tenfold between 1965 and 1980, whereas the consumer price index increased over this period by about a factor of 1.5.[46] The essential cause of this price escalation is an example of what Barbara Adam discussed as a "timescape" arising from the collision of social and natural time under capitalist modernity.[47] Specifically, the supply of old-growth is naturally unresponsive ("inelastic") to price and demand increases because it takes 175 to 250 years to regenerate true old-growth Douglas-fir.[48]

Yet in response to price escalation and diminishing supplies of available old-growth, firms began to experiment with alternative sources of veneer logs outside the Douglas-fir region. Thus, in 1950 the first plywood mill east of the Cascade Mountains was built in Quesnel, British Columbia, followed in 1952 by a Potlatch Timber Company mill in Lewiston, Idaho. Each mill made use of alternative sources of fiber, including inland Douglas-fir but also western larch, red fir, and spruce. Contradicting the conventional wisdom in the industry uniquely linking large logs and veneer production, log diameters used for peeling veneer in these mills were in the fourteen-inch diameter range, not much larger than the cores left behind in mills further to the west.[49] Moreover, by the early 1960s investment in technologies used to manufacture plywood from southern pine had commenced, including the establishment of a plywood mill in Diboll Texas in 1963-64.[50] Yellow pine offered several attractions, including cheaper stumpage and increasing abundance as regenerated forests matured in the region. Moreover, yellow pine timber grew closer to the expanding housing markets of the southern states.

It was in this context that Georgia-Pacific acted to reverse its earlier westward march. In 1963 Georgia-Pacific purchased timberlands and a lumber mill in Fordyce, Arkansas, with the intention of manufacturing plywood there. Using new adhesive formulations and drying processes specifically designed for the cheaper yellow pine, Georgia-Pacific converted its Fordyce facility to plywood manufacture. Georgia-Pacific's breakthrough, along with continued aggressive timberland acquisitions, became the basis of the company's return to the American South as one of the industry's integrated giants and helped propel a new geography of plywood production with two main poles, one in the Pacific Northwest, the other in the Southeast. By the mid-1970s, southern yellow pine was supplying more than one-third of the nation's veneer, or peeler, logs (see Figure 3.7).

Remanufacturing Regional Advantages: Reconstituted Wood Products

At the time of this writing, Georgia-Pacific remained the leading producer of softwood plywood and other wood panels and is one of the largest forest-products companies in the world.[51] Yet the stories of plywood, and of Georgia-Pacific in particular, have broader implications. Douglas-fir is still a plywood industry staple, and as of this writing, Oregon is still the nation's leading softwood plywood—and lumber—manufacturing state.[52] However, in developing new techniques designed to break the singular link between large-diameter Douglas-fir and softwood plywood, the Georgia-Pacific story, and that of the industrial geography and ecology of plywood production more generally, is highly germane to subsequent trends in solid wood-commodity manufacture in the Douglas-fir region. In fact, plywood, as a product manufactured from whole logs (not unlike lumber), yet processed using energy-intensive processes, high temperatures, and chemical resins, may be seen as the first example of industrial appropriation in solid wood-products manufacture; that is, the first composite solid wood commodity and a forerunner to a suite of newer, reconstituted wood commodities.

Similar to plywood, a number of new solid wood products have been developed during the past three decades. Although there is tremendous diversity in these products and all aspects of their manufacture, what they share in common is a significant transformation in the relationships among products, processes, raw materials, and industrial geographies. A broad trend may be seen in the manufacture of these new solid wood products involving first the reduction of whole logs down to wood fractions, including particles and even fiber, followed by reconstitution or recombination of the wood using highly material and energy-intensive processes. The resulting reconstituted or composite wood products, although diverse, share in significant measure a degree of liberation from the constraints entailed by whole-log processing and the natural, species-specific characteristics of logs, constraints that are central to lumber production (see chapter 4), and witnessed to a lesser degree in plywood. Reconstituted wood products thereby comprise a form of industrial appropriation in their reliance on manufactured inputs (principally chemical resins) and high-energy processes as substitutes for natural binding agents (particularly lignin), allowing capital to significantly rework the relationships between raw material characteristics on the one hand and products and processes on the other. At the same time, these composites also share in common the production of new environmental risks stemming principally from their reliance on chemical adhesives. Although liberating capital from certain natural constraints, or more accurately by transforming these constraints, the move toward reconstituted wood products also comes with new industrial ecologies associated with the chemical adhesives used in binding together various wood fractions.

Reconstituted solid wood products offer ways of circumventing some of the endemic heterogeneity and idiosyncrasies involved in using logs as raw materials. The key issue to consider is that the less log reduction and reconstitution that takes place in the production of a wood commodity, the more commodity production is constrained by the inherent characteristics of the original raw material. Although perhaps overstating the extent to which raw material constraints are being wholly overcome by the new reconstituted wood products, Julie Graham and Kevin St. Martin nevertheless identified the dynamic by which these constraints are being transformed or remade:

> The *natural* weaknesses of wood, including knots and splits, are counteracted in products that involve the recomposition of wood fragments. These engineered products are highly predictable in strength and performance, since they no longer depend on the *idiosyncratic growing conditions of individual trees.*[53] (emphasis added)

Nathan Rosenberg reinforced this view, arguing that raw material dependence in the wood-products industry has factored centrally into the pattern of recent technological innovations. He noted particularly the importance of raw material heterogeneity in the wood-products sector, making it a special case among industries more generally, even among other natural resource sectors. He wrote,

> If the industry is not unique, it is at least at the extreme end of a spectrum of possibilities with respect to variability or heterogeneity in physical characteristics of its primary raw material. This heterogeneity is based on the fact that wood is an organic material with a remarkable degree of natural diversity.[54]

The new reconstituted wood products help alleviate the problems in manufacture associated with such idiosyncrasies, opening up new possibilities for tailoring the characteristics of commodities to their intended uses. Thus, "Because wood properties vary among species, between trees of the same species, and between pieces of the same tree, [traditional] solid wood products cannot match reconstituted wood in the range of properties that can be controlled in processing."[55]

Although some of the idiosyncrasies and heterogeneity of timber as raw material are endemic to all wood, some of them are species specific, and others are evident between trees of the same species (as noted by Youngquist), including those differentiating old-growth and young-growth. It seems quite significant then that the age of the new reconstituted wood products coincides with exhaustion of America's last frontier of old-growth timber, namely, the Douglas-fir region. Although stumpage prices in the region rose,[56] the quality of available sawtimber also dropped, making it difficult to obtain logs with the characteristics suited to high-quality structural wood products—in short, giving rise to a particular source of uncontrolled heterogeneity and unpredictability in log supply.[57] Although Douglas-fir, as I have noted, is still widely perceived to be the best source tree for softwood sawtimber, converting old-growth to young-growth was nevertheless accompanied by a notable decline in wood quality from a commodity production standpoint. As noted succinctly by David Pease, publisher of the magazine *Forest Industries,*

> Managed second-growth timber is generally characterized by smaller diameters, higher costs on a unit basis, and reduced quality. The smaller diameter of second-growth trees make this wood more costly to handle. . . . In addition, managed second-growth trees grow faster than non-plantation forests, reaching harvestable size at an age when the tree contains a high proportion of juvenile wood. . . . [Juvenile wood] is inferior to mature wood in many ways that are important to wood products conversion. . . . These factors adversely affect the mechanical and dimensional stability of solid wood products.[58]

The climate of high stumpage prices and raw material transitions precipitated by the liquidation of old-growth is therefore implicated in a series of shifts in the industrial geography and ecology of the forest sector, changes that are apparent within and across commodity types and within and outside of the Douglas-fir region. Plywood was the first instance. Parallels may be witnessed too in Douglas-fir lumber manufacture, which has shifted from a focus on higher end, custom grades ("selects") manufactured from the highest quality and largest timber toward more standard commodity grades manufactured for heavy and light framing in construction. Data compiled by James Wigand of the U.S. Forest Service's Pacific Northwest Research Station (see Figure 3.8) capture this shift.[59] Between 1971 and 1995, the proportion of Douglas-fir lumber manufactured in Douglas-fir region mills classified as either "C selects" or "D select and shop" (the two highest quality classifications designated for custom cuts of typically large-diameter clear lumber) dropped from a combined 15.6 percent to a combined 0.8 percent of the total. Over the same period, the proportion manufactured in the heavy- and light-framing categories increased from a combined 56.1 percent to a combined 79.2 percent. This change in manufacturing tendencies is not due to market shifts; price data show that over the same period, the ratio of C select prices to heavy- and light-framing prices actually increased somewhat. Yet if sagging market demand for high-end lumber was responsible for its declining share of production, one would expect a decrease in the ratio of the price of select to framing grades. Rather, the shift is due to the transition from old-growth to young-growth, as noted by David Pease.

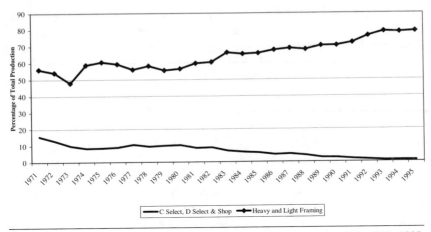

Fig. 3.8 Proportion of Douglas-fir Lumber Types, Pacific Northwest Coast Mills, 1971–1995. Source: Weigand, 1998.

Yet in addition to changes within commodity lines such as plywood and lumber, decisive inroads have been made by a number of wholly new commodities in response to the emerging scarcity of old-growth.[60] These new, reconstituted commodities include glued-laminated lumber and laminated veneer lumber (LVL), both of which have increasingly competed with conventional dimensional lumber; oriented strand board (OSB); and a range of hybrid products.[61]

Glued-laminated lumber, or glulam, is based on a technology invented in Europe at the turn of the century and first used in North America to build a demonstration structure at the U.S. Forest Service's Forest Products Laboratory in Madison, Wisconsin.[62] However, glulam experienced a more recent renaissance in North America as a substitute for large beams, the likes of which were once made from old-growth logs, including Douglas-fir. Glulam is produced from individual pieces of lumber that are finger-joined together to form desired lengths. These pieces are glued, and the resulting boards are pressed and dried, offering a substitute for large, long, and structurally sound lumber cut as a single piece from large-diameter logs.[63]

LVL was first developed in 1970 and is also used as a substitute for traditional dimensional lumber. But as its name suggests, LVL is manufactured from veneer strips, peeled from logs as in plywood manufacture, with layers glued together using adhesives and pressed under high heat and pressure. The only real differences between LVL and plywood are that LVL has more layers to match the thickness of typical lumber products; it is trimmed to strips to match lumber dimensions, and the veneer layers are oriented in parallel, rather than orthogonally, to mimic natural wood grain.[64]

Other products have been developed for the structural panels market, competing with plywood. Of these, waferboard and OSB are the most significant. Waferboard was first developed in Canada during the 1960s and is produced by first reducing logs to small flakes not unlike the wood chips from which pulp is manufactured. These flakes are dried and coated with wax and powdered adhesives; separate inner and facing layers are formed, combined, and then pressed under high heat and pressure. The resulting board is lighter and cheaper than plywood but is also not typically as strong. OSB was developed after waferboard, but it has made deeper inroads as a substitute for structural plywood. Producing OSB is not unlike manufacturing waferboard, with the notable exception that the wood chips are broken across the grain into strands prior to being reconstituted. These strands can then be oriented (thus the name) in parallel in each layer, with offsetting layers orthogonal to one another, conferring engineering advantages similar to plywood but with greater precision.[65] OSB production has grown exponentially in recent decades, with total U.S. output rivaling that of softwood plywood.[66]

Numerous hybrid products are also now manufactured from combinations of reconstituted wood products. For example, composite I-joists produced using a combination of LVL and OSB comprise about 17 percent of the market for floor beams in U.S. residential construction, substituting for beams traditionally manufactured from whole logs, including large-diameter Douglas-fir. One of the features of these and other engineered structural wood products is that they can be designed for highly specific uses; for example, offering higher strength and density in some parts of the beam than others based on the particular conditions of their use. This is one of the bases of industry claims that engineered lumber products are more materially efficient.[67]

Yet all of these products allow greater flexibility in the use of raw materials by being less tied to logs of specific dimensions and to particular species than is the case with conventional structural wood commodities. They do this in part by relying on high temperature processes and a range of chemical adhesives to reduce and remanufacture (reconstitute) wood-based products. Although the specific ways in which wood is remade is highly particular to specific firms, processes, and commodity lines, there is a common thread that may be conceptualized along a spectrum of appropriation in manufacturing wood products, with simple whole-log conversion at one end and complete wood reduction and reassembly from fiber at the other (see Figure 3.9).[68] Lumber comprises the archetypal commodity produced simply from whole-log conversion, whereas wood-based pulp and paper manufacture occupies the other end.

Of course, modern wood-based pulp and paper production is not a new process, dating back to at least the middle of the nineteenth century.[69] What is new, however, is the range of solid wood-commodity production processes that are in many ways kindred with, and foreshadowed by, plywood. Moving from

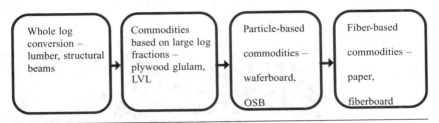

| Whole log conversion – lumber, structural beams | Commodities based on large log fractions – plywood glulam, LVL | Particle-based commodities – waferboard, OSB | Fiber-based commodities – paper, fiberboard |

Fig. 3.9 Spectrum of Appropriation in Wood Products Manufacture.

left to right on the spectrum entails greater degrees of wood reduction.[70] It also involves an increase in industrial appropriation, because (generally) more and more capital intensive processes and more and more nonwood inputs are required in moving between lumber and reconstituted products. Moreover, along this spectrum, there are distinct clusters of production technologies. The first cluster is based on the use of large log fractions, including veneer stripping and thus conventional plywood, along with LVL and glulam. The second, intermediate category applies to products manufactured from wood that has been reduced to particles but not to raw fiber. This includes waferboard and OSB. The third category belongs to fiber-based products, including paper, but also other new solid wood products such as fiberboards. Although the spectrum indicates a decreasing level of dependence on heterogeneous whole-log attributes moving from left to right, this direction also indicates the overall direction of innovation over time, away from whole-log processes and toward greater industrial appropriation.

New Geographies   In part because the supply of forest resources is geographically differentiated, the coevolving relationship between raw materials and solid wood commodities has geographic dimensions, including implications for the Douglas-fir region. It is certainly clear, for example, that the substitution of OSB for structural plywood is further eroding the region's position in structural panel production. OSB can be manufactured from relatively low-grade, and generally less-specific, raw materials, including, for example, abundant, cheap, and heretofore little-used aspen.[71] Not surprisingly, investment in OSB capacity has been directed to a number of nontraditional wood-products producing states and to those with substantial supplies of low-grade hardwoods. This includes northern and intermountain western states, as well as southern states such as Tennessee and West Virginia. In fact, southern states account for the largest share of installed production capacity in the OSB industry, with almost 70 percent of the nation's total.[72] Growth in OSB's production in areas outside the Douglas-fir region is thus the single greatest contributor to a geographic shift in the aggregate softwood structural panel market since the early 1980s, the evidence of which can be seen in Figure 3.10. Between 1983 and 2000, while

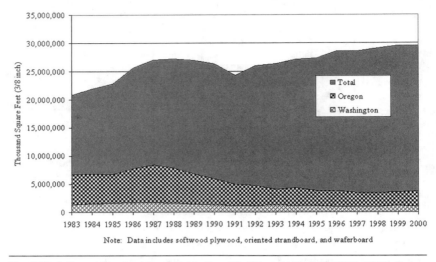

Note: Data includes softwood plywood, oriented strandboard, and waferboard

Fig. 3.10 Softwood Structural Panel Production, OR, WA, and U.S. Total, 1983–2000. Source: Warren (2002) and other years.

aggregate softwood structural panel production increased from fewer than twenty-one billion square feet (3/8-inch basis) to almost thirty billion square feet, production in Oregon and Washington combined declined from more than eight billion square feet to fewer than five billion square feet.

This is not to say that the development of reconstituted wood products has been exclusively at the expense of the Douglas-fir region, however. Indeed, some products have tended to reinforce the natural advantage of Douglas-fir, because they still rely on some of the species' unique attributes described previously; appropriation is by no means about the replacement of natural production but rather about its industrialization and reconfiguration. For example, engineered lumber products such as glulam and LVL no longer rely exclusively on large trees, but they are still dependent on fiber with a combination of strength and flexibility. The glulam process allows for large structural lumber to be manufactured from smaller pieces by gluing and pressing them together; moreover, the gluing process makes the resulting structural members stronger on average than similar sized pieces manufactured from a single piece of wood because of the strengthening effect of the laminates.[73] Even still, the process requires relatively high-grade lumber for fiber supply; not surprisingly, with its ample Douglas-fir forests, Oregon has ten glulam mills, more than any other state. Similarly, the LVL process, as the name indicates, is based on conventional veneer stripping, for which Douglas-fir remains well suited even as young-growth. Oregon also led all states in 1999 with nine LVL mills.[74] The overall importance of the region to the new reconstituted wood products is reflected by 1997 census of manufacturing data that point to Oregon as the leading state in reconstituted and engineered wood products, with California and Washington also significant producing states. In particular, in the classification for reconstituted wood products (including waferboard, particleboard, medium density fiberboard, and hardboard), Oregon led all states with twenty-eight facilities, followed by California with twenty-three; Washington ranked thirteenth with nine facilities. In addition, Oregon led all states in employment and value added in this classification. In the classification for engineered wood member (except truss) manufacturing, which includes LVL, glulam, and engineered joists and beams, Oregon also led all states with fifteen of a total of fifty-three facilities, followed by Washington and Louisiana with three each.[75]

Environmental Impacts of the New Solid Wood Products   If the dynamic of appropriation in solid wood-products manufacture results in new industrial geographies, it is also based on new industrial ecologies that entail more than mere changes in raw material demands. Indeed, somewhat ironically, if reconstituted wood products rely less on old-growth timber, and thus take some of the pressure off remaining old-growth forests, they also bring with them a series of environmental and health concerns. These concerns emanate from aspects of the very production processes that make composite or reconstituted wood products possible, notably their manufacture under typically high heat and pressure conditions and the use of synthetic organic resins and glues. Here too, plywood is an important forerunner to the trend.

High temperature and pressure processes that are required to produce most of the reconstituted products, including plywood, require large inputs of high-quality energy. High pressures may be generated by electricity, but the electricity has to come from somewhere, and electricity generation is among the most serious sources of a range of environmental pollution in the contemporary United States, including air pollutants such as particulates, oxides of nitrogen and sulfur, and carbon dioxide, as well as water pollution and the various destructive effects associated with extracting the oil, natural gas, coal, and uranium used to fuel power generators. Although the concerns surrounding use of these fuels in electricity generation cannot be ascribed to reconstituted wood products alone, a comprehensive life cycle assessment[76] of the industrial ecology of these products requires that the ecological footprint of each phase in the commodity chain be considered, including upstream energy requirements. Similarly, high pressure and high heat processes are enabled in many production facilities by on-site boilers. These are typically natural gas fired or fueled by wood wastes such as chips and sawdust, but both of these fuel cycles produce pollutants of various kinds. Natural gas is implicated in climate change because its combustion produces carbon dioxide emissions, but also because methane is a powerful greenhouse gas when leaked during extraction, distribution, and storage.

Methane combustion also produces NO_x, a significant air pollutant. Wood waste combustion is arguably less of a concern vis-à-vis climate change if the uptake of carbon by growing trees offsets the release of carbon dioxide from combustion. However, wood waste combustion also produces NO_x, a range of hydrocarbon compounds, and very high levels of particulates.

Serious environmental and health impacts also stem from the various adhesives used to bind wood fractions together in the manufacture of reconstituted wood products. The very use of these adhesives is one of the links joining the reconstituted wood products together, because such use is a form of industrial appropriation directed at displacing lignin, the substance that binds wood fibers together in trees, with manufactured alternatives applied during production. Industrial adhesives, by allowing wood fractions to be disassembled and put back together, partially offset the problems and limits of potentially scarce and heterogeneous log supplies, as noted previously. Although a diversity of adhesives is used within and across particular commodity lines, most are synthetic organic polymers, and in structural wood products in particular, most are thermosetting or heat-transformed varieties. The most common are thermosetting formaldehyde compounds, including urea-formaldehydes, phenol-formaldehydes, resorcinol-formaldehydes, and melamine formaldehydes.[77]

The use of organic adhesives to bind reconstituted wood products together originates with plywood. Until the 1930s adhesives used to manufacture plywood in the United States were of natural origin, derived from plant and animal extracts. Although some naturally derived adhesives are still in use, the rising costs of such extracts in the 1920s and 1930s, compared with increasingly cheap and readily available substitutes derived from fossil fuel sources, led to a shift to synthetic organics in plywood manufacture, foreshadowing the use of synthetic organic adhesives in reconstituted wood products more generally.[78] Because of this shift, and the onset of a wider range of composite wood products, petrochemical adhesive use in the manufacture of particleboard, OSB, and plywood alone has grown to more than one billion pounds (454 million kilograms) per year in the United States.[79]

Yet concerns surrounding the occupational and environmental health risks of these adhesives are long standing and primarily associated with emission of volatile organic compounds, including phenols and formaldehyde gas. It bears noting first that synthetics are derived primarily from petroleum and natural gas and are thus inherently nonrenewable and associated with a range of noncombustion environmental impacts of the petrochemical commodity chain, including air and water pollution, and a virtual pantheon of toxic contamination effects. Much of the early concern specifically surrounding adhesives in the wood industry was directed to the dangers posed by formaldehyde gas, due to occupational exposures in reconstituted wood-products manufacturing facilities and exposures to formaldehyde gas given off from finished wood products.[80]

Formaldehyde is a noxious, colorless gas that is an irritant to the eyes and the respiratory tract in levels higher than 0.1 parts per million of air. It is implicated in asthma and allergic reactions, is a suspected carcinogen, and is a congressionally designated hazardous air pollutant in the United States.[81] Given these characteristics and the ubiquity of the substance in wood adhesives, it is not surprising that workers in plywood, and later OSB and particleboard plants, complained of the noxious smell of formaldehyde, as well as of irritation of the eyes, nose, mouth, and throat, along with chronic breathing problems. There is a shortage of good data available to indicate exposure levels in earlier years; however, a conservative estimate would be that formaldehyde gas levels in reconstituted wood-products plants prior to the mid-1970s were routinely in excess of one to five parts per million, or ten to fifty times levels now considered acceptable.[82] Research on the health effects among workers as a result of chronic, long-term exposure to formaldehyde gas in the workplace indicates serious and lasting effects, particularly, and as one would expect, among those exposed to higher levels during earlier years. These health effects include allergies, persistent breathing problems (including asthma), and elevated cancer levels.[83]

As a result of concerted efforts by workers and their unions to bring these problems to light and to force companies to adopt measures to control them, numerous technical changes have been introduced to reduce occupational exposures to formaldehyde and to reduce emissions more generally in the wood-products industry.[84] However, there are remaining concerns about formaldehyde exposure levels among workers and respiratory and dermal (i.e., through the skin) exposures to other dangerous gases given off by wood adhesives, including phenol. These risks are more acute in areas of mills where products are glued, pressed, and dried, particularly if workers are stationed next to or must directly handle products without benefit of tight controls on their exposure to gasses from the adhesives—controls that can be quite difficult to achieve if workers need to handle products; for example, in aligning layers of veneer during plywood production.[85]

Concerns have also been raised about some products giving off formaldehyde in postproduction. This problem has also been largely addressed by reducing the ratio of formaldehyde in the adhesives and by curing wood products to seal them and prevent off-gassing. Moreover, the rate at which formaldehyde is released in postproduction tends to decay rapidly, reducing concerns about indoor air quality in homes and offices.[86] However, the combination of long-standing occupational health concerns with those surrounding exposure to gasses given off in postproduction reinforce a central point regarding reconstituted wood products: synthetic organic adhesives used to produce them are sources of dangerous toxic pollutants.

This fact was recognized in the Clean Air Act Amendments of 1990 when the Lumber and Wood Products sector was targeted for new controls on air emissions. Specifically, Congress and the Environmental Protection Agency (EPA) have recognized a class of wood products—composites—that share in common the use of a set of synthetic organic adhesives that give rise to a range of hazardous air pollutants designated for control under the Clean Air Act. As such, the EPA has been working for more than a decade on the development of new standards for regulating composite wood-products facilities under the auspices of maximum achievable control technology regulations to be promulgated by the EPA. This campaign has a rather long, convoluted history of bureaucratic wrangling and industry resistance behind it, resulting in considerable delays to the implementation of new standards.[87] However, the campaign also highlights two important themes relevant to this discussion. First, the EPA has recognized and designated a class of products and processes, included under the title "Plywood and Composite Wood Products," sharing in common many of the attributes named in this chapter; specifically, they share in varying degrees the manufacture of solid wood products from a combination of wood fractions and adhesive resins. The specific products in the EPA's amalgam include plywood, waferboard, and OSB; particleboard; medium density fiberboard; hardboard and fiberboard; and engineered wood products. Not all of these are structural wood products but all of the reconstituted solid wood products discussed in this chapter are included in this categorization. Second, and more important, the EPA's campaign has generated considerable knowledge about the production of in excess of thirty hazardous air pollutants hazardous to human health and the environment, including formaldehyde, methyl diphenyl diisocyanate, xylenes, toluene, phenol, and others. All of these substances are given off in various proportions by specific combinations of production facilities, products, and adhesives in the reconstituted wood-products sector.[88]

Conclusion

The environmental impacts of reconstituted wood products is a complex and complicated story that deserves a more complete treatment than I can give here. My point is to highlight potentially serious problems associated with this direction of innovation. I suspect that this facet of reconstituted wood products will continue to be an important political and policy issue, one that may rise to the level of a new political ecological crisis on par with the politics of old-growth in time. Certainly, although the EPA's campaign to develop maximum achievable control technologies applicable to reconstituted wood-products facilities will reduce the emission of known hazardous air pollutants, there will continue to be legitimate pressure placed on the industry to switch back to more environmentally benign adhesives derived from renewable plant and animal sources.[89] But it is important to appreciate the systemic character of the processes and problems involved in this story, as an appropriation-like process supplants aspects of natural wood (and the heterogeneous characteristics of logs) in piece-meal fashion. Although the direction of innovation toward reconstituted wood products comprises an example of industrial appropriation and a response to the ecological-economic crisis posed by exhaustion of old-growth, offsetting one source of ecological crisis seems to be giving rise to another.

More broadly, I have tried in this chapter to examine a particular facet of the fictitiousness of nature as a commodity by examining a broad industrial ecology arising from the interactive or dialectical relationship between social and ecological production in the Douglas-fir region. The industrial geography of raw material processing and conversion in the wood-products industry is clearly influenced by the sector's nature-centered character, with regional differences in forest types, resource stocks, and biological productivity all profoundly influencing the dynamics of regional specialization in the Douglas-fir region. Large, old-growth Douglas-fir helped give rise to a solid wood-products specialization in the coastal Pacific Northwest, one particularly specialized in the production of structural lumber and plywood. This solid wood-products advantage is in some measure reproduced by the unique characteristics of young-growth as well, and thus by no means is the solid wood orientation of the region entirely an old-growth effect. Nevertheless, the exhaustion of old-growth has helped propel new industrial ecologies, as firms have increasingly pursued strategies of industrial appropriation that reconfigure relations between commodities and raw material in solid wood-commodity manufacture. Using high-energy processes and chemical adhesives, a range of reconstituted wood products allow more diversity in raw material use (in terms of log size and species) by breaking logs down into wood fractions and reassembling them with glues.

Again, I stress that this is by no means a story of capital's conquest over natural limits. Rather it is about an evolving dialectical relationship in which limits are transformed and reconfigured. Historically and geographically evolving institutions, biophysical constraints, and production technologies interact dynamically in ways that are translated into new industrial and geographical outcomes.[90] At the same time, the environmental impacts of raw material dependence and conversion are reworked by industrial appropriation in ways that generate their own problems and limits. These dynamic changes require careful analysis of the way that the fictitious commodity problem is reformulated within and across regions, at different historical junctures.

4

Geographies of Scale
and Scope in Lumbering

Raw Material Dependence and Economies of Scale and Scope

The physical properties and growth characteristics of wood . . . would seem to impose certain limitations on the productivity gains that can be achieved. Wood does not come in uniform and standardized pieces and therefore does not lend itself readily to automatic handling as homogeneous materials do. Trees are scattered over wide areas which makes the assembly of large quantities for highly mechanized handling difficult, if not impossible. . . . In the mill each log is studied separately to determine the method of log breakdown that will yield the most quality lumber.[1]

In *Scale and Scope: The Dynamics of Industrial Capitalism,* Alfred Chandler argued that economies of scale and economies of scope were critical factors that allowed certain U.S. manufacturing firms to grow large and to capture significant market shares. As these firms grew large, their corresponding industries became concentrated into oligopolies, and even monopolies. He argued further that a distinct form of firm organization arose as an institutional answer to capturing and managing economies of scale and scope. This model comprised vertically and horizontally integrated firms practicing what Chandler called "managerial capitalism." These firms and their sectors came to represent, according to Chandler, the basis of a uniquely American form of twentieth-century capitalism.[2]

How representative Chandler's paragons of corporate managerialism were and are remains in question.[3] Yet it is also fair to say that Chandler had much more to say about big firms and concentrated industries than he had to say about small firms and fragmented industries. Whether exceptional, or just

different, Chandler largely ignored sectors where bigness remained elusive, arguing simply that their production technologies did not allow economies of scale
(or presumably economies of scope) to develop sufficiently that managerial
capitalism was appropriate, required, or very successful.[4]

In this respect, it is interesting to look at the history and geography of industrial organization in the lumber industry. As Chandler noted, lumbering is
one of the manufacturing sectors in which large, managerial firms were relatively rare and slow to emerge. In this chapter, I seek to explain why bigness has
remained (relatively) elusive in the lumber sector and, conversely, how and
why bigness has been achieved where and when it has. Although it is true that
economies of scale—or more accurately the lack of economies of scale—are
critical to the explanation, an examination of the lumber industry reveals that
its relative failure to give rise to large, vertically integrated managerial firms
does not entirely confirm Chandler's thinking. Specifically, the challenges to
bigness in the lumber industry did not and do not rely exclusively on the development of technologies of rapid throughput per se, but revolve instead around
access to sufficient timber supplies. In this chapter, as a contemplation on the
link between raw material access and economies of scale and scope, intended as a
complement to the previous chapter, I round out the discussion of how economics and ecology combine and collide in nature-based commodity production.

Chandler and Bigness

Although organizational innovations in the development of managerial capitalism were critical to achieving bigness, in Chandler's view these developments
were subordinate to or driven by technology-based economies of scale that
emerged as part of the second industrial revolution. In other words, the emergence of a corporate, managerial capitalism allowed the capture and management of economies of scale that originated in production.[5] As Chandler put it,

> As a result of the regularity, increased volume, and greater speed of the
> flows of goods and materials made possible by the new transportation
> and communication systems, new and improved processes of production
> developed that for the first time in history enjoyed substantial economies
> of scale and scope. Large manufacturing works applying the new technol-
> ogies could produce at lower unit costs than could the smaller works.[6]

He attributed the ultimate basis of these new high-speed production possibilities to key technological innovations, not least the extensive availability and
use of high-quality energy—particularly fossil fuels and electricity—to power
mass production.

In parallel, new economies of scope allowed firms to lower unit costs by also
dedicating newfound capacities for rapid production to products linked by similar production processes and technologies. As Chandler described it,

The economies of joint production, or scope, also brought significant cost reduction. Here the cost advantage came from making a number of products in the same production unit from much the same raw and semi-finished materials and *by the same intermediate processes*. The increase in the number of products made simultaneously *in the same factory* reduced the unit cost of each individual unit.[7] (emphasis added)

But although Chandler noted that numerous manufacturing sectors—including oil, chemicals, tobacco, and, significantly, the pulp and paper sector—developed the technologies and organizational capacities to "go big," he recognized but did not dwell on those that did not. These include the apparel, textile, furniture, printing, and publishing sectors, and of course the lumber industry. One might expect that, if Chandler's theory of bigness holds, its converse would also be true: industries and firms not characterized by bigness must have been unable to develop the technological and organizational capacities for bigness, and in particular must not have been able to develop high-quality and high-quantity energy processes to achieve rapid throughout.

The lumber industry, however, does not conform to this expectation. Although large quantities of high-quality energy—specifically steam and then electricity—have indeed been harnessed to industrial lumber manufacture, and although these technologies were critical in the emergence of larger mills and industrial consolidation during the twentieth century, lumber has remained relatively fragmented and has seen comparatively sluggish productivity growth when measured against other industries. It is my contention in this chapter that this is due to the industry's essential nature; that is, a raw material gathering and conversion industry. In particular, the influence of an extensive geography combined with the heterogeneity of logs as a raw material input have constrained the development of economies of scale in lumber manufacture per se. Moreover, where bigness has been achieved in lumbering, the acquisition of large quantities of timberlands has been the key to bigness, not technology.

Aside from the theoretical significance of this question, concentration and economies of scale in the lumber industry, and in the forest sector more generally, is one of considerable importance. From capital's perspective, a fragmented sectoral structure, particularly prior to World War II, presented the industry with some of its biggest political and economic challenges. Because there were so many producers, and with such divergences of interest between large and small operators, industry lobbying and political coordination suffered from a lack of cohesion. In fact, it was sometimes almost impossible for the lumber interests to speak with a single voice, as evident in repeated and failed attempts to control overproduction, itself a problem tied to minimal economies of scale and thus low capital requirements as a condition for entry into the business.[8] Conversely, industrial consolidation is one of the primary concerns of economic regulation and policy, important to concerns about the concentration of economic power and influence and critical to the dynamics of labor markets and efficiency-related job losses; efficient industries are in fact defined in large measure by how little labor they require per unit of output. Because scale economies allow greater efficiencies, they translate into less employment per unit of output. Thus, concentration or its absence is a key concern, all the more so in the lumber sector because of its historic significance to rural communities (a topic I explore in more depth in chapter 6). Concentration in the lumber industry also takes on an added significance because, as I argue here, bigness in the sector has primarily revolved around the guarantee of access to sufficient quantities of raw material and around the capture of economies of scope that exist in the production of diverse products from a heterogeneous raw material base. Thus, it has been through consolidation of timber supply, a public policy concern, that concentration has been achieved in lumbering.

Lumber and the Limits to Bigness

Despite gradual increases in scale economies in lumber manufacture, bigness and the development of significant market power have remained elusive in the U.S. lumber industry and in the Douglas-fir region—at least by Chandlerian standards. Entrepreneurial and even familial capitalism have remained remarkably significant in lumbering. In fact, these have long been noted as distinguishing features of the trade. The 1913 U.S. Bureau of Corporations mammoth report on the lumber industry, although documenting rapid accumulation of U.S. timberlands after the turn of the century, noted at the same time that the lumber industry was extremely fragmented and dominated by small-scale producers. Specifically, the report stated, "The *nature* of the sawmill business makes the output of the largest mill very small compared with the total of the country"[9] (emphasis added) and "To enlarge a mill beyond a capacity of 20 or 25 million feet a year is to duplicate mechanical units, with small or doubtful advantage in manufacture."[10] In 1909, the peak year of U.S. aggregate lumber production with more than forty-five thousand operational lumber mills in the United States, more than half manufactured less than five hundred thousand board feet per year.[11] Such mills accounted for 13 percent of U.S. lumber output. Conversely, the aggregate output of the nation's thirty-six largest sawmills (i.e., those cutting in excess of fifty million board feet per year) accounted for less than 6 percent of the nation's total. In Oregon, where lumbering was just beginning a long boom, there were already at least six hundred different lumber mills in 1909.[12]

Vernon Jensen noted that by 1929, some large mills were emerging. By that time, Jenson noted, more than 120 mills in the United States were manufacturing in excess of fifty million board feet per year, accounting as a group for 26 percent of all production. Yet still, there remained more than twenty thousand lumber mills in the nation, and mills cutting less than twenty-five million board feet per year cut more than half of the country's lumber.[13] Noting the lack of economies of scale, Jensen wrote, "Although small mills are usually inefficient, a good medium-sized mill is just as efficient as a good large one, and the large mills do not enjoy any significant advantage in terms of productive efficiency."[14] In 1947 the four-firm concentration index for U.S. lumber manufacturing was only 5 percent, dwarfed by the kinds of consolidated manufacturing industries interrogated by Chandler.[15]

At the same time, the lumber industry was characterized by a decided lack of Chandler's managerial capitalism. Jensen wrote,

> The structure of ownership of the lumber industry has conformed to older patterns than has been the case in many other industries. Individual ownership has been widespread. . . . *Considering all units, the corporate form of business organization has been utilized by a relatively small number.*[16] (emphasis added)

This is not to suggest that the lumber industry has never seen increases in economies of scale and increases in economic concentration, however uneven and relatively slow these increases have been. Moreover, somewhat true to Chandler's model, energy technologies and, in particular, power supply to sawmills have been integral to what development of rapid throughput has been possible in lumbering and thus to the development of greater economies of scale. A report commissioned by the Federal Works Agency's Work Projects Administration on mechanization in the lumber industry in fact noted that the widespread introduction of steam power to lumber manufacture by the late nineteenth century allowed lumber production to take on "the characteristics of the processes of a mass-production industry" with attendant indicators of a trend toward consolidation.[17] Yet the report goes on to note that, despite the move toward electrification in larger sawmills by the 1920s, the momentum toward bigger mills had reversed, with declining average mill capacity and production.[18] Thereafter, with the increasing penetration of electricity in lumber milling, and attendant increases in the flexibility of logging operations through the use of mechanized, diesel-powered skidders, tractors, and logging trucks, tendencies shifted once more toward increasing average mill size and industry consolidation; between 1929 and 1958, although total U.S. lumber production was essentially unchanged, the total number of mills fell by a further 25 percent.[19]

In the Douglas-fir region of western Oregon—by this time the most important lumber-producing region in the nation—expanding output during and after World War II led to an initial increase in the number of mills, as small, flexible mills persisted. However, beginning in the late 1940s, a marked shift toward the dominance of larger mills took place, indicated by an upswing in

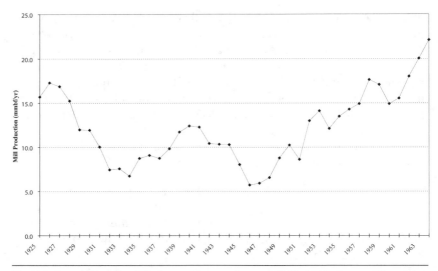

Fig. 4.1 Average Lumber Mill Output, Western Oregon, 1925–1964. Source: West Coast Lumbermen's Association (1964) and Western Wood Products Association (1998) and various other years.

average mill capacity (see Figure 4.1), driven by mill electrification and improved economies of log transport.[20] As the average size of lumber mills began to rise, some truly large producers emerged with significant and increasing market shares. Thus between 1947 and 1972, the four-firm concentration index for lumber manufacture in the United States rose from 4.8 percent to 18 percent, whereas the eight-firm concentration index rose from 6.4 percent to 23 percent.[21] The Douglas-fir region showed the highest levels of concentration, with the four-firm index reaching 23.9 percent by 1960. The largest producer in the region was Weyerhaeuser, accounting for 14.6 percent of regional output, followed by Georgia-Pacific with 4.3 percent, the U.S. Plywood Corporation at 2.7 percent, and Pope and Talbot at 2.2 percent.[22]

This trend of gradual and somewhat lurching increases in lumber industry consolidation has continued up to the present day. Thus, between 1968 and 1994, although aggregate Oregon lumber production remained essentially unchanged, the total number of sawmills declined from three hundred to eighty-nine. This attrition has been almost exclusively at the expense of small mills, indicating the continuing increase of economies of scale in the industry. During this interval, the number of sawmills capable of producing in excess of 120 thousand board feet of lumber per eight-hour shift (the largest-rated category) remained essentially constant, rising from fifty-nine to sixty mills.[23] With such consolidation, combined with essentially stagnant or slightly declining aggregate lumber production, employment in the industry has declined by nearly half over this period, and the social implications of this long-term employment decline have been significant, even devastating in some mill towns (see chapter 6).[24]

These longer term data indicate that, however uneven, the tendency has been toward larger sawmills and industry consolidation, markers of increasing economies of scale. Yet, in a relative sense, the lumber industry nationally and in the Douglas-fir region has remained a relatively fragmented sector. The four-firm concentration index for lumber actually fell slightly during the 1970s and 1980s, and as of the 1997 Census of Manufacturing, stood at 16.8 percent.[25] This is a very low degree of concentration when measured against other manufacturing sectors, ranking among the least concentrated of all industries listed in the census. What is intriguing is that, although they are not exceptionally concentrated, the softwood plywood and the pulp and paper sectors present contrasts to sawmilling with higher four-firm concentration indexes; softwood veneer and plywood manufacturing was at almost 50 percent, whereas the pulp sector was at almost 59 percent.[26] Even the top fifty lumber companies accounted for less than half of total value added in U.S. lumber manufacture in 1997.[27]

Moreover, the degree of concentration in the lumber industry (nationally and regionally) is particularly low compared with current and historic levels in the sectors highlighted by Chandler as paragons of American managerial capitalism—petroleum products, tobacco, chemicals, and others. In these sectors, single-plant and multiplant economies of scale and scope allowed firms to construct large individual facilities and networks of facilities. By contrast, the size of individual lumber mills remains small compared to industry capacity. For example, the largest lumber mill in Oregon in the late 1990s was the Seneca Sawmill of Eugene, among the most technologically advanced lumber mills in the nation. In 1998 the mill employed 265 people, had a daily capacity of 750 thousand board feet per eight-hour shift, and produced a total of 275 million board feet of lumber.[28] Yet although this mill's production dwarfed that of most of its contemporaries in the region, its output accounted for only 5 percent of Oregon's lumber production in 1998 and less than 1 percent of the nation's softwood lumber output.

Not only are individual mills small by the Chandlerian standards of American manufacturing but the ownership of many facilities remains familial, entrepreneurial, and largely independent. The aforementioned Seneca Sawmill, despite being the state's largest lumber mill, was independently owned and operated. This type of firm structure also characterized other significant mills and lumber firms in the state during the late 1990s, including, for example, (1) the Roseburg Forest Products Company of Dillard (near Roseburg), established in the 1930s and one of the largest firms in the state's wood-products sector, yet independently owned and operated by the Ford family of Roseburg; (2) the Roseboro Lumber Company of Springfield, one of the largest and most enduring lumber companies of central Oregon, also independently owned; and (3) Rough 'n' Ready Lumber of southwestern Oregon, featuring a relatively large mill with modern technologies, 150 employees, and a substantial lumber export trade, and independently owned and operated by the Krauss family since 1922.[29] Far from being anomalies, independent firms whose ownership is held by a single entrepreneur or a small group of investors has actually remained typical of the Douglas-fir region's lumber sector. Again taking data from Oregon, although large, integrated, and publicly traded corporations with Chandlerian managerial structures (e.g., Weyerhaeuser, Willamette Industries, and Georgia-Pacific)[30] do operate lumber mills in the state, only about one-third of all Oregon lumber mills either are owned by publicly traded firms or are part of firms with multiple lumber production facilities.[31]

The lumber sector also has been a somewhat laggard sector in relation to other manufacturing industries in its rate of productivity increase—crudely, a measure of improvement in output per unit input (e.g., capital, labor, or all inputs).[32] This should not be surprising, given a direct relationship between the advance of economies of scale and productivity improvement over time.[33] From a historical perspective, the lumber industry has ranked at or near the bottom of all U.S. manufacturing in terms of its rate of productivity advance, with total factor productivity improving in the lumber sector at an annual average rate of only 1.0 percent between 1899 and 1953. This compares with an average of 2.0 percent for all U.S. manufacturing industries over this same period. Over the same period, labor productivity in U.S. lumber production increased at an average rate of 1.2 percent per year, also among the lowest in manufacturing and substantially lower than the manufacturing average of 2.2 percent per year.[34]

In more recent years, the rate of productivity improvement in U.S. manufacturing has been somewhat more sporadic, including instances of industry declines during the 1970s. By contrast, between 1970 and 1979, labor productivity in lumber manufacturing grew at an average rate of 1.2 percent per year, slightly higher than historic levels and higher than numerous other manufacturing sectors. Moreover, during the 1980s, productivity improved in U.S. lumber manufacturing at a reasonably rapid 2.9 percent per year. This too is higher than rates achieved in numerous U.S. manufacturing sectors during the decade.[35] Yet, these recent upswings notwithstanding, the longer term trend is apparent: productivity growth in the lumber sector has been remarkable largely for its sluggish pace.

The Problem of Raw Material Dependence

U.S. and Douglas-fir region lumber production have never conformed with Chandlerian standards of managerial capitalism. This is apparently due to comparatively low economies of scale and productivity improvement. Yet a visit to virtually any contemporary lumber mill will confirm that high-quality energy supply and rapid throughput technologies are no strangers to modern lumber manufacture, the onset of which were documented more than sixty years ago by the Federal Works Agency's Work Projects Administration.[36] Why then is lumber different from Chandler's paragons of bigness? To answer this question, one must not look to production technologies per se but instead consider the sector's character as a raw material gathering and conversion industry. Dependence on geographically dispersed, heterogeneous timber supply with physical characteristics that are (largely) ecologically determined and only somewhat malleable creates problems that have historically impeded the development of significant scale economies and the advance of productivity in lumber manufacture. Essentially, the problem is the industry's quasi-extractive character, relying as it does on the conversion of a slow-growing, extensive resource into commodities. It follows from this that, where bigness in lumbering has been achieved, it has historically come through control of timber rather than through scale economies in industrial throughput and through associated economies of scope in processing heterogeneous raw materials into a range of commodities.

The significance of lumbering's nature-centered character to the development of economies of scale, to the advance of productivity, and more generally to the industry's shifting geography was in fact recognized as a defining feature by the Federal Works Agency's Work Projects Administration's study of mechanization in the industry. The report compares lumber manufacture to extractive industries, such as mining, noting,

> In extractive industries technology is faced with continually increasing physical difficulties as operations are continued at the same location; these must be overcome if the productivity of labor is to be maintained or increased. To the extent that the lumber industry has employed an extractive technique, it has faced obstacles to increased productivity analogous to those encountered by mining industries. *If there is no accompanying change in the techniques employed, labor requirements are higher per thousand board feet of lumber realized in the logging and milling of smaller timber, and they are likewise increased as logging operations extend further from the mill into less accessible territory or as the cutting of less dense stands of timber is undertaken.*[37] (emphasis added)

Refining these insights, one can consider this problem of raw material dependence and the challenges it presents to lumber manufacture by considering two constraints on economies of scale and productivity advance: (1) in securing access to adequate timber supplies, including their harvest and transport, and (2) in the conversion of round logs to rectangular pieces.

Scale Economies and the Economic Geography of Extraction

The geography of timber supply is highly extensive. The density of standing timber in the forest is generally low relative to the raw material demands of a mill, particularly a large, modern one, and because trees grow relatively slowly, there is a limit on the amount of timber that can be supplied within any particular area. This gives lumbering an inescapably extractive character at a local scale of resolution. To secure a continuous supply of raw materials, mills must source timber from a wide area. This problem has only become worse as mills have become more efficient, capable of more rapid throughput due not least to larger and higher quality energy supplies and, more recently, to highly automated and computerized log-handling and processing equipment—which, according to Chandler, are the driving forces behind bigness. This in turn means that greater scale economies in mills translate directly into more extensive spatial requirements to ensure adequate timber supply; scale economies and the "timbershed" of a mill are directly related and can be in tension with one another.

Consider, for example, that an average sawmill in Oregon consumed about twenty-five million board feet (Scribner) of logs in 1994. Assuming sawtimber density of fifteen thousand board feet per acre of timberlands, this would mean that the mill needs to draw timber from an area of about 1,700 acres of timberlands to operate for one year.[38] Now, assuming a sixty-year growing cycle, the mill would need to own or have access to about one hundred thousand acres of timberlands. Considering that ownerships are typically fragmented, the problems of an extensive geography of timber supply for lumber milling are made even more acute. Compounding this, however, are penalties to hauling logs from logging sites to mills. The ratio of the volume of lumber produced to the volume of wood consumed in sawmills has increased in recent years, but even efficient mills achieve only about 50 percent volume recovery.[39] This means that transporting logs to mills entails pulling about twice the volume of wood that ends up as lumber; the rest comprises what is known in the industry as "residuals" (e.g., chips, sawdust, bark). These residuals have value as inputs for other production processes, but their value per volume is certainly lower than lumber. Moreover, lumber must be dried—by air or in a kiln—before being distributed to wholesalers or consumers.[40] This imposes a further penalty on long log hauls, because water is heavy and its weight is a pure loss. Thus, the geography of collection and transport costs constrain typical log haul distances for a lumber mill to perhaps fifty to one hundred miles, with some variation depending on market conditions.[41] One of the central design constraints on a mill's production capacity is therefore how much timber can be supplied to it on a continuous basis from within this radius. These are the spatial economic foundations of the notion of a timbershed or "working circle," the long-standing recognition in the industry that mills compete not only in an aggregate market based on lumber prices but also in a very localized market for timber supply.[42]

One way for lumber capital to solve this problem is to be mobile, one of the key factors favoring smaller mills; larger, more capital-intensive facilities were all that much more difficult and expensive to move or decommission. As stated by the Federal Works Agency's Work Projects Administration's report on mechanization in the lumber industry,

> The small mill is able to operate closer to the timber and can draw on stands of smaller volume than are required for the successful operation of large, stationary mills. To do so, however, the small mill must move frequently; it attains the requisite mobility by sacrificing much of the labor-saving machinery of the large, stationary mills.[43]

Though this quote specifically pertains to very small, highly mobile "peckerwood mills" typical of the late nineteenth century, the dilemma or trade-off between scale economies and timber supply is manifest at a broader scale in the historical geography of American lumbering, and particularly prior to the onset of industrial plantation forestry in the mid-twentieth century, when lumbering was an exclusively extractive industry. Expanding and relatively inelastic domestic demand for wood (particularly lumber), long cycles of forest renewal, and huge regional accumulations of standing timber in various forested parts of the United States propelled an industry that confronted forests that capital had no role in producing. As long as sufficient forest stocks existed elsewhere, prices were too low for firms to justify treating the raw material in any particular location as anything other than an extractive, nonrenewable resource. Even under the influence of regulatory and social pressures to regenerate forests, firms have until relatively recently treated reforestation expenses as a cost of doing business, a purely regulatory burden, as opposed to a form of production and investment (see chapter 5). As former U.S. chief forester William B. Greeley put it in the 1920s, "It is written in the immutable law of commerce that industries seek their cheapest source of raw material."[44] Thus, prior to its arrival in the Douglas-fir region, as industry exhausted forests in any particular region, capital moved on; during this extractive phase, the nature upon which capital relied was truly external or "found" in a material sense. In its wake, the lumber industry left a landscape of devastation.

This is not to say that the timing and character of shifts in the historical geography of the U.S. lumber industry are due solely to the availability of timber. Rather, institutional, political, and regulatory factors, including, for example, the railroad land grants, the availability of cheap and abundant land courtesy of federal disposal policies, and a westward expansion of the United States propelled by the doctrine of Manifest Destiny, all helped propel regional shifts in the industry's geography, leading to its arrival in the Pacific Northwest. At any given time, even within the same regional context, trees can be extractive and renewable resources; which label applies is largely a question of politics, economics, history, and geography. Moreover, with time, industrial plantation forestry has increasingly taken hold of forest growth, moving from extraction to cultivation. Yet the rise of the Douglas-fir region's lumber industry came on the heels of dramatic expansions and declines in lumber production, first in the Northeast, then the Great Lakes, and finally in the Southeast. As one region's production peaked and sagged, it was supplanted, even in local markets, by production from the next frontier in a wave of essentially extractive forestry.[45]

In fact, prior to and underlying the arrival of lumbering in the Douglas-fir region during the mid-nineteenth century, a shifting nomadic character was arguably its defining feature, with accompanying fears of timber famine and related community social and economic instability the most pressing political issues facing American forestry.[46] The first forests subject to large-scale exploitation were those of the Northeast. Sawmills were established in New York and Maine by the middle of the seventeenth century, and community sawmills soon became fixtures in towns with nearby supplies of sawtimber.[47] In 1839, for example, the Northeast states accounted for two-thirds of the nation's increasingly industrial lumber production. Yet, within a few short years, as the forests of Maine, Vermont, Pennsylvania, and New York were liquidated, the geographic center of lumber milling had moved on, to the white pine forests of the Great Lake states and the yellow pine forests of the Southeast.[48] Chicago's emergence as the hub of the American Midwest was critically underpinned by the city's centrality to the lumber trade. Expanding lumber output from mills in Wisconsin, Minnesota, and northern Michigan was shipped by boat and barge down rivers and lakes to be sold in Chicago's great wholesale yards, transforming the city into what William Cronon termed "the single greatest lumber market in the world" by the 1850s.[49]

In turn, lumber production in the Lake states peaked in absolute terms, and as a proportion of the nation's total (about 35 percent) during the late 1880s and early 1890s. Thereafter, limited and falling regional timber supplies and buoyant demand led lumber interests from the northern states to turn their attention to the extensive pine forests of the Southeast. Lumber production in the South soared to more than fifteen billion board feet by 1912, with the region accounting for more than one-third of the nation's supply. Yet here, too, cutting exceeded growth, leading to rapid exhaustion of forest stands. An estimated ninety million acres of cutover lands had been created by 1920, only a third of which was reforested.[50] Resource depletion coupled with the availability of ample timber in the Northwest undercut the region's dominance, and the industry turned increasingly to the West. Massive stands of timber in the region, coupled with the ruinous exhaustion of forest resources in other regions of the country, meant that a long boom in Pacific Northwest lumbering came when the region's forests were estimated to contain perhaps half of the nation's standing timber.[51] Much of this massive inventory was concentrated in the Douglas-fir region, a

120- to 150-mile-wide strip between the Cascade summit and the Pacific Coast (see Figure 1.1).

The first mechanical (steam- and water-powered) sawmills were actually established on the Pacific Coast in the 1820s and 1830s. The Hudson's Bay Company built a mill at Fort Vancouver (Washington) across the Columbia River from what is now Portland in 1827, whereas Oregon's first sawmill was constructed on the lower Willamette River (near what is now Oregon City) in 1842 by the Island Milling Company.[52] By 1850 there were thirty-six mills in Oregon and Washington combined, and more than two hundred by 1880. However, these mills served mainly local markets, and the region's lumber boom did not really take off until after the turn of the century (see Table 4.1).

When it came, explosive growth was concentrated first in northern California and in Washington. Puget Sound quickly developed as the center of the Northwest's lumber industry, with much of the production serving the rapidly expanding California market. Washington emerged as the nation's leading lumber-producing state by 1909, increasingly serving midwestern and eastern markets (in addition to California) with lumber shipped along recently completed transcontinental railroads.[53] Led by thousands of small firms and several large integrated companies such as Weyerhaeuser, the lumber industry began to tap Oregon's Douglas-fir forests in earnest during the 1920s and 1930s. The state surpassed Washington as the nation's leading lumber producer in the late 1930s, with annual production exceeding 5 billion board feet by 1940, and peaking in 1955 at almost 9.2 billion board feet, close to one-third of the nation's total at the time.

With the onset of the lumber boom in Oregon, the interregional dynamics of the lumber industry within the United States changed; quite simply, there were no more frontiers of untapped, old-growth sawtimber, at least not in the forty-eight contiguous states. Perhaps not coincidentally, scale economies in lumbering took a decidedly upward turn in the mid-twentieth century as larger, less mobile mills became more prevalent and lumber capital began to develop new ways of sourcing timber to mills on an ongoing basis. This includes important new technologies in logging such as the chain saw, the truck, and mechanized

Table 4.1 Sawmills in Oregon and Washington, 1850–1920

Year	>10	5–10	1–5	0.5–1	<0.5	Total Mills
1850	0	0	5	18	13	36
1860	2	5	9	17	117	150
1880	7	7	32	29	122	197
1900	53	42	200	123	259	677
1920	246	81	372	167	377	1243

Source: Mitchell, 1988.

diesel skidders and tractors, all making logging more flexible and easier to relocate and allowing longer distance transport to mills (no doubt aided by the
expansion of federal and state highway systems).

Yet if the pronounced historical geography of interregional shifting in lumbering had reached its limit, a geography of extraction in lumbering persisted
within the Douglas-fir region. Sustained-yield forestry doctrine notwithstanding (see chapter 6), the cycle of boom and bust enacted on a national scale was
repeated at a more local scale of resolution, as mills were established and local,
high-quality timber stocks were exhausted. Some glimpse of this turbulence
can be gleaned by examining Figures 4.2 and 4.3 and the data in Table 4.2.
Figure 4.2 shows lumber production in the nineteen counties of western Oregon (the state's portion of the Douglas-fir region) from 1925 to 1990.[54] Note
the wide fluctuations from year to year, testament to the limits of sustained-
yield forestry in smoothing lumber production swings. Figure 4.3 is a rather
busy graph showing thirteen lumber production trend lines over the same period, ten for individual counties and three compiled from aggregations of contiguous counties necessary to ensure consistency in reporting throughout the
period in question.

These data are the most disaggregate available, aside from individual mill
output, and they show several things. First, as shown in Figures 4.2 and 4.3, it is
apparent that some broad parallels exist between the aggregate and disaggregate data—as indeed one would expect, because the state total is only the aggregate of individual counties, and the western counties of Oregon are by far

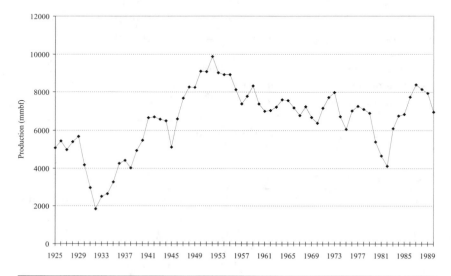

Fig. 4.2 Lumber Production in Western Oregon, 1925–1990. Source: West Coast Lumbermen's
Association (1964), Western Wood Products Association (1998), and various other years.

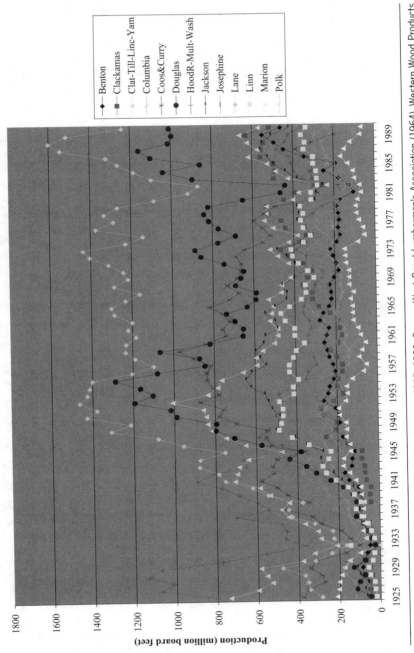

Fig. 4.3 Lumber Production in Western Oregon Counties, 1925–1990. Source: West Coast Lumbermen's Association (1964), Western Wood Products Association (1998), and various other years.

Table 4.2 Correlation Matrix of Lumber Production Trends in Western Oregon Counties, 1925-1990

	Benton	Clackamas	Clat-Till-Linc-Yam	Columbia	Coos & Curry	Douglas	HoodR-Mult-Wash	Jackson	Josephine	Lane	Linn	Marion	Polk
Benton	1.00												
Clackamas	0.70	1.00											
Clat-Till-Linc-Yam	0.31	0.00	1.00										
Columbia	0.13	0.03	0.44	1.00									
Coos&Curry	0.25	0.03	0.47	0.31	1.00								
Douglas	0.77	0.71	0.46	0.06	0.55	1.00							
HoodR-Mult-Wash	-0.42	-0.59	0.27	0.43	-0.31	-0.57	1.00						
Jackson	0.75	0.58	0.46	0.12	0.70	0.94	-0.57	1.00					
Josephine	0.64	0.60	0.44	0.09	0.69	0.93	-0.64	0.93	1.00				
Lane	0.78	0.75	0.41	0.22	0.60	0.94	-0.60	0.93	0.93	1.00			
Linn	0.64	0.55	0.44	0.12	0.72	0.86	-0.64	0.92	0.94	0.92	1.00		
Marion	-0.07	-0.12	0.56	0.69	0.11	0.02	0.54	0.03	0.02	0.05	-0.02	1.00	
Polk	0.57	0.53	0.63	0.48	0.34	0.64	-0.06	0.60	0.58	0.73	0.62	0.44	1.00

Note: Clat-Till-Linc-Yam represents an aggregation of data for Clatsop, Tillamook, Lincoln, and Yamhill counties; HoodR-Mult-Wash is an aggregation of data for Hood River, Multnomah, and Washington counties; Coos and Curry combines data for these two counties.

Source: These aggregations are necessary because of the way the data are originally reported in Western Wood Products Association (1998) and various other years.

and away the most significant lumber producers during this period. State-level output generally shows an upward trend following the Depression in the early 1930s, peaking in the mid-1950s, oscillating but generally trending downward through the early 1980s, rebounding in the late 1980s, and beginning to drop again as the spotted owl crisis came to a head. These fluctuations have many underlying causes, with lumber production closely tied to the housing market and overall macroeconomic conditions (indicated by deep dips in the early 1930s and 1980s coinciding with serious, generalized economic crises). Trends for at least some of the individual counties and small aggregates of contiguous counties show broad parallels with the overall state totals (e.g., see Lane County). Figure 4.3 also shows broad parallels between some pairs of individual counties and aggregates, which might also be expected given broad economic tendencies, not least the increasing intensity of investment and production in Douglas-fir lumbering beginning in the 1920s and extending into the 1950s. What is surprising about the data in Figure 4.3 in this context, however, is the divergence of trends between many of the pairs of counties (and aggregates). Some of the trend lines do not track well at all. Table 4.2 is a matrix of pairwise correlations between county trends, each number representing the correlation between trends in one county and another (or a small aggregate). Although some pairs of counties do track very closely ($r > .90$), others are not well correlated at all ($r < .20$), and still others actually show negative correlations. Why? The answer is that lumbering moved around the Douglas-fir region, in this case within the nineteen counties of western Oregon, just as it had in the nation more generally. Although counties that are close to or even contiguous with one another tend to track closely (e.g., Jackson and Douglas counties, or Lane and Douglas counties), others that are further apart (e.g., Jackson and Clackamas counties, or Josephine and Columbia counties) do not. This reflects the historical geography of pronounced booms and busts in Douglas-fir lumbering, with escalating production first in coastal counties and those around the Portland area later eclipsed by the interior and more southern counties.[55] This is apparent, for example, by looking at the aggregate of Hood River, Multnomah, and Washington counties (all around Portland), positively correlated only with Marion County just to the south, as lumbering in these counties peaked in the late 1920s. Likewise, the aggregate of Clatsop, Lincoln, Marion, and Yamhill counties, all broadly in the coastal Northwest of the state, show weak correlations with the other counties (except perhaps with nearby Marion and Polk counties), again, as lumbering peaked in these areas earlier than in the rest of the region.

The point here is that the historical geography of lumbering has been characterized at multiple scales of analysis by cycles of local timber extraction, as capital exhausts nearby timber supplies and moves on. This essentially extractive geography has been a profoundly significant phenomenon from a political and ecological standpoint, because it has propelled wide swings in employment and economic development at regional and community levels (see chapter 6) and has led to forest depletion and profound ecological transformations. The incentive to clear-cut around a lumber mill, to minimize the distance over which logs must be hauled, is also a critical factor reinforcing this method of forest harvest, which is in turn one of the most controversial aspects of industrial forestry.[56] However, this geography of extraction is also part of the dynamic underlying lumber's exception to Chandler's managerial model of corporate capitalism. The extensive spatial requirements of supplying timber on an ongoing basis have factored centrally in the industry's historically migratory tendencies and have also comprised a direct challenge to the construction of larger mills.

Yet economies of scale in lumbering also are constrained by the heterogeneity of trees and logs as raw material. Though organized in assembly-line fashion, lumber mills are in reality disassembly lines, reducing whole logs into pieces of finished lumber and waste materials (chips, sawdust, and bark). The production line in a lumber mill is in essence a series of decisions about how to convert each log into products, and how to sort the products based on size, species, and quality. Consider, for example, the succinct and still essentially accurate description of the lumber manufacturing process offered by the Federal Works Agency's Work Project Administration's report on mechanization in lumbering:

> At the mill the logs are hoisted, dragged, or conveyed to the log deck in the mill. From the deck the log is shifted onto a carriage which travels on rails and conveys the logs past the headsaw. The headsaw reduces the [log] to cants (large dimensions for later reduction by resaws or gang saws) or, by successive operations, to boards. The "sawyer," who is the key employee of the sawmill, stands beside the headsaw and directs the workers on the carriage in turning and setting the log for cutting. All boards from the headsaw, gang saw, or resaw which have not been cut to uniform width are passed through an edger where the bark is cut off and the width made more standard. The boards then pass to trimmers where the ends are squared, inferior portions removed, and standard lengths made. After they have been trimmed, the boards are graded, sorted, and stacked.[57]

Contemporary sawmills are more mechanized, typically including a computer to help make decisions on the recovery of value from each log and also to increase the precision of cutting. Specific jobs, including, for example, workers turning cants by hand on the carriage, have been replaced by machines. Yet, in essence, lumbering remains a craft of decision making on how to reduce a log into a collection of pieces whose aggregate value is maximized.

These decisions are based on market criteria but are a reaction to the heterogeneity of logs coming into the mill.[58] Logs vary by species, diameter, wood density, the frequency of knots and other flaws, straightness, and so forth. As I discussed in the last chapter, differences between species and age classes of logs are central considerations in regional commodity specialization and to decisions about how to recover the most value (or volume) from each log, within and across commodity sectors. This includes how to reduce a round log to lumber and achieve the greatest return. Although expertise is accumulated by workers (and by computers) from one log to the next, there are limits in the extent to which decisions about one log affect the decisions about another, because each is in some way unique, and this in turn limits the degree to which the production process can be rationalized and standardized. For these reasons, the head sawyer is widely acknowledged to be the most important person in determining the profitability of a sawmill. Moreover, because the production process is, despite the aforementioned changes in technologies, a relatively simple matter of mechanically reducing logs to lumber, the costs of building even larger mills is low compared with the costs of building production facilities in more technologically involved sectors.

Finally, all of these issues and problems are made more acute by the conversion of old-growth to young-growth forests in the Douglas-fir region. This conversion—the substitution of one type of resource for another as discussed in the previous chapter—has exacerbated the challenges to increasing economies of scale and the advance of productivity. Specifically, timber densities drop in moving from old-growth to young-growth forests, as do yields of lumber as the diameter of logs decreases (recall Figure 3.3). Thus, smaller logs mean that, all other things being equal, a mill needs access to more area of young-growth timberlands to sustain an equal level of production, pays a higher premium for transport, and must at the same time process more logs per unit of output than is the case for old-growth timberlands. Thus, the Federal Works Agency's Work Projects Administration report on mechanization in the lumber industry, though it did not deal with transition from old-growth, was nevertheless able to identify succinctly the relationship between economies of scale and productivity in lumbering on the one hand and the raw material centered character of the industry on the other:

An examination of the relationship between unit labor requirements and technological developments in the lumber industry must begin with a recognition of two facts: (i) that unit labor requirements, measured by the number of man-hours required for the production of 1,000 board feet of lumber, increase with a decrease in the size, quality, and accessibility of the timber available for cutting and (ii) that the lumber industry has operated for the most part on the principle of clear-cutting first the most accessible stands of large size, high-quality timber, so that there has been a steady decrease in the accessibility of available timber, or in its size or quality, or in both as operations continued within a particular region.

Because of these conditions, a consistent process of adjustment has been necessary to accomplish a reduction or even the maintenance of the level of unit labor requirements.[59] (emphasis added)

The Keys to Bigness I: Timberland Ownership

If lumber as a whole diverges from the paragons of Chandlerian managerial capitalism for reasons having to do with securing access to and processing timber, it is also the case that certain lumber-producing firms that have been able to develop into large, complex, and integrated corporations. What are the keys to their ability to achieve bigness? Numerous factors explain the growth and consolidation of large vertically and horizontally integrated firms in the lumber sector, and in solid wood products more generally, some common to many firms in the sector and some specific to particular firms. But raw material issues are once again central. Specifically, if securing access to adequate timber combined with processing a heterogeneous raw material present challenges to bigness, it is the successful confrontation with these challenges through (1) control of large areas of timberlands and (2) product and process diversification driven by economies of scope in log processing that link the largest, corporate lumber-producing entities.

Given the raw material intensity of lumber and wood-products manufacture, the need to secure access to large quantities of fiber is an obvious prerequisite to achieving bigness. At the same time, because of the relatively low-tech character of mechanically reducing logs to lumber, smaller mills have generally been competitive in the lumber sector, and barriers to entry in the form of large fixed capital outlays for mill construction do not exist to the degree that they do in, say, pulp and paper manufacture. However, acquiring large quantities of contiguous timberlands addresses the trade-off between building larger, more efficient sawmills and ensuring adequate raw material supply, while comprising a defense against entry. Moreover, because timber is in limited supply, both in aggregate and in local timbersheds, and because ownership of private timber is exclusive, ownership of large blocks of timber can be and has been used as a strategy to achieve and protect market shares. Speculative acquisitions have allowed some firms to obtain timber at low costs and use their resulting raw material cost advantages to achieve significant market shares in wood-products manufacture, a dynamic not unlike that noted by Barham in the context of the aluminum industry.[60]

The relationship between bigness and the control of raw material has long been recognized in the lumber industry. The 1913 U.S. Bureau of Corporations report on the lumber industry noted,

If there is any condition in the industry to prevent the introduction of competition, that condition becomes the basis upon which those controlling it may combine. Such a potential condition is present in the ownership of the timber. Those who control the standing timber can control the manufacture of lumber; any permanently effective combination in the manufacture of lumber or in the wholesale trade must almost certainly be a combination among timber owners.[61]

Though published during a current of strong antitrust sentiment in American public opinion,[62] the report also came out soon after, and was partly inspired by, the most significant single purchase of timberlands in the history of the United States. In December 1899, in a transaction of some questionable legality under the rules governing railroad land grant sales, Frederick Weyerhaeuser of the Weyerhaeuser Timber Company of Minnesota negotiated the purchase of 900,000 acres of land grant Douglas-fir timberlands in Washington and Oregon from the Northern Pacific Railroad, headed by Weyerhaeuser's St. Paul neighbor James J. Hill. The price was $5.4 million, or $6.00 per acre, for some of the best quality timberlands left in the United States at the time—best quality because of their high productivity and also because these lands held massive volumes of old-growth. Weyerhaeuser then bought a further 380,000 acres in Oregon and Washington in August 1901, some at $6.00 per acre and some at $5.00 per acre.[63] Reflecting the significance of controlling timberlands to competitive positioning in the lumber industry, Frederick Weyerhaeuser told George Long, then manager of the company's West Coast operations, simply, "The less [timber] we sell and the more we buy, the more money we make."[64]

The Weyerhaeuser purchase was by far the largest and most dramatic acquisition of Douglas-fir region timberlands of the era. But it highlighted the significance of such acquisitions in a highly competitive industry and a more general rush to lock up Douglas-fir timberlands. Widespread speculation of a timber famine at the time, fueled by the extensive deforestation of timberlands in the Lake states, and in the Southeast, no doubt propelled a green gold rush of sorts to the nation's last timber frontier.[65] At the same time, however, lumber operators had learned through experiences in other regions that the key to a strong position in an industry plagued by overproduction and wide price fluctuations was upstream integration into timber ownership. Intense efforts to acquire timberlands—and the ease with which federal land disposal grants were abused—quickly made timber ownership in the coastal Northwest more concentrated than in any other part of the country.[66] By 1913 sixteen of the largest Douglas-fir timberland owners controlled more than 40 percent of all private timberlands in region.[67]

This is not to say that private timberland ownership allowed price and market manipulations in the lumber industry typical of monopolistic and oligopolistic industry structures, even if, as Walter Mead demonstrated, local control of timbersheds—both public and private—is no stranger to the Douglas-fir region.

Rather, limited economies of scale in manufacturing, large amounts of timber remaining in small ownerships, and the increasing availability of federal timber from vast public forests in the region made oligopolistic practices almost impossible.[68] The real importance of timberland ownership is that it has consistently proved critical to those firms that have managed to "get big" and stay big in the lumber industry, and in solid wood products more generally. Weyerhaeuser was the first to demonstrate this link, scaling up its rate of lumber production during the 1920s and emerging as the nation's largest lumber producer during the postwar. The company has maintained this position up to the present.[69] Weyerhaeuser's ability to achieve bigness through raw material acquisition was mirrored by the Georgia-Pacific Corporation's rise to prominence in the Douglas-fir region based initially on a series of timberland purchases in Oregon and Washington during the 1950s and 1960s, as noted in the previous chapter. Georgia-Pacific became most dominant in softwood plywood production in the region, and it remains the nation's largest manufacturer of structural wood panels. Yet the firm rapidly became the second-largest producer of Douglas-fir lumber after Weyerhaeuser, and it remains the number two producer of lumber in North America.[70] Critical to Georgia-Pacific's rapid rise to bigness, and to the firm's growth into one of the largest forest products firms in the world, has been and continues to be control of timberlands and the firm's capacity to diversify production from these lands.

The importance of timber acquisition as a strategy to achieving bigness is based not only on technical aspects of the relationship between the geography and heterogeneity of timber supply on the one hand and economies of scale in production on the other. Rather, the significance of holding timber has been augmented by regulatory considerations. In particular, the acquisition of timberlands as a competitive strategy by large firms such as Georgia-Pacific was encouraged by key changes in federal rules regarding the financing of timber acquisitions during the mid-twentieth century, changes that in fact helped blur the lines between nature and capital. Specifically, in 1953 the Federal Reserve for the first time allowed banks to loan money against the value of standing timber, essentially creating a credit system based on timber as a capital asset. The Federal Reserve's decision built on changes introduced under the Revenue Act of 1943, altering section 631(a) of the Internal Revenue Code to allow integrated firms to count as a capital gain the net value of cut timber at the time of harvest. By contrast, timber sold through market transactions was taxed at a higher rate, thereby creating an obvious advantage to vertical integration. These changes are key when one considers the amount of money that must be locked up in timberlands and, in the absence of credit financing on their value, the responsibility of timberland owners to raise sufficient capital to acquire them. Not only did the Federal Reserve allow firms to count nature as a form of capital for the purposes of securing financing, it also opened a door for the credit system to spread and manage risks associated with holding and growing timber—risks from fire, storms, or disease outbreaks that can devastate forest resources. In this instance, then, finance capital, aided by Federal Reserve rules, expanded the circuits of accumulation by taking on some of the problems posed by nature as time and nature as risk. This is a prime example of the timescapes formed by collisions between biological and social time under capitalist modernity, as discussed by Barbara Adam, yet also indicates a distinct parallel to the ways in which credit finance acted to address natural risks, vulnerabilities, and time lags in California agriculture, as discussed brilliantly by George Henderson.[71] Not surprisingly, major timberland acquisitions and mergers accelerated in the wake of these changes, as firms sought to integrate upstream into timberland ownership, led in substantial measure by Georgia-Pacific; timberland holdings of the twelve largest U.S. forest products companies more than tripled between 1950 and 1970.[72] Moreover, control of timberlands by larger, more integrated firms has become an even more pronounced aspect of firm strategies in the Douglas-fir region than in the United States as a whole. Although in the Douglas-fir region 35 percent of all timberlands and two-thirds of all private ownership is held by industrial landowners, only 14 percent of all U.S. timberlands and 20 percent of all private timberlands are held in industrial ownerships.[73] In Oregon most of the state's industrial timberlands are held by a group of only twenty owners.[74]

Whether timberland ownership is overly concentrated, or is becoming so, is an important political question, one that factors centrally into discussions regarding the links between forest policy goals (employment, community development, etc.) and forest tenure. However, this is not the main thrust of my discussion here. Rather, the point is that upstream integration into timberlands is a prerequisite to achieving bigness in the lumber industry. This has proved to be the case in the postwar lumber industry, and it remains the case in the Douglas-fir region. The succinct observation made by the U.S. Bureau of Corporations in 1913 remains essentially true: no enduring combination or large corporation in the solid wood-products industry in the Northwest is without its own significant raw material base. Of all the firms owning multiple lumber mills in Oregon, for example, almost all are significant owners of timberlands. This is not to say no firms attempt other strategies. One firm I encountered during my fieldwork, for example, is WTD Industries, a firm that at the time operated six softwood lumber mills and one finger-joined lumber mill in Oregon and was the fourth-largest softwood lumber producer in the United States. Yet WTD relied on open market purchases for approximately 95 percent of its raw material needs. How enduring this strategy is, however, remains to be seen. WTD was formed only in 1984 and has already experienced severe financial difficulties, filing for Chapter 11 bankruptcy in 1991 and reporting a net loss of $12.1 million in 1998.[75]

WTD notwithstanding, not only do large, complex corporate firms own substantial timberlands but also even most of the enduring midsize familial or entrepreneurial lumber firms have retained substantial timberlands. Moreover, most industry observers agree that the importance of vertical integration into timberland ownership has only increased in recent years, as federal timber sale programs in the region have been scaled back in the wake of the spotted owl crisis. This has prompted further restructuring of private timberland ownership, a process in which major firms wishing to remain active in the region have sought to add to or consolidate their landholdings. Unquestionably the most prominent example of this is Willamette Industries. Willamette was founded in Dallas, Oregon, in 1906. The modern version of the company was formed from a merger among several independent firms in 1967, and first publicly traded in 1968.[76] By the 1980s, Willamette Industries had become one of the largest solid wood products producers in the U.S., yet despite being large and diversified, remained approximately 50 percent reliant on federal timber sales to feed its mills. But in August of 1991, the firm moved to consolidate its position in the face of restrictions on federal timber sales, buying forty-five thousand acres of Oregon timberlands from Bohemia Inc. One of the critical aspects of the decision by Bohemia to sell, and Willamette to buy, was that neither company was self-sufficient in timberlands.[77] Subsequently, in March 1996, Willamette agreed to pay $1.59 billion to Hanson PLC for five hundred thousand acres of timberlands in Oregon and Washington. The purchase doubled the company's timberland inventory, vaulting it to a rank as the tenth-largest timberland owner in the United States, and making the firm self-sufficient in Douglas-fir region timberlands.[78] Yet the financial strain of these restructurings left the firm vulnerable, and it has since been swallowed by Weyerhaeuser, decisively consolidating the latter firm's stranglehold as the leading private holder of timberlands in the Douglas-fir region and its largest lumber manufacturer.

The Keys to Bigness II: Economies of Scope

Vertical integration upstream to control timberlands is key to achieving bigness among those managerial firms that have emerged and endured in the Douglas-fir lumber industry. Yet these firms are more than vertically integrated lumber manufacturers. Indeed, another important aspect of their bigness is that they are diversified and active in the production of a variety of wood commodities. Here again, raw materials are a crucial consideration. Specifically, Chandler's model of economies of scope is inadequate to explain the advantages of diversification in wood-products manufacture because these advantages do not come from exclusively technological or process continuities across commodity lines. Rather, economies of scope in wood-products manufacture originate also in the diversity of the raw material base, a diversity that returns maximum value to firms able to process logs of different species and qualities into different kinds of wood products.

Log heterogeneity is a key problem in lumber mills, as noted previously, even within a given forest region such as the Douglas-fir region. Yet in a lumber mill, heterogeneity of log types is largely restricted to those logs suitable for lumber manufacture—sawlogs. A much broader diversity of logs comes from most forest stands because of the diversity of species and ages and the wide variation in the specific characteristics of each tree. In Oregon's Douglas-fir region, for instance, although about 75 percent of the volume of all timber harvested in 1995 was Douglas-fir, the remainder came from a diversity of species, including spruce, pine, hemlock, cedar, and others, all with their own unique biophysical characteristics (recall Figure 3.4). At the same time, different wood products have different raw material requirements that mesh more or less well with different kinds of trees. Thus, log flow from a single logging unit (and from a contiguous ownership more generally) is typically directed along numerous distinct pathways or "commodity chains" to production facilities of different kinds, including sawmills, plywood mills, pulp and paper facilities, and the newer fabricated solid wood-products manufacturers; to each go, respectively, sawlogs, peelers or veneer logs, and pulp logs. In the Pacific Northwest, as noted in the previous chapter, sawlogs dominate, accounting for 70 percent of the total harvest in 2001 (see Figure 3.6). However, 17 percent of logs were allocated to plywood mills as veneer logs primarily because their physical characteristics were better suited to this commodity line. The segmentation of log types into different commodity chains is reflected by the existence of formally distinct log markets, with their own posted prices, for logs of different kinds, including sawlogs of various qualities, peeler logs of different grades, and chip logs. Although these markets indicate that firms can and do transfer logs to mills by way of market transactions, larger, diversified firms sort these logs internally, sending them from logging operations to mills of different kinds controlled by the same parent company. By diversifying wood-products facilities, firms can develop a set of material pathways from raw logs to finished products that suits the mix of log types (species, age class, wood densities, etc.) coming off their lands in a given area.

Thus, in the Douglas-fir region, some of the region's largest landowning and lumber-producing firms also have been its largest producers of other commodities, including plywood.[79] Large, managerial firms operating lumber and plywood facilities in Oregon in 1997, for example, included Boise Cascade (four lumber mills, seven plywood mills), Willamette Industries (five lumber mills, three plywood mills), and Weyerhaeuser (two lumber mills, one plywood mill). It is also true that many of the integrated medium-sized and familial firms, though hardly Chandlerian in their structures, have followed this path to economies of scope, including Roseboro Lumber Company, Roseburg Forest

Products Company, Superior Lumber Company, and Sun Studs Inc., all of which operate at least one lumber mill and one plywood mill in Oregon.[80]

As noted in the previous chapter, the Northwest is not a leader in pulp and paper production, in some measure because of regional commodity specialization linked to forest types. For this reason, most of the fiber going to pulp and paper manufacture in the Douglas-fir region comes from residual wood chips produced as a by-product of solid wood-products facilities. However, here too there are economies of scope from using whole trees in commodity production by processing residuals and by using lower grade pulp logs for making pulp and paper. As a result, horizontal integration between solid wood-products facilities and pulp and paper facilities under the umbrella of a single parent firm is common to several of the largest firms active in the region. This is epitomized by Boise Cascade, Georgia-Pacific, Willamette Industries, and Weyerhaeuser,[81] firms whose facilities account for about half of the state's total pulping capacity and whose strength in timberland ownership, sawmills, veneer and plywood mills, and other types of processing facilities rounds out their dominant positions in the region.

The key to all of this is that firms have been able to grow big through diversification, capturing economies of scope based on control of a heterogeneous raw material. Economies of scope in the industry, to the extent that they exist, do not originate solely in process technologies—indeed, there are few such links between lumbering and pulp and paper manufacture, as indicated by the fact that facilities producing different commodities tend to be quite independent of one another, with production lines in entirely separate facilities, often located far apart but also set apart in different buildings even when located at the same site. For these facilities, and their parent firms, process technologies for producing different wood products are distinct. Chandler's technology-based notion of economies of scope simply does not apply.[82]

Conclusion

In this chapter, intended as a complement to the previous one, I extended a discussion of the ways that the disjuncture between social and ecological production in nature-based commodity manufacture carries with it implications for the organization of raw materials sectors, including the wood-products industry and, specifically in this discussion, the lumber sector. My argument, drawing on Chandler's celebrated account of the rise of twentieth-century managerial capitalism differs from Chandler's by considering lumber as a case study that tests the converse of his hypothesis about the basis of economies of scale and scope. Specifically, if the key to bigness is rapid throughput enabled by a suite of technological innovations in factory production (including high-energy processes) that enable large economies of scale and scope, then the absence of bigness must be accompanied by the absence of these high-speed, high-quality energy processes. However, examining the lumber sector, with its relatively slow march toward the capture of scale economies, relatively low concentration indices, and historically slow rates of productivity growth, we find that the limit to bigness cannot be explained by the absence of rapid throughput processes per se but rather is based on the problems posed by raw material dependence. On the one hand these problems have to do with sawmills confronting a conflict between the advance of scale economies due to the development of high-quality energy (e.g., steam- and electricity-powered mills) and high-speed throughput and the need to secure adequate access to timberlands to feed these processes continuously. On the other hand, limits to process speed in the mills are also due in significant measure to raw material dependence and, specifically, to the heterogeneity of logs as a raw material. Thus, although high-speed processes are indeed key to the achievement of scale economies, and the absence of rapid throughput does seem to be a factor in the lumber sector's relatively low-scale economies, this absence is fundamentally related to the fact that lumber manufacture is essentially a disassembly process reliant on spatially extensive, slow-growing, and heterogeneous raw material inputs. Strengthening and complementing this insight is the fact that in the Douglas-fir region the few wood-products firms that have achieved a level of bigness comparable to Chandler's paragons of the corporate form are characterized by extensive timber holdings that can sustain their mills and various facilities and, in addition, are characterized by generally diverse types of production facilities and commodity lines that take advantage of economies of scope that also stem from heterogeneous log inputs.

I stress once again that by no means do I posit a static relationship between raw material demands and characteristics on the one hand and the institutional and social organization of commodity production on the other. Neither do I posit a kind of environmental determinism. My goal, in this chapter and in the last, is to explore a kind of codetermination or conjoined materiality between social and ecological production in nature-based commodity manufacture, interrogating the ways in which the fundamentally fictitious character of nature as commodity (i.e., its to some degree ecologically produced character) leads to important ways in which, at particular junctures, capital may be subject to a kind of ecoregulation.[83] In the next chapter I explore this notion in greater depth by examining the historical political economy of tree improvement and the ways in which the particular problems of trying to grow and customize the growth of forest trees have helped influence the peculiar political economy of industrial tree improvement in the Douglas-fir region.

5

Toward Organic Machines

The Historical Political Economy
of Douglas-Fir Tree Improvement

In volume 2 of *Capital: A Critique of Political Economy,* Karl Marx made one of his few, scattered, insights of direct relevance to the commodification of nature, dismissing the very notion of a truly capitalist forestry. He wrote,

> The long production time (which comprises a relatively small period of working time) and the great length of the periods of turnover entailed make forestry an industry of little attraction to private and therefore capitalist enterprise.[1]

In essence, Marx identified in forestry an example of what Barbara Adam, writing more recently, termed a "timescape of modernity" originating in the collision between biological time and social (in this case capitalist) time. For Marx, this collision presented seemingly insurmountable obstacles to capital, dissuading the reproduction of forest trees as a form of commodity per se. Yet history has shown and continues to demonstrate that Marx was only partly right in this admittedly casual observation. Although capital has indeed been slow to directly embrace industrial forestry as a proprietary undertaking, the history of intensive silviculture and tree improvement in particular bears witness to an assault on the boundaries between biophysical nature and capital in the industrial cultivation of forest trees.

In this chapter, I trace the shift from extraction to cultivation in Pacific Slope forestry. Specifically, I examine attempts to subordinate the reproductive biology of trees to the dictates of capital accumulation as an example of nature's capitalization[2] in the very material sense of this term. Viewed through this lens, nature is increasingly made (or, more accurately, remade) as opposed to found (as

113

under earlier extractive forestry regimes); "socially produced" by industry, the state, and science in Neil Smith's jarringly prescient framing.[3] Through nature's social production in tree improvement, the challenges posed by the reproductive biology of Douglas-fir region forest trees—including the problem of biological time—are increasingly confronted and transformed, allowing capital to circulate less and less around nature and more and more through it.[4] And as forestry capital, state agencies, and science together attempt to grow trees that are bigger and better, and grow them faster, biological productivity becomes the object of capitalist accumulation strategies, and the subsumption of nature by capital shifts from the formal to the real.[5] In short, nature is converted, albeit unevenly, into a form of commodity per se.

The most dramatic and clear evidence of these tendencies—in technological terms and in terms of the enclosure of genetic sequences as forms of private property—is apparent in the application of new biotechnologies in forestry. In particular, trees as commodities in full are suggested by the seemingly imminent prospect of proprietary genetically engineered (GE) tree varieties to be deployed in commercial forest plantations—subject, that is, to the resolution of a suite of important social and environmental contradictions and challenges. As I argue in this chapter, however, more than ninety years of research and applied science in forest genetics and tree improvement precede these more recent and dramatic developments. Any attempt to confront, explain, or critique the new biotechnology and the commodification of forest trees that seems now imminent must embed these projects in a longer term trajectory of political economy, ecology, and science. One of my aims in this chapter is to do exactly this and, at the same time, to explore another facet of nature-based commodity production, the commodification of nature per se.

My second aim of this chapter is to give the lie to narratives that would represent the shift to industrial cultivation and socially produced forests—through either conventional breeding or now genetic engineering—as simply a story of the triumph of capital and the market over nature's limits. For if Marx was mistaken in thinking capital had no interest in forestry as a private undertaking, he was right that the interface between capital and nature in the cultivation of forest trees is problematic. In response to a highly specific set of biological obstacles and challenges confronting industrial forestry and intensive tree breeding, complemented by the absence of clear legal provision for proprietary control of improved varieties (i.e., until recent changes accompanying GE organisms), firms have chosen to pool their risks and resources by cooperating and enlisting science and the state (or both at once through regional research universities) to do what private capital would not or could not do alone. The specific biological problems in question include reproductive delays, barriers to mass propagation of certain species of improved forest trees (including Douglas-fir), slow maturation rates, and breeding zones fragmented by provenance or seed source concerns. Thus, if capital has increasingly targeted and transformed the bio-logical basis of forest growth through industrial tree improvement, it is equally true that biophysical nature is inscribed into the very institutional organization of nature's social production; nature's social production is in this respect a dialec-tical conversation between social and environmental change.[6] Moreover, any notion of a "final" victory of capital over nature coincident with industrial culti-vation of commodified trees is misguided. Significant and unpredictable ecolog-ical risks accompany GE trees. At the same time, considerable social and political opposition to the genetic-engineering project persists based in part on such risks. These factors may yet impede the successful commodification of forest trees.

I begin this chapter with a discussion of some of the myriad challenges or obstacles to the industrialization of Douglas-fir reproduction, issues that delayed the onset of industrial tree improvement in the region and that have shaped the morphology of nature's real subsumption. I then review developments in tree-improvement operations through the 1980s, during which tree improve-ment became more closely tied to public science through silvicultural research cooperatives at state universities in the region. In the last section of the chapter, I discuss the prospects of more complete commodification of improved trees coincident with the emergence of forestry biotechnology, including genetic engineering. I stress that the success of this project is contingent on state and broader social sanction for commercial biotechnology and, ultimately, on the attendant environmental and social implications of trees socially produced using the new biotechnologies.

Delays and the Timescapes of Douglas-Fir Tree Improvement

Contemporary industrial tree improvement along the Pacific Slope dates to experiments in the geography of Douglas-fir breeding and environmental adap-tation initiated in 1913 by U.S. Forest Service scientist Thornton Munger.[7] Using test plots scattered between the crest of the Cascade Mountains and the coast, Munger helped to pioneer American research on seed source or provenance in forest trees in seeking to understand how genetic variation in Douglas-fir populations changed across space in response to locally specific environmental conditions such as drought, cold, heat intensity, and soil chemistry.[8] Provenance testing of the sort conducted by Munger had originated in European experi-mental forestry programs during the late nineteenth century, and research had shown that any successful industrial planting program needed to match seed sources with environmental conditions where the seeds (and later seedlings) would be grown; the alternative was to risk crop failure.[9]

Despite the unquestionable significance of Munger's work as a precursor to applied forest tree genetics and breeding along the Pacific Slope, tree improvement per se did not begin in the region for another forty years, and then only on an experimental basis. One reason for this delay is that for decades after Munger's work, there was little impetus for capital to undertake industrial forestry in a region still stocked with large accumulations of old-growth timber and in a political and regulatory context devoid of significant pressure to force the issue until into the 1940s. Forestry was, after all, an industry whose particular version of the "spatial fix"[10] involved an incessant search for new sources of raw material, manifest as a march from the Northeast, to the Midwest and Southeast, and finally to the Northwest.[11] Once the industry arrived on the Pacific Slope, liquidation of old-growth forest resources proceeded as it had elsewhere, and it was only after evidence of depletion became a concern within and outside the industry that industrial reforestation became a reality, punctuated in 1941 by Oregon's passage of the nation's first legislation governing forest practices on private lands.[12] That same year, Weyerhaeuser established the nation's first industrial tree farm near Grays Harbor, Washington, while the West Coast Lumbermen's Association, led by William Greeley, established a cooperative nonprofit nursery—the region's first—to supply five million seedling trees per year for reforestation purposes; seven years later, the Industrial Forestry Association (IFA) was formed as a collaborative effort among large industrial operators to coordinate Douglas-fir nursery and planting activities in the region.[13] All of these developments underpinned an increasingly social production of forests, propelling the expansion of industrial reforestation in the region (see Figure 5.1). And overwhelmingly, expanded reforestation efforts focused on regenerating the region's staple tree, Douglas-fir.

Initially, expanded Douglas-fir seeding and planting efforts made use of unimproved stock; that is, they relied exclusively on seeds collected from wild

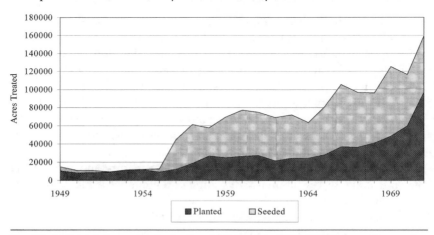

Fig. 5.1 Artificial Reforestation in the Douglas-fir Region, 1949–1971. Source: Hagensteing, 1973.

forest trees without any attempt at systematic tree breeding. Despite the foundation for tree improvement laid by Munger, a much longer European lineage of research and experimentation on forest and fruit tree variation and breeding dating to the eighteenth century, and expanding industrial reforestation programs, capital had little initial appetite in the region for applied forest genetics.[14] This points to a second source of delay; namely, a suite of scientific uncertainties and biological challenges that impeded industrial tree improvement, many of which continued to be challenges for decades after industrial efforts commenced. These include not only the provenance question but also challenges to controlled breeding, impediments to the mass production of desirable genotypes, and reproductive delays and long maturation rates. Some of these problems, in various guises, are endemic to the industrial cultivation of any forest trees, whereas others are highly specific to Douglas-fir. But as a set, they comprise an excellent example of what Boyd, Prudham, and Schurman referred to as the "problem of nature"; that is, the contradiction-ridden and uneven processes by which biological nature is formally subsumed in natural resource sectors, even those nominally renewable ones.[15]

Consider first the issue of time, and specifically biological time. Adam noted that social theory has given too little notice to the formation of distinctive timescapes formed from the collision of social and natural time under capitalist modernity. As she wrote specifically in relation to GE food debates, "A timescape analysis is not concerned to establish what time is but rather what we do with it and how time enters our system of values."[16] This theme is prominent in the agrarian literature, where observers have noted the significance of, for example, seasonal crop cycles and animal gestation periods that produce built-in delays to the turnover of capital.[17] Such challenges are not dead ends for capital; instead, they shape and constrain the specific avenues by which capital, aided in various ways by science (state supported and otherwise) takes hold of and transforms biological time and biological organisms in the production of commodities from natural inputs and in the commodification of nature per se. Examples include all manner of intensive breeding and cultivation schemes aimed at industrializing organisms, from fish to foul.[18]

Although taking hold of and compressing biological time is one of the goals of tree improvement (insofar as the endeavor selects for accelerated growth as a central goal of almost all commercial efforts), the collision of social and biological time in tree improvement presents major problems, not least because of delays and lags larger than anything encountered in agriculture. With a commercial rotation age along the Pacific Coast on the order of sixty to eighty years, Douglas-fir is particularly problematic, making socially produced Douglas-fir a distinct form of fixed capital. For example, perhaps the most significant manifestation of long growing times is that uncertainties persist to this day regarding rates of returns to investment from tree improvement because of the sheer novelty of the undertaking and the long waiting periods for sound empirical data to accumulate. This is reflected, for example, in research contemporaneous with or subsequent to the commencement of industrial tree-improvement efforts along the Pacific Slope on fundamental questions such as the degree to which growth rates in Douglas-fir are genetically controlled.[19] Observed results from tree improvement have tended to reinforce early expectations of significant gains in height and volume to be realized at rotation age.[20] Yet uncertainty remains regarding the correspondence between observed traits in young trees and the realization of gains at commercial maturity; as the cliché goes, only time will tell.[21]

Time is also an issue and of the essence in breeding. Delays are introduced in waiting for seedlings to grow large enough to provide a basis for their evaluation, and retained varieties must be allowed to reach sexual maturity before they can be crossed with other selected varieties to produce improved seed; Douglas-fir, under normal conditions, take twelve to fifteen years to produce flowers. Though ameliorated more recently by techniques to stimulate early flowering,[22] this was a key rationale for the development of tree-improvement techniques employing immediate collection of commercial quantities of seed from selected parent trees; waiting to cross the progeny of such trees in so-called full-sib crosses was beyond the time horizons of participating firms, particularly in the absence of secure knowledge of the expected gains from tree improvement.[23]

Yet time is not the only problem. There is also a need to regulate gene flow in the production of commercially improved varieties of trees, with significant implications in the event that this fails. Controlled breeding in the production of crosses can be highly problematic indeed, not least because male and female flowers are ubiquitous on conifers (including Douglas-fir) creating large numbers of possible cross-fertilization points. This, combined with the physical architecture of the trees, makes manual fertilization difficult and laborious.[24] Moreover, seed orchards used to produce improved seed are located in forest settings among wild trees. Although this best matches environmental conditions in the orchards to the sites where progeny will be cultivated, it also creates risks that uncontrolled crosses will be produced by pollen drift from the surrounding forest. This has proved a stubborn problem indeed, with concerns that crosses are not adequately controlled persisting to the present day.[25]

It also bears noting that the integration of biological discoveries into the realm of applied tree improvement on an industrial scale has been slow and uneven, an example of the shifting interface between science and capital, basic and applied research, and nature and commodity.[26] Specifically, scientific understanding of forest genetics, including issues surrounding provenance, was accumulated and made commercially useful only at a gradual pace, with specific obstacles standing in the way of the emergence of industrial tree-improvement programs. For example, relatively basic research on Douglas-fir provenance and breeding, commenced by Munger, continued well into the 1960s.[27] More-over, problems emerged in translating research into operational tree breeding. One of the most difficult of these problems, common to work with most conifer species (including Douglas-fir), is resistance to clonal propagation. Without easy cloning options, reproducing even one individual of a desired genotype in seed orchards to provide seed stock for reforestation remained problematic for some time; yet, initially, at least, use of seeds harvested from desired trees was considered unreliable for controlled breeding in seed production. One facet of the cloning problem was addressed in conifer breeding by Syrach Larsen in his research in Denmark during the 1930s. Larsen pioneered the grafting of "plus" or "elite" tree scions (cuttings) onto established host root stocks for the purposes of controlled experimentation. The trees resulting from these grafts expressed the genotype of the scion, not the root stock, and thus Larsen was able to create orchards of selected or plus tree-breeding stock used in the production of im-proved seed stock.[28] Yet Larsen's work was useful only for reproducing one clone at a time in seed orchards; producing large quantities of improved seed would work only in the long run using cross-pollination of the grafted (cloned) parent stock in seed orchards. In effect, the natural obstacle to mass propagation of desirable genotypes remained, impeding the use of hybrids in commercial tree improvement (the darling of early experimental forest tree improvement),[29] while also preventing the mass production of improved genotypes pending the advent of new techniques in clonal propagation and tissue culture.[30]

Although the problem of mass producing improved genotypes has impeded the development of economies of scale in breeding, so too has provenance pre-sented a major hurdle, particularly in the Douglas-fir region. Numerous special-ized traits in populations of Douglas-fir evolve based on widely ranging climatic conditions in the topographically uneven coastal area of Oregon and Washing-ton. This includes, for example, adaptation to extremes of drought and cold.[31] Under these circumstances, ensuring that like breeds with like in tree improve-ment requires the use of many distinct seed source zones. Determining the precise boundaries of such zones is often as much guesswork as science, and it has remained a source of debate in the region since the first establishment of industrial Douglas-fir tree improvement.[32] However, a cautious approach first developed by Roy Silen, the architect of cooperative Douglas-fir tree improve-ment, underpins the use of seventy-five separate breeding zones along the Pacific Slope, with forty-nine in western Oregon alone (see Figure 5.2).[33] The resulting fragmentation in breeding limits the area from which genotypes may be selected and crossed and thus further inhibits the capture of scale economies in breeding. Capturing the most gain possible within each zone—particularly because breed-ing zones typically span multiple landownerships, both public and private— virtually requires that breeding programs be shared undertakings, and that genetic resources be shared as well.

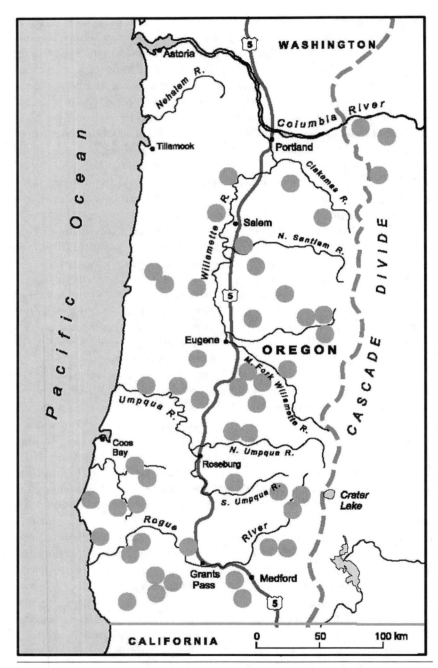

Fig. 5.2 Oregon Seed Zones Used in Douglas-fir Breeding by the Northwest Tree Improvement Research Cooperative. Source: Adapted from Bordelon, 1988, by the Cartography Office, Department of Geography, University of Toronto.

Because of this series of biological challenges confronting the capitalization of forest trees by means of industrial breeding, a pattern of predominantly cooperative and joint public-private organization has prevailed in Pacific Slope tree improvement. These patterns of institutional organization are remarkable in a capitalist economy, particularly considering that improved trees are the next generation of wood fiber in the region and thus play a large role in future industrial profitability. Despite the potential competitive advantages to be gained in developing proprietary breeding programs, with one major exception (Weyerhaeuser, see below), firms in the Pacific Slope region have tended to cooperate extensively on several levels, including sharing genetic resources as "common property" while also relying extensively on university scientists and state institutions and funding. On the one hand, such extensive cooperation and open research and development structures reflects an ethos common to factions of capital, science, and the state dedicated to taming—or as two forest scientists put it, "domesticating"—nature (i.e., wild trees) for the purposes of more efficient commodity production.[34] But at the same time, this institutional pattern also indicates the reluctance of firms (on the whole, and as of yet) to embrace tree improvement as an exclusive and proprietary undertaking; that is, as a basis of capitalist competition. Reflecting back after thirteen years of cooperative Douglas-fir tree improvement using Silen's "Progressive System," and noting many of the technical problems, uncertainties, and delays endemic to the undertaking, Silen and Joe Wheat bluntly stated, "Under no circumstances could any of the owners have carried such a comprehensive program alone."[35]

In short, Pacific Slope tree improvement comprises an example of the "inscription of ecological processes" onto the morphology of nature's appropriation suggested by Enrique Leff. Yet, as I argue in discussing more recent incursions of forestry biotechnology into Pacific Slope tree improvement, a central contradiction of this cooperative and open strategy is that it inhibits competition between firms on the basis of proprietary tree improvement. Specifically, sharing genetic stock creates coordination problems, undermines the development of tree improvement tailored to highly specific firm strategies, and impedes the emergence of breeding programs with the kind of exclusive control over plant germplasm that would allow firms to compete on the basis of their own, fully commodified tree varieties. However, with the technological potentials unleashed by biotechnology, including GE forest trees and their attendant implications for the exclusive control over higher life-forms as commodities in and of themselves, tree improvement may well be on the cusp of following agriculture's lead in the commodification of life-forms as an avenue for nature's real subsumption. In the remainder of the chapter, I explore in more detail the history of industrial tree improvement in the region that has brought it to this point.

Cooperative Tree Improvement, 1956–95

Experiments and Industrial Cooperation   Biological and political economic delays aside, experimental and eventually operational industrial tree improvement drew momentum from increasing industrial reforestation efforts along the Pacific Slope during the 1950s. Having spearheaded the establishment and certification of numerous industrial tree farms, in 1954 the IFA hired the region's first industrial forest geneticist, John W. Duffield, to coordinate experiments in Douglas-fir tree improvement involving multiple forest landowners.[36] Duffield had obtained his Ph.D. from the University of California–Berkeley, and had worked at the U.S. Forest Service's Institute for Forest Genetics in Albany, California, which by the 1930s had become a leading center for research on forest tree improvement and where considerable early research focused on tree improvement through hybridizing conifer (particularly pine) species.[37] Yet in 1956, when the IFA invested $40,000 in a research facility at Nisqually, Washington, Duffield initiated a program based on Larsen's model, using grafted parent stock for the establishment of seed orchards that would produce standard, not hybrid, crosses. The appeal of this approach was not hard to fathom; although hybrids might show great vigor, conventional sexual crosses were the cheapest way to produce the large quantities of improved seed that would be necessary to sustain expanding industrial reforestation in the region. Under Duffield's guidance, collaborating private companies selected plus trees, collected scions, and used grafts to establish three IFA seed orchards, two located in western Oregon and one in western Washington.[38]

The IFA experimental program became the basis for the region's first applied industrial tree improvement cooperative, established in Vernonia, Oregon, in 1967. Although largely a private sector initiative involving collaboration between three main industrial partners (Crown Zellerbach, Longview Fiber, and International Paper), the Vernonia cooperative also established a long-standing pattern of public science contributions in Pacific Slope tree improvement, drawing on key contributions from the Oregon State Department of Forestry and the U.S. Forest Service.[39] In particular, the Vernonia cooperative was assisted by Silen of the U.S. Forest Service's Forest Science Laboratory at Oregon State University and his "Progressive System" of tree improvement.[40] Silen's system, though like Duffield's in relying on sexually reproduced and reproductive parent and seed stock, departed from the IFA experimental program by eliminating grafting and seed orchards from the first generation of seed production. There were two reasons for this reversal. First, experiments with Douglas-fir suffered from widespread failures of the grafted parent stock as the result of incompatibility between root stock and scions;[41] this problem has since been largely addressed. At the same time, firms were also concerned with the delays introduced in applied versions of Larsen's model, because grafted parent stock had to be allowed to reach sexual maturity and evaluated before a second selection process produced parent stock intended for the production of commercial quantities of seed. Silen reasoned that genetic gains from wild, wind-pollinated seed collected from "superior trees" in the forest could certainly do no worse than random seed dispersal from natural regeneration but would also save time when compared with rigorous testing in seed orchards prior to commercial seed production. Thus, under Silen's system, tests of progeny were conducted on wild seed to more carefully refine the selection process, and seed orchards came into play only in producing a second generation of seed, allowing commercial seed production and rigorous tree improvement to proceed in parallel.[42]

As more and more firms with land in western Oregon and Washington expressed interest, Silen worked with Duffield's successor at the IFA, Joe Wheat, to expand the Vernonia cooperative into a network forming the backbone of Douglas-fir tree improvement in the region. By 1979 the network had grown to include thirty-three companies and approximately 2.5 million hectares of timberlands distributed across nineteen local cooperatives involving public and private forest landowners.[43] This network was renamed the Northwest Tree Improvement Co-operative (NWTIC) in 1985, and in 2001 it involved thirty public and private members, with seventy-five separate breeding zones and twenty-two local tree improvement cooperatives producing improved seed from sexual crosses of grafted Douglas-fir in seed orchards (see Figure 5.3).[44]

Fig. 5.3 Grafted Douglas-fir in a Pacific Slope Seed Orchard. Source: Photo by the author.

The fragmentation evident in this proliferation of local cooperatives and distinct breeding zones emerged specifically from Silen's attempts to account for provenance and ensure that like breeds with like. According to his specifications, breeding zones were defined initially as ecologically similar areas on the order of sixty thousand hectares, with elevation ranges of fewer than three hundred meters, although the size of breeding zones has since expanded somewhat.[45] The program has worked by having landowners within each breeding zone share in the collection of seed and in the selection and testing of potential parent trees to establish seed orchards, also cooperatively managed, with the goal of capturing the greatest genetic gain across the largest possible area while ensuring provenance is retained. Based on this local, cooperative model, the NWTIC was meeting 80 percent of the seed requirements of cooperative members in the region by the late 1980s and has produced enough improved seed to reforest in excess of three million hectares of forestland with socially produced trees.[46] The network has become by far the single greatest source of improved Douglas-fir varieties and includes all of the major industrial landowners in the Douglas-fir region, with the lone exception of Weyerhaeuser, whose nonparticipation I specifically discuss below.

Cooperative Research and Public Science   Industrial tree improvement in the Douglas-fir region also has been reinforced since approximately 1980 by cooperative, public-private research institutions in the region. Specifically, research cooperatives housed at Northwest research universities, including Oregon State University in Corvallis, the University of Washington in Seattle, and Washington State University in Pullman, have emerged as key partners in tree improvement. Moreover, the institutional character of these research cooperatives reinforces tendencies established under the auspices of the NWTIC. That is, research cooperatives are characterized by collaboration between private capitalist firms and strong support from the state, in this case coming in the form of funds, personnel, and facilities. And all of this is aimed at more deeply enlisting science to the tree-improvement project; research cooperatives have emerged as the key avenues of scientific research in regional tree improvement and intensive silviculture and are in this respect vital to the social (re)production of timber resources by means of increasingly intensive plantation forestry in the region.

Silvicultural research cooperatives vary somewhat in the ways in which they are organized, yet at the same time exhibit some key common characteristics. These characteristics include cooperative relations between private and public members, extensive assistance in funding and staffing from the universities, and largely informal management of innovations and information, including (until recently) common and relatively informal property regimes over plant varieties.[47] In addition, cooperatives are based on partnerships between private forest-products companies and timberland owners, as well as public forest management agencies (including, most significantly, the Forest Service, the Bureau of Land Management, and state forestry departments). Other members may also include commercial interests whose business draws them to particular cooperatives: for example, forest tree nurseries. Another trait shared by the research cooperatives is that they are led by one or more academic scientists, typically a forestry professor in one of the region's main state universities. These researchers manage the cooperatives and undertake much of the actual research, assisted by technicians and graduate students. Experiments are carried out in laboratory conditions and on experimental plots on land belonging to the host university, to cooperative members, or both.[48] Data and the results of research are made available to all members, and to some extent more widely through annual reports. Funding for the research cooperatives is drawn from a combination of annual member dues (in fixed amounts, or prorated according to their acreage); university contributions of facilities and equipment, staff, and direct financial inputs; and external grants and contracts from agencies such as the National Science Foundation and the U.S. Department of Agriculture.

In exchange for their dues, member firms gain some control over cooperative governance, including over the cooperative research agenda. Decision making is typically conducted by means of periodic meetings at which members and cooperative staff debate and vote on research direction. Perhaps not surprisingly, even though they are heavily funded and staffed by state contributions, the overall research direction of these cooperatives shows a decidedly applied and commercial orientation, and, specifically, it emphasizes harnessing, taming, domesticating, and subordinating the biological underpinnings of forest reproduction to the dictates of capital.

In tree improvement, the most important cooperative in the Pacific Slope region is the Pacific Northwest Tree Improvement Research Co-operative (PNWTIRC) at Oregon State University. The PNWTIRC was founded in 1983 to provide research support for the network of Douglas-fir tree improvement conducted under the NWTIC. That is, whereas the NWTIC coordinates applied cooperative tree improvement, the PNWTIRC acts as its research arm, providing technical assistance through its experimental programs.[49] The organization's membership is made up of most of the region's largest integrated forest-products companies, together with public forestland management agencies.[50] The research undertaken by the PNWTIRC is also generally representative of the commercial orientation of cooperative research, again, no surprise given the membership, funding, and governance structures of such cooperatives. For example, major emphasis in the PNWTIRC has been directed at the early selection of Douglas-fir for cold and drought resistance, identifying the genetic basis of these traits and the relationship between these traits and growth characteristics with more direct commercial advantages (e.g., wood density).[51] Emphasis on these issues is driven by a desire to breed trees across wider geographic ranges to increase the economic efficiency of tree improvement and to continue to capture genetic gain in successive generations of improved trees. The cooperative is also working on techniques to more tightly control the lineage of crosses produced in the seed orchards, motivated by uncertainties in gene flow that plague open, wind pollination.[52]

High-Yield Forestry   Taken together, applied and research initiatives in Douglas-fir tree improvement along the Pacific Slope feature extensive coop-eration among private firms and public agencies, including scientific contribu-tions from academics at major state-funded research universities. This, I have argued, is evidence of the inscription of the distinct timescapes and biological challenges of tree improvement into the particular institutional organization of nature's formal subsumption by means of tree improvement. Yet there is an important exception to this prevailing organization and division of labor in Pacific Slope tree improvement. In 1969 Weyerhaeuser initiated an integrated approach to intensive reforestation seeking to match chemical inputs and other management techniques with the use of improved seed. By the early 1970s, under the auspices of what the company called "High Yield Forestry," Weyerhaeuser had commenced an aggressive program of forest intensification, including the creation of a significant and proprietary research facility in Centralia, Washington. According to one former company geneticist,

> They took this big audio-visual presentation to the Board of Directors meeting in Dallas back in whatever year it was, '70 or '71. The Board of Directors bought off on it big time. Money was no object . . . the whole High Yield Forestry Program, all the research, was staffing up so rapidly that . . . in Centralia [Washington] at one time I think we had probably 30 Ph.D. scientists at that one center, in every discipline you can imagine. We had a total staff of over 100 working out of Centralia.[53]

Like other forestry firms embarking on tree improvement, Weyerhaeuser faced uncertainty and delays in tree improvement. However, following the model of the NWTIC, the firm aggressively pursued progeny testing and applied tree improvement in parallel, not in sequence.[54] Yet unlike the NWTIC, which took time to expand, Weyerhaeuser scaled up its efforts very quickly to meet all of its reforestation needs in western Oregon and Washington with socially produced seed. As a result of its rapid mobilization, Weyerhaeuser is widely acknowledged to be ahead of other firms in the region in intensive silviculture and, in par-ticular, in the quality and advanced stage of growth of a next generation of improved forest trees on its lands. Indeed, a first harvest from high-yield trees may come sometime within the next decade.[55]

However, aside from the scale and ambition of Weyerhaeuser's program, High Yield forestry stands out because it is a much more proprietary form of intensive forestry research and development model than is typical of the region's other firms and public agencies. This is not to say that Weyerhaeuser avoids involvement in the research cooperatives. Indeed, most of the cooperatives with research interests applicable to Weyerhaeuser's lands and operations do claim the company as a member. But the High Yield forestry program stands out as a proprietary effort in a region and an industry characterized by more cooperative approaches—for now at least. Yet I argue that Weyerhaeuser is the exception that proves the rule. Weyerhaeuser is by far the largest landowning firm in the region with approximately two million acres of timberlands.[56] More than half of this land was purchased in the 1899 transaction negotiated between Frederick Weyerhaeuser and James J. Hill and a second purchase in August 1901.[57] Because of the way they were obtained, these lands are more contiguous than most industrial holdings. Also, being primarily located in western Washington, these lands are generally less topographically uneven than in western Oregon, with correspondingly less fragmented provenance and thus larger breeding zones. Large areas of top-quality timberlands allow for greater economies of scale in industrial tree improvement, further enhanced by relatively more homogeneous land located in more contiguous configurations. Thus although Weyerhaeuser has unquestionably taken a different approach to tree improvement, this difference is further evidence of the inscription of biophysical nature in the tree-improvement project.

Tendencies in the Political Economy of American Forestry

If Douglas-fir region tree improvement is any indication, Marx was right to note that forestry is a problem for capital. In a predominantly capitalist economy and industry, Douglas-fir tree improvement in research and applied settings is characterized by extensive cooperation among firms and relies heavily on publicly supported science. These comprise strategies for confronting biological constraints that dissuade more aggressive, proprietary forms of capital investment in the social reproduction of forest trees—Weyerhaeuser excepted, that is. This being the case, patterns in Pacific Slope tree improvement reflect and reinforce long-standing themes in the political economy of American forestry. Indeed, despite persistent anxieties about timber famine and the exhaustion of old-growth forests in the history of American forestry debates,[58] enthusiasm for privately funded, in-house forestry research and development has been and remains tepid among forestry capital. It has instead been left to the state and cooperative institutions to assume the largest responsibility for harnessing science to the intensification and rationalization of forest growth.

According to the National Science Foundation, private research and development expenditures in the lumber, wood products, and furniture industries averaged 0.7 percent of sales from 1984 through 1994, whereas expenditures in the paper and allied products industry averaged 0.9 percent of sales over the same period.[59] These figures are quite low compared with an overall average rate of about 3.1 percent of sales in manufacturing industries during the same period (see Table 5.1). More specifically, private research and development expenditures in the forest-products sector are heavily skewed toward investment in forest-products research rather than forestry. Paul Ellefson estimated that four or five companies dominate private investment in forestry (one of which is Weyerhaeuser), and that more than 90 percent of private research and development investment in the forest sector is oriented toward forest-products research, not growing trees. Conversely, although total state support accounts for only 5 percent of forest-sector research and development expenditures, state expenditures account for about 85 percent of research funding for research in forestry per se.[60]

The division of labor is evident; although industry invests in commodity and process innovations, the state provides the vast majority of funds for what has traditionally been considered more basic, that is, biology-based, research. This is underpinned by a history of federal and state support dating back almost one hundred years and channeled through various federal laboratories managed under the auspices of the U.S. Forest Service (e.g., the Pacific Northwest Forest and Range Experiment Station in Portland) as well as the state land grant

Table 5.1 Company and Other (nonfederal) Research and Development Funds as a Percentage of Net Sales in Wood-Based Industries and All Manufacturing Sectors, 1984-94

	All Manufacturing	Lumber, Wood Products, and Furniture	Paper and Allied Products
1984	2.6	0.7	0.8
1985	3.0	0.8	0.8
1986	3.2	0.6	0.7
1987	3.1	0.6	0.6
1988	3.1	0.6	0.8
1989	3.1	0.6	0.8
1990	3.1	0.6	1.0
1991	3.2	0.9	1.1
1992	3.3	0.9	1.0
1993	3.1	0.7	1.1
1994	2.9	0.6	1.0
Average	3.1	0.7	0.9

Source: National Science Foundation, 1996.

universities. The origins of these programs cannot be dissociated from state efforts to ensure forest reproduction in the face of industry cut-and-run tendencies (see chapter 6), yet such public efforts were ever propelled by a simultaneous desire to perpetuate forest commodity production; that is, to reproduce nature to sustain capital accumulation.[61] There are, in this, strong parallels between the political economy of American forestry research and development and what Jack Kloppenburg so effectively documented in American agriculture.[62] That is, a division of labor has developed whereby the state does what capital will not, laying the foundations for an eventual commodification of nature by undertaking the basic biological research that is either too far removed from immediate commercialization to interest capital or not fully appropriable on a proprietary basis for a myriad reasons. These reasons include, significantly, biological obstacles such as the self-reproducing character of plants and, in the case of forest trees such as Douglas-fir, resistance to clonal and thus mass propagation.

In this context, cooperative institutions such as the PNWTIRC, many of them also housed at state-supported research universities, have comprised a specific and essential foundation for operational and research dimensions of U.S. forestry innovation. Although relatively little research—academic or otherwise—has been done on such institutions, an American Forestry Council survey conducted in 1987 identified a total of fifty-one cooperative forestry institutions housed at research universities and drawing membership from a range of private firms and public agencies.[63] North Carolina State and Oregon State were the leading universities in terms of the number and size of their cooperatives, yet other important host institutions included the University of Washington, the University of Florida, the University of Maine, Texas A&M, and Virginia Polytechnic Institute and State University. The significance of these institutions should not be underestimated, again keeping in mind a context in which more proprietary and private forms of investment in forestry have been extremely limited.

The key theme here is that although the entire purpose of industrial forestry—including tree improvement—is to intensify and rationalize forest tree growth, the biophysical challenges of doing so have inspired cooperative institutional strategies, extensively assisted by state-supported science. As Ellefson plainly yet unremarkably stated in his review of forest-sector research and development in the United States, "Industrial forestry research efforts . . . must face the realities of long pay-back periods for investments in forestry research, high risks and uncertain consequences of research investments."[64] That is, forestry research must confront the challenge of a reproductive biology that obeys its own laws and rhythms, twinned with scientific uncertainties regarding how these laws and rhythms may be subordinated and rationalized.

These observations are in certain respects quite consistent with broader themes in the political economy of science, technological innovation, and capital accumulation. For one thing, as Kloppenburg nicely argued, it is essentially ideological claptrap to see science and capital as separate realms, even when science is manifest in its so-called basic guise. Rather, science (increasingly supported by state largesse over the past century or so) has long been an arena for undertaking what capital cannot or will not do on its own. There are several reasons for this. One reason is that there is a widespread belief, again essentially ideological, that innovation in capitalist societies is the result of the diffusion of ideas from an essentially autonomous science into the realm of economics and applied technology. Yet, as Nathan Rosenberg argued, there is little basis in historical fact for the sweeping character of this view; rather, science is just as well understood as the realm in which systematic exploration of phenomena follows the application of these in the realm of technology. In other words, science is in important ways driven by the needs of and discoveries within the realm of applied technology, not the other way around. In this respect, and as Rosenberg wrote specifically, it is just as accurate to see that "we can learn a great deal about the activities of scientists—even those of scientists engaged in basic research—by starting our inquiry in the realm of technology."[65]

Second, the ideological belief that what is good for capital is good for society is widespread—in more familiar terms, what is good for GM is good for America. Thus, quite simply, the state and science should serve capital's need for innovation. There are those who would and do unreflexively subordinate science and the state to capital in exactly these terms; that is, advocate for the state and science to engage in the provisioning of public goods, with these in turn understood simply as those things that are difficult or impossible at any particular juncture to make available as capitalist commodities. The implicit argument here is simply that anything that can be provided in the form of a capitalist commodity should be.[66] The political struggle for control over state spending on science, as Kloppenburg also demonstrated in the context of American agriculture, often has been waged on exactly these terms, and as Robbins and, more recently, Rajala demonstrated, the same may be said for forestry.[67] One is thus reminded of Engel's prescient observation regarding the procession of cognate sciences toward closer integration with industrial innovation: from mechanical sciences, to physics, to chemistry, and then to biology.[68] Cooperative and state-supported research and development in the area of intensive silviculture may in these respects be understood as key strategies for enlisting the biological basis of forest productivity to the dictates of industry, laying the groundwork for the real subsumption of nature.[69] Or, as the Forestry Research Task Force of the U.S. Department of Agriculture and the Association of State Universities and Land Grant Colleges put it in framing a research agenda for state-supported science in forestry, and specifically tree improvement, in 1967,

Today's forest practice is based almost entirely on wild forest trees. Unlike crop plants, trees have not undergone centuries of selection and breeding to make them more useful to man [sic]. There is strong evidence that through application of genetic principles we can produce stock that grow twice as fast as the parent stock, that resistance to most major destructive pests can be bred into trees, that specified wood properties can be produced at will. . . . It should be feasible to develop straighter form, fewer limbs and resistance to climatic extremes.[70]

That it was in the public interest to do this, that is, enlist the reproductive biology of forest trees to the logic of more efficient commodity production, apparently did not warrant comment.

Forest Tree Genetic Engineering

Yet if cooperative- and state-funded scientific and applied tree improvement confront a suite of challenges to subordinating the reproductive biology of forest trees to capital, these strategies are at the same time characterized by serious tensions and contradictions undermining the exclusive appropriation and commodification of growth and reproductive processes and of tree varieties themselves. Specifically, relatively open property regimes governing tree varieties and tree-improvement techniques are essential to cooperation between firms and between public and private actors in tree-improvement efforts such as the NWTIC; yet these regimes by definition do not allow individual firms to take hold of forest tree varieties and the biology of their reproduction as bases of exclusive, proprietary competition and innovation. As yet, forest tree improvement has not witnessed even the degree of plant commodification seen in agriculture and fruit tree breeding even prior to age of the new biotechnology.

However, just as biotechnology has led to accelerated commodification of plant varieties and processes in agriculture and fruit tree breeding, so too in tree improvement may the advent of new biotechnology applications lead eventually to a more complete form of commodification of forest trees.[71] The new biotechnologies present a suite of new possibilities ranging from interspecific genetic transfers by means of genetic engineering to new tissue culture techniques that open up the door to mass production of specific plant varieties.[72] All of these allow new ways of confronting and taking hold of plant biology, enabling nature's real subsumption. Given these possibilities, there is every reason to suspect the potential for biotechnology to have a significant commercial impact on tree improvement and for these technological changes to catalyze a social reorganization in the project of intensifying forest growth.[73] Tendencies in the Douglas-fir region, as I discuss, do suggest these possibilities.

And yet this potential, and the tensions it portends for cooperative and publicly supported tree improvement, are still barely evident. For instance, there are as of this writing no GE forest trees in commercial plantations in the Douglas-fir region or in the United States more generally. Reflecting the ongoing significance of specific biological properties in shaping and constraining the appropriation of forest tree reproduction, work with at least some forest trees has been slowed by the fact that their genomes are relatively complex and take longer to map and also by the difficulties propagating some species from shoots or tissue. Thus, although the first successful regeneration of a transgenic tree (a poplar) occurred in 1987, the first successful experimental regeneration of a nontransgenic conifer—specifically Norway spruce (*Picea abies*)—took place in 1985; the first successful regeneration of a transgenic conifer (a spruce) was accomplished in 1993.[74] Work with Douglas-fir has been relatively slow in these respects because the species has a relatively complex genome, and, as I have noted previously, it is also notoriously difficult to propagate from cuttings and tissue. According to one forest geneticist, "Producing a transgenic plant with Douglas-fir is technically very possible. Clonally propagating that [plant] and using it on a broad scale is much more expensive and technically challenging."[75]

In this respect, as in proprietary tree improvement, Weyerhaeuser stands apart, having pursued a technology called somatic embryogenesis aimed at the mass production of varieites of improved Douglas-fir.[76] However, Weyerhaeuser's efforts notwithstanding, far more intensive focus on the introduction of forest tree biotechnology and genetic engineering in forest tree breeding has been directed at poplar varieties, even in primarily softwood forest regions such as the Pacific Slope. This is due in part to certain attributes of the genus, including high natural growth rates and relatively simple genomic structures. Moreover, and perhaps most important, poplars tend to be very easy to clone using planted cuttings.[77] Here too, then, is another example of the interaction of the ecological and the economic, as the particular traits of species help direct specific trajectories of scientific and technological development in the introduction of forest tree biotechnology. That is, if conventional forest tree breeding and biotechnology take hold of reproductive biology, they hardly do so on a biological tabula rasa but rather exploit the particularities of intra- and interspecific variation.

The Tree Genetic Engineering Research Co-Operative   Direct evidence of the potential significance of biotechnology for Pacific Slope tree improvement is manifest in one of the more recent research cooperatives established in the region, the Tree Genetic Engineering Research Co-operative (TGERC), founded at Oregon State University in 1994. As its name might suggest, the explicit goals of the cooperative's founders were to develop GE varieties of forest trees through cooperative research and, through this research, to become the source of the first successful commercially deployed GE trees.[78]

TGERC research involves the production of GE varieties of hybrid black (*Populus trichocarpa*) and eastern (*Populus deltoides*) cottonwood, hybrids now used in a relatively small cumulative acreage of forest tree plantations located primarily in eastern Oregon and Washington and developed to supplement fiber supply to regional pulp and paper mills. Significantly, these plantations are already among the most intensive forest tree cultivation systems anywhere; trees are planted at a density of approximately 250 stems per hectare and harvested after only six to eight years of growth at heights of approximately twenty-five meters. Although actual rates of biological productivity for the plantations are not made available by firms, even a rough guess based on these figures indicates rates far in excess of conifer plantations further to the west. Moreover, reflecting the ease with which they are propagated from shoots, each block of trees in the plantations is composed of a single variety selected for its advantageous growth properties and cloned by planting cuttings or shoots in even rows under close chemical and hydrological control by means of a drip irrigation system (see Figure 5.4).[79]

Fig. 5.4 Boise Cascade Hybrid Cottonwood Plantation in Southeastern Washington. In the foreground are rows of new growth, with older trees growing in a block in the distance. The trees are grown on a 7-year rotation and harvested at a height of 75 feet. Source: Photo by the author.

At the TGERC, researchers have successfully developed prototypes of GE cottonwoods that are in field trials in Oregon under regulatory review by the Animal and Plant Health Inspection Service of the U.S. Department of Agriculture and by the Environmental Protection Agency.[80] This TGERC research has to date pursued three principal avenues: engineered sexual sterility, engineered insect resistance, and engineered herbicide resistance.[81] Although introducing the latter two traits is of direct commercial interest, attempts to produce sterility in the trees is intended to control the spread of introduced gene constructs from any engineered varieties of the trees into wild populations of cottonwoods and related species, a key regulatory concern.[82]

The TGERC research on engineered insect resistance makes use of an increasingly common technique in GE food crops involving the production of *Bacillus thuringiensis* (Bt) toxicity. The insertion of a gene extracted from a common soil bacterium from which the toxin derives its name results in a plant that produces Bt toxin—a common agricultural pesticide—in its own tissues.[83] Bt pesticide is widely used to control cottonwood leaf beetles in conventional cottonwood plantations; thus, the chief commercial advantage of Bt cottonwoods is that they promise to eliminate expensive aerial spraying. However, this could have also have ecological benefits if it reduces the dispersion of Bt into the surrounding environment.[84] One of the problems with Bt toxicity, however, is that it is likely to accelerate selection pressure in pests for resistance to Bt toxicity, because unlike aerial spraying, Bt toxicity in plant tissues means constant exposure of the pests to the toxin.

Engineered glyphosate resistant or so-called Roundup-Ready cottonwoods is the third major research emphasis at the TGERC.[85] These trees have a gene construct inserted into their DNA that makes them tolerant to the application of the herbicide glyphosate, a broad spectrum weed killer. Prototypes of these Roundup-Ready cottonwoods are in TGERC field trials and will be of interest in existing plantations because glyphosate is toxic to conventionally bred hybrid cottonwoods and is therefore of limited use in the cultivation of non-GE varieties.[86] Yet glyphosate is among the most widely used, broad-spectrum herbicides in current use and has a relatively low ecological impact because it breaks down relatively quickly in the environment. Research to date also indicates low levels of toxicity to aquatic organisms and mammals. The introduction of glyphosate use in cottonwood plantations will be attractive to firms now cultivating these trees who are prevented from using it on conventional cottonwoods.

In many ways the TGERC builds on and extends the cooperative tradition of tree-improvement research and development in the Pacific Slope region. Based at Oregon State University in Corvallis, the TGERC is a joint undertaking relying on state and private sector inputs and contributions of research staff and facilities from the university. Members originally included major owners and operators of cottonwood plantations in the region and interested public agencies, including the Department of Energy and Oregon State University.[87] Subsequent additions included forest-products giants International Paper, Georgia-Pacific, Weyerhaeuser, and Westvaco, who, though without significant investments in cottonwood operations in the region, expressed interest in potential applications of the cooperative's research in GE forestry. In 1999 the TGERC entered its second five-year research plan and claimed as members Aracruz Cellulose, Alberta Pacific, the Department of Energy, International Paper, the National Science Foundation, Oregon State University, Potlatch, Westvaco, and Weyerhaeuser.[88]

Yet, in other ways, the TGERC is clearly set apart. Monsanto and Mycogen have also been involved as associate members, for reasons that are not difficult to discern; the cooperative has formal license agreements covering gene constructs owned wholly by these firms. The TGERC Bt cottonwoods contain proprietary gene constructs (owned by one of either Mycogen or Monsanto), while Roundup-Ready cottonwoods, as their name suggests, are tied directly to Monsanto, the producer of the herbicide Roundup, the number-one selling brand of glyphosate. In each case, commercial use of GE varieties produced by the TGERC could open potential revenue streams from licensing fees for Monsanto and Mycogen covering cultivated, proprietary (i.e., commodified) varieties of the trees. Moreover, under the auspices of the Oregon State University Office of Technology Management, formal patent claims have been filed with the U.S. Patent Office under the ownership of the TGERC and Oregon State University for a range of products and processes, including transgenic trees produced by the cooperative. This is a first for tree improvement in the region, and it opens up avenues for other potential licensing arrangements covering GE varieties or, at the very least, new property rights issues that will affect the social control of plant varieties and the commercial diffusion of TGERC research.

Conclusion

Whether developments under the auspices of the TGERC and the advent of genetic engineering in forestry more generally will lead to a more proprietary form of tree improvement, and an increasing role for private science in forestry, as Mark Sagoff predicted, remains to be seen.[89] However, the TGERC experience already demonstrates the extent to which genetic engineering in tree improvement is being accompanied by the creation of more formal and exclusive property rights, including more strictly controlled forms of technology transfer from public science to applied tree improvement. This tendency toward formalism may signify the emergence of a more complete commodification of improved forest trees, involving some combination of more exclusive, bilateral partnerships between academic science and capital, and more entirely exclusive endeavors by individual or multiple firms in the production of proprietary forest tree varieties.[90] Yet whatever the economic and ecological outcomes, forays into GE trees represent an extension of the project of enlisting forestry science to industrial capital and, in this respect, draw a direct lineage to Thornton Munger's experiments on Douglas-fir provenance more than ninety years ago. Any appraisal of tree improvement, conventional or otherwise, must confront this history and, specifically, the social relations and processes underpinning nature's real subsumption in intensive plantation forestry.

More broadly, tree improvement, including the TGERC's incursion into genetic engineering, is one aspect of the shifting interface between capital and nature in the long transition from timber extraction to forest cultivation in the Douglas-fir region. In this sense, capital—with no small amount of help from science and the state—may be seen to have responded to the exhaustion of one type of resource (old-growth) with the production of another (genetically improved trees). This history comprises a lens on the evolving interface between capital and biophysical nature in the political ecology of Douglas-fir forestry, but at the same time, the tree-improvement story represents a more specific case study of the formal subsumption of biophysical nature by capital.[91] If the TGERC is any indication, the industry may be headed toward the commodification of forest tree varieties.

Yet this story is clearly not simply one of the Promethean conquest of nature's limits by industrial and scientific innovation. For one thing, there is little in this story to support naked, market triumphalism. Rather, if trees are to become commodities along the Pacific Slope, they will do so in no small measure because of assistance provided by the state in applied and research settings. And as I have tried to argue, the history of cooperation and collaboration among private firms, and involving science and the state, should be interpreted in significant measure as a set of institutional strategies for confronting biophysical challenges. If biophysical nature cannot be understood as a set of fixed constraints or limits to social action, neither can it be approached as the passive recipient of social action. Political economy and environmental change are instead locked in dialectical conversation.

Moreover, if the march toward tree improvement and, perhaps, proprietary tree varieties is a response to ecological crisis in the form of the exhaustion of old-growth, the development of commodified trees, though seemingly imminent, may engender new crises of its own. There are, for instance, genuine ecological risks associated with commercial deployment of GE trees. There seem to be good reasons to worry that introduced genes will spread to wild populations of poplar, and possibly to other species of plants.[92] In fact, despite a generally lower level of public controversy, GE trees arguably bring greater risks that engineered genes will spread into nontarget populations than do engineered agricultural crops. This is because industrially cultivated trees are more closely related to noncultivated or wild trees than are most crops given the short history of forest tree cultivation and the primarily wild genetic resources used in tree improvement.[93]

These concerns are not merely technical. Rather, they are political and regulatory issues. The ecological risks associated with commercial deployments of GE trees are the subject of regulatory scrutiny in test plots now under evaluation.[94] The contingency of regulatory approval based on these tests attests to the requirement that commercial use of GE trees and their more complete commodification is ultimately subject to social sanction, and thus to potential obstacles within and outside of the formal regulatory system. Social opposition to GE trees, whether related to environmental concerns or to the implications of biotechnology for the social control of forests and forestry, could very well threaten the ultimate legitimacy of the entire project of producing GE forest trees. Underscoring this potential, in March 2001, members of a group identifying itself as "concerned Oregon State University students and alumni" destroyed approximately 1,200 GE TGERC prototypes and other hybrid cottonwoods at testing sites near Corvallis and outside Klamath Falls, Oregon. In a subsequent press release, the group declared its opposition to genetic engineering of forest trees at Oregon State University, stating, "The test plots of *Populus* genus trees (poplars and cottonwoods) at these places were independently assessed and found to be a dangerous experiment of unknown genetic consequences."[95]

Other incidents in recent years involving protests against genetic engineering in forestry reinforce the basic fact that that public acceptance of GE trees can by no means be ensured.[96] The actual probability that engineered gene constructs will escape to nontarget populations and the broader biological implications of tree improvement for genetic diversity, when combined with the politicization of these issues, bear all the hallmarks of O'Connor's ecological crisis: part objective, material transformation of nature, yet inescapably politically interpreted, defined, and constructed.[97] Thus in no sense can the social production of forest trees, culminating in commodified, GE varieties, be considered an escape from the contradiction and crisis-prone character of nature's commodification, but it is instead best understood as a reconfiguration of the political and ecological dimensions of these tendencies.

This in turn opens the door to broader considerations concerning the tree-improvement project. If, as Neil Smith asserted in his controversial excursis, the history of capitalism is in part a history of nature's increasingly social production, then, as he also argued, the challenge is not to deny nature's social production or to engage in romantic wishes for some return to an original nature. Rather, the challenge is to reimagine a politics of nature's social production, to ask what sorts of nature are desirable and who should decide. Here, it bears reiterating that the entire tree-improvement project, whether involving conventional breeding or the new biotechnologies, has been predicated on a willingness on the part of the state to underwrite the integration of forestry science with industrial forestry. With relatively little fanfare, the silvicultural research cooperatives critical to various aspects of industrial forestry in the United States have blurred meaningful distinctions between public and private science, conferring as they do exclusive access to research results and significant control of research governance to private, capitalist firms. This is a troubling trend that is consistent with a much broader shift in academic-industrial relations in biological research in the United States since at least the early 1970s; that is, in the age of recombinant DNA.[98]

This only serves to further underscore that all reinventions of nature, including those achieved by means of the new biotechnology, entail simultaneously technical, scientific, and sociopolitical projects, each with specific histories and geographies.[99] Far from inevitable, these projects are ultimately contingent. Thus, any critique of biotechnology or genetic engineering and, for that matter, any socially produced nature needs to confront the underlying social processes that propel these outcomes. The alternative is to reify new natures as mere "things" and debate their meanings in a vacuum, rather than against the backdrop of social processes and social relations.[100] Specific objections to GE trees, whether based on assessments of ecological risk, the ethics of manipulating and controlling life itself, or concerns about proprietary control of life-forms, must, like objections to industrial forestry more generally, consider the social origins of these projects. In this case, this includes almost one hundred years of science and capital working together in tree improvement and, more broadly, the project of capitalist liquidation of old-growth in the Douglas-fir region. In the next chapter, I explore more directly questions surrounding the politics and social regulation of this project of liquidation.

6

Timber and Town
The Rise and Fall of Sustained-Yield Regulation in Oregon's Illinois Valley

In the spring of 1983, a small band of militant environmental activists buried themselves in the ground up to their necks to halt extension of a logging road in the Siskiyou National Forest of southwestern Oregon. The protesters, most of them young men from nearby back-to-the-land hippie communities, were members of Earth First!, the vanguard of an emerging direct action campaign attempting to halt industrial logging in the old-growth forests of the Pacific Slope.[1] Though it took place in a remote location far from Oregon's media, population, and political centers, the protest was nevertheless of considerable significance, not least in pointing to the emergence of a political movement that would remake the regulation of industrial forestry in the region and beyond. Among the first salvos in an escalating campaign of civil disobedience, the protest highlighted not only opposition to industrial logging but also the declining political legitimacy of a policy regime that had underpinned management of federal forests for most of the post–World War II era. This policy regime, based most centrally on the idea of maximum sustained-yield forest management, entailed the application of forest science to yield maximum harvests of timber volume every year in nondeclining amounts.[2] Despite entrenchment of the sustained-yield doctrine in American forest policy dating to the late 1930s, and its apparent success in regulating capitalist appropriation of public forests, in the 1980s escalating opposition to sustained-yield orthodoxy led to its downfall. When the northern spotted owl was listed as a threatened species under the Endangered Species Act in June 1990, a bubbling crisis boiled over. The listing catalyzed displacement of the regime of sustained-yield forest management, to be replaced by the New Forestry and the rise of ecosystem management.

The rise and fall of sustained-yield orthodoxy as an approach to regulating forest commodity production along the Pacific Slope is the subject of this chapter. I argue that sustained-yield management must be understood as a regulatory response to the twin Polanyian fictions of nature and labor as commodities and the ways these fictions became manifest during the earlier, more liberal era of regulating land-based capital accumulation in the U.S. Northwest. Sustained-yield forest management doctrine arose in the context of a generalized crisis in American capitalism during the Depression and in response to the specific crises of American forestry. These crises were caused by the social and economic disruptions of cut-and-run forestry on the one hand and the threat of resource exhaustion suggested by a growing inventory of denuded, cutover lands during the early twentieth century on the other hand.[3] As a response to these crises, sustained-yield regulation reinforced generally state interventionist and science-based regulatory tendencies typical of the New Deal, becoming the linchpin of federal attempts to secure ecological and socioeconomic renewal in the forest sector. Sustained yield would address the fictions of nature as commodity by ensuring the reproduction of forests in the image of the *Normalbaum* or normal forest, an organic machine prescribing even-aged, rationalized forests intended to produce maximum flows of timber volume.[4] At the same time, sustained yield would address the fictions of labor apparent in social tensions arising from a highly volatile and geographically variable forest industry by means of these same constant harvest volumes, guaranteed to capital in exchange for providing income and employment in rural, forest dependent communities.[5] The result was what David Harvey called a "structured coherence" of capital accumulation, regulation, and social life.[6]

Yet the principles of sustained-yield doctrine guiding postwar federal forest management ultimately failed to redress these twin fictions. The ecological contradictions of sustained-yield regulation were grounded in problematic social constructions of forests as mere assemblages of timber volumes, belying the diverse and complex ecological communities that depend on forests for their survival and, critically, the key differences between the ecology of old-growth and young-growth forests. These contradictions became politicized in protests against industrial logging in old-growth forests in the Northwest; for example, the 1983 incident in Oregon's Siskiyou region and also through myriad lawsuits filed by environmental and citizen groups, culminating in the spotted owl crisis (see epilogue). Yet though the ecological contradictions of sustained-yield regulation exploded into open political conflict in the late 1980s, they must be understood as outgrowths of the structured coherence of sustained yield, based as it was not only on problematic ecological visions of forests but also on fundamentally naive, albeit largely implicit assumptions about the stability of industrial capitalism. Specifically, the dissolution of petty commodity production in favor of industrial, factory-style forest-products manufacture, long-term employment attrition driven by gradual yet inexorable productivity increases in the wood-products sector, and cyclical industrial downturns punctuated by the devastating recession of the early 1980s all gave the lie to any notion that constant and even rising annual harvest volumes could ever be sufficient to provide stable income and employment levels in forest dependent communities. Long before anyone ever dreamed that spotted owls could be the canary in the coal mine of Pacific Slope forests, the social contract between communities, capital, and the state forged by sustained-yield regulation was moribund. To explore these ideas in greater depth, I develop in this chapter a case study of southwestern Oregon's Illinois Valley.[7] Historically one of the most forest dependent areas of the Pacific Slope, the Illinois Valley was heavily reliant on federal forests—and sustained-yield doctrine ostensibly guaranteeing nondeclining annual timber volumes—to sustain its economy for much of the postwar period. Emerging as one of the most chronically depressed and troubled regions in Oregon during the 1980s and into the 1990s, the Illinois Valley case study throws into sharp relief the contradictions of sustained-yield social regulation and, in particular, the failure of this regulatory model to address the twin fictions of nature and labor as commodities.

The Illinois Valley

The Illinois Valley is a small, remote, and rural area located in the extreme southwest corner of Oregon, within Josephine County and perched near the California border (see Figure 6.1). The valley is named for its main watercourse, the Illinois River. Waters in the Illinois River eventually join the larger Rogue River, wind through the surrounding Siskiyou Mountains, and empty into the Pacific Ocean near Gold Beach. The mountains, part of the Klamath formation linking the Cascade and the Coast ranges, are not particularly high, reaching peak elevations on the order of two thousand meters. But they form a jumbled and densely packed cluster of ridges and summits that cut off the valley from the coast and the interior, lending it a sense of isolation. The valley's population in the 1990 census was about eight thousand,[8] although local residents told me they believed it was closer to twelve thousand.[9] This population is widely dispersed, with the largest settlement being Cave Junction, a town of about thirteen hundred.

Although wet like most parts of the Pacific Slope, the Illinois Valley gets most of its precipitation during long, cool winters. By contrast, summers are very hot and very dry. Microclimates abound, and the area is known for its incredible diversity of biota; for example, there are an estimated thirty-five hundred plant species in the Klamath-Siskiyou region. Most of the landscape that has not been cleared for farming and settlements is heavily forested, and the area is home to an astounding thirty-eight different species of conifer trees.[10]

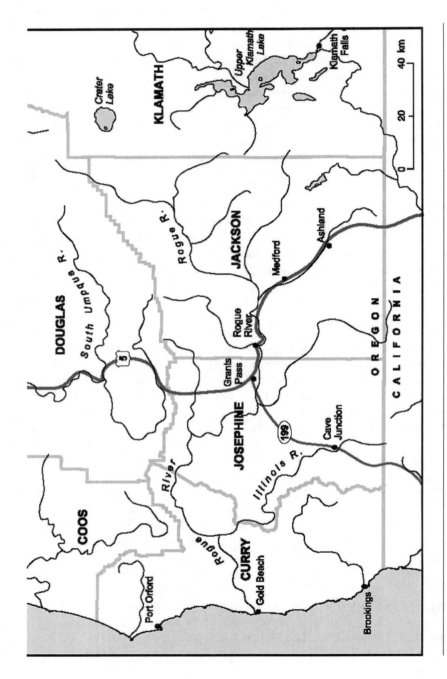

Fig. 6.1 The Illinois Valley and Southwestern Oregon. Map by the Cartography Office, Department of Geography, University of Toronto.

Species of commercial significance include ponderosa pine (*Pinus ponderosa*), Jeffrey pine (*Pinus jeffreyi*), sugar pine (*Pinus lambertiana*), lodgepole pine (*Pinus contorta*), and, of course, Douglas-fir. Not surprisingly, the area has long been subject to intense logging pressure, concentrated in most recent decades on the upslopes now marked by extensive networks of logging roads and clear-cuts (see Figure 6.2). Federal lands are extremely significant, particularly in the upland, more forested areas, and are administered by either the U.S. Forest Service or the Bureau of Land Management (BLM). Together, these two agencies control almost 60 percent of the total land area in Josephine County, and a higher proportion of the county's timberlands (see Table 6.1); in the Illinois Valley, federal lands comprise as much as 80 percent of the timberlands. With so much land in the area under federal control, and logging and lumbering so important to the local economy (at least historically), the rise and fall of sustained-yield management is central to the recent history of the Illinois Valley.

Liberal Extraction

Early Nature-Centered Production: Farming and Mining

The Illinois Valley and its environs have been caught up in and shaped by nature-based extraction and capital accumulation since the arrival of white settlers to

Fig. 6.2 Logged hillsides in the Illinois Valley, Southwestern Oregon. Photo by author.

Table 6.1 Land Tenure in Southwest Oregon Counties

County	Bureau of Land Management	U.S. Forest Service	State	County	Private
Coos	16	5.4	6.2	2.1	70.3
Curry	6.5	53.4	1.1	0.2	38.8
Douglas	20.1	28	1.5	0.9	49.5
Jackson	23.9	22.2	0.3	0.5	53.1
Josephine	30	28.5	0.8	3.1	37.6

Source: Dicken and Dicken, 1979.

the area in the 1840s. Long before sustained-yield forest regulation underpinned the postwar industrialization of the local forest sector, more liberal forms of extraction and resource-based accumulation in the mining, agriculture, and forest sectors propelled the local economy. And in many ways, the twin fictions of labor and nature as commodities that were evident locally and at wider scales during this period set the stage for sustained-yield regulation as part of a wider New Deal project of reform.

Settlers began to enter southwestern Oregon in substantial numbers during the late 1840s, following the establishment of the Applegate Trail in 1846 linking the Willamette Valley with established wagon trails by way of the Klamath Mountains.[11] Homesteading was reinforced by congressional passage of the Oregon Donation Land Act of 1851, part of a suite of federal legislation aimed at disposing of vast and newly acquired federal lands in the western states and (in principle at least) at establishing a social fabric of Jeffersonian independent farmers and petty commodity producers; in short, a quintessentially liberal dream to be achieved with federal largesse in the transfer of public land to private hands. Even if dramatically interventionist in one sense, the ethos in homesteading was hardly to regulate land use but rather to give it away as quickly as possible. By 1852 the wave of settlers drawn by available land had reached the Rogue River country east of the Illinois Valley, near the present route of Interstate 5.[12] Most of them were farmers who quickly established agrarian homesteads, and the first crop of Rogue Valley wheat was harvested that same year.[13]

Although homesteading brought settlers to the general area, it was the lure of gold that drew migrants specifically to the Illinois Valley, also reinforced by laissez-faire federal policies on land claims by (primarily) white settlers.[14] As pressures on panning sites in the Sierra Nevada mounted during the California gold rush, more and more gold seekers moved north into Oregon, including into the valleys of the Rogue and Illinois.[15] By 1851 gold miners had established the first non-Indian community in the Illinois Valley, naming it Sailor Diggings after a crew of sailors who abandoned ship in Crescent City to seek gold in the

Siskiyous. Renamed Waldo, the town became the first seat for the newly established Josephine County in 1856.[16]

Land-based production dominated primarily by Euro-American agrarian settlers and miners became the basis of the local economy. With the notable exceptions of large work crews—including Chinese labor gangs—deployed to build irrigation networks for mining and to extend the railroad system through southern Oregon, this early land-based production was typically of a small scale and even petty commodity variety. In fact, Pomeroy noted that by 1870 some degree of in-migration to Oregon and the rest of the Northwest was composed of settlers self-consciously spurning the more monopolistic tendencies already apparent in control of land in California, and farms in the Northwest, though productive, were typically small, though hardly disinterested in commodity production.[17] Thus, as recently as 1930, the census reported more than half of the agricultural workforce in Josephine County was composed of independent farmers working their own land. The census also reported that independent farmers outnumbered farm laborers by more than two to one, and that one-quarter of all farm laborers were unpaid family members.

Early mining was also primarily small scale, and atomistically organized, although some operations did achieve an industrial character as the complexity of separating ore from parent materials increased. Liberal social regulation of these activities was further reenforced by the opening of federal lands to individual claims under the 1872 Mining Law, enabling small-scale operators to set up independent and exclusive claims on public land and to work them alone using simple panning techniques. With increasingly sophisticated methods of hydraulic mining used to improve efficiencies and yields, however, the scale of environmental disruption in mining and the complexity of social organization increased. Hydraulic mining diverted increasingly large quantities of water from local streams to move ore into and through the sluice boxes, and the use of water cannons for this purpose became widespread to quickly section hillsides of loose aggregate material.[18] Groups of laborers were required to undertake larger hydraulic operations, building canals to funnel large quantities of water to the mining sites and operating the larger scale mining equipment. Evidence of these operations may still be seen in the Illinois Valley in the form of cutaway hillsides and snaking networks of ditches that directed water from the surrounding slopes.

Further increases in the scale of operations and their environmental transformations were tied to the introduction of hard-rock mining. For example, the Greenback gold mine in Grants Pass employed as many as three hundred people at its peak in the 1920s. Sporadic mining for gold, copper, bronze, and chrome remained a facet of the extractive economy of the Illinois Valley, and it was still quite central to the area well into the 1940s, largely in the form of larger underground or shaft mines.[19] But most operations eventually panned out, and many would-be miners moved into and out of the area, or stayed and took up other vocations.

Early Nature-Centered Production: Railroads, Land Grants, and Lumber

Reinforcing its commitment to federal disposal, Congress established a massive land grant in 1866 to support the construction of a rail link from Portland to Sacramento, and thereby to San Francisco. The grant was turned over to railroad mogul Ben Holladay by the Oregon legislature in 1868, eventually leading to formation of the Oregon and California (O&C) Railroad. In all, a total of 3.7 million acres of land were made available as alternating 640-acre parcels within a twenty-mile wide strip on either side of the proposed route through Oregon.[20] The clear intent of the O&C railroad grant was that the grant lands be transferred to private ownership in the area to further reinforce homesteading.[21] However, very little O&C land was actually transferred to settlers.[22] Instead, by 1898 the O&C grant had fallen under the control of the Southern Pacific, and from the company's perspective, a requisite price of $2.50 per acre for sale to homesteaders was simply too low for the sale of richly forested timberlands. It was more profitable if the land could first be cleared of its timber and then sold. This fraudulent ethos contributed significantly to an emerging contemporaneous problem of concentrated private timberland landownership in the Douglas-fir region.[23]

Yet, if it impeded homesteading, the railroad did spark development of a timber and lumbering sector in southwestern Oregon nonetheless. For one thing, the actual construction of the railroad required large amounts of wood, primarily for ties and trestles. According to Richard White, construction of railroad lines accounted for as much as 20 to 25 percent of the nation's timber harvest between 1870 and 1900.[24] Augmenting this direct demand, when the rail line linked the southwestern Oregon line to the California line in 1887, local lumber could be shipped north to Portland and south to Sacramento and San Francisco, providing key outlets for the landlocked timber. These links proved critical to the expansion of the Oregon lumber industry, including subsequent to the 1906 San Francisco earthquake when lumber from southwest Oregon factored centrally in the city's reconstruction.[25] By 1929 the Oregon lumber sector accounted for a remarkable 62 percent of total wage employment and 44 percent of total value added in the state.[26] And expansion of the industry, with its extensive, raw material dependent geography, helped propel rapid population growth in rural, timbered parts of the state, particularly in the western Douglas-fir counties.

These processes eventually proved critical to the development of Josephine County's lumber sector, although the county lagged somewhat behind its neighbors. Sandwiched between the coast and the main north-south rail line,

Josephine County's isolation slowed its growth. Thus, although the county's population grew steadily during the first four decades of the twentieth century, it did so less quickly than in surrounding counties, and slower than in the state overall. Moreover, the Illinois Valley's remoteness was even further reinforced when the main railroad line bypassed it to the east.[27] Thus, although other counties saw earlier booms, up to 1940, agriculture accounted for almost one-third of Josephine County's total employment, whereas mining employment ran second (see Table 6.2).

Only with the expansion of national lumber demand leading up to World War II did the Oregon lumber boom arrive in the state's more remote locations, including Josephine County and the Illinois Valley (see Figure 6.3). Josephine County's total population and workforce more than doubled between 1940 and 1950 (see Table 6.2), propelled by expansion in the local wood-products sector, which grew by more than seven times over the decade. By 1950, more than 90 percent of all manufacturing employment and more than one-quarter of the county's overall workforce were accounted for by the wood-products sector.

Still, throughout this early wave of expansion, a predominantly laissez-faire model of social regulation continued to prevail over forest practices. Up to this point, private lands continued to sustain local extraction in the form of grant lands sold to independent farmers and lumbermen, or in the form of lands retained by the railroads. As yet, the state did not regulate forest practices, and the growing inventory of federal lands in the area did not yet provide raw material for industrial expansion. And although large conglomerate railroad companies played a prominent role in the control of timberlands in southwestern Oregon, the actual cutting of timber and milling of lumber was predominantly carried out by small-scale operations. These small operations reinforced tendencies in the social organization of early agriculture and mining and reflected the vision of yeoman capitalism in the West that had been used to legitimate federal land disposal policies. Thus, between 1935 and 1951 the number of operating lumber mills in Josephine County expanded from eighteen to sixty-four, whereas lumber production increased from twenty-four million board feet to more than three hundred million board feet (see Figure 6.4); yet typical mills employed on the order of only eight to ten men. One of the extraordinary

Table 6.2 Changing Employment Structure, Josephine County, Oregon, 1940-60

Year	Total	Agriculture	Mining	Wood Products	Manufacturing
1940	5263	1642	473	300	533
1950	9381	1641	113	2381	2623
1960	9593	832	44	2323	2708

Source: U.S. Bureau of the Census, Census of Population, various years.

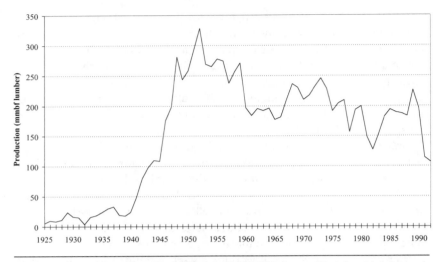

Fig. 6.3 Josephine County Lumber Production, 1925–1992. Source: West Coast Lumbermen's Association, (1964), Western Wood Products Association (1998), and various other years.

aspects of these mills, in addition to their small scale, was their flexibility and portability, typical of the industry more generally at the time (see chapter 4). Small mills were powered by four- and six-cylinder diesel engines and drew logs from a one- to two-mile radius. Only the highest quality logs were used; everything else was left behind. And when an area was logged over, the mill was simply dismantled and moved to another location, where the operation began again. Private timber was obtained through the outright purchase of land or through purchasing the cutting rights from independent owners. Many mills purchased small parcels, cut the timber, and left the land to go back to the county in lieu of paying taxes on it.

Moreover, in addition to their mobility, most small mills were operated for only seven to nine months of the year because of severe winter weather. During the off season, workers drew on what other employment they could; collected unemployment insurance; and hunted, trapped, and collected firewood to supplement their cash incomes.[28]

One such mill was operated by a longtime resident of the Illinois Valley who moved to the area in 1949 from Arkansas, where he was born and raised. He remembers the small mill he bought in August 1949 for $4,000 as one of fifty-one operating in the valley at the time:[29]

My first mill was a little mill that I had down here at O'Brien. I bought that in 1949, and started it in August. I had eight men, and one mule. We cut about 10 or 12 thousand feet a day. And I paid them all not by the

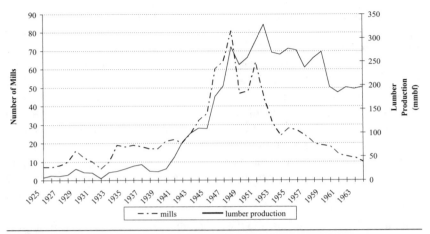

Fig. 6.4 Lumber Mills and Lumber Output, Josephine County, Oregon, 1925–1964. Source: West Coast Lumbermen's Association (1964), Western Wood Products Association (1998), and various other years.

hour, but by the thousand. I paid my sawyer, who was the main ramrod of the thing, a dollar and a half a thousand. The log turner and the edger I paid a dollar and a quarter a thousand, and the rest got a dollar a thousand. We'd move into a place like here, and cut the timber, and push the slabs and the sawdust over the canyon. We'd work there for, oh, maybe two or three weeks, and move. It took a couple of days to move. They all worked for nothing then, when we moved. And I'll tell you, they didn't make a hell of a lot of money. But I didn't do too bad. I know I could figure that if I paid about a thousand dollars in wages this week, well that's about what I made. I made just about what all the eight men made put together.[30]

Far from anomalous, mill operations such as these were typical during the Illinois Valley lumber boom. If there were fifty mills in the area at the time, there were perhaps only five larger and more permanent installations in the valley capable of cutting twenty to thirty million board feet of lumber per day.[31] As a result, the face of capitalism in the local lumber sector was still largely yeoman and of an almost petty commodity character. Workers outnumbered owners, but by factors of ten, not hundreds. Class distinctions were in significant ways blurred by the fact that owners of small mills worked closely alongside their employees, as in the previous example. Moreover, boundaries between workers and owners were further erased as many of the former tried their hands at entrepreneurship (again, epitomized in the previous example), returning to wage labor if things did not work out.

This liberal, almost petty commodity form of land-based extraction developed against a backdrop of relatively laissez-faire social regulation over forest practices and timber supply. Although the state played a minor role in providing raw material from public lands prior to the Depression and the New Deal, federal forest policy primarily reflected a cooperative approach to regulating private forestry. Under this paradigm, federal initiatives largely took on the character of voluntary programs of coordinated federal research and funding, relying on incentives rather than command and control regulation vis-à-vis forest practices. These policies reflected compromises between opposing forces, with industry aggressively attempting to avoid direct federal regulation of private lands, and reformers such as Gifford Pinchot and Ferdinand Silcox arguing for much more aggressive intervention by the federal state.[32] Perhaps no forest legislation epitomized the era of compromise and conciliation preceding the advent of sustained-yield management and the escalation of federal timber sale programs more than the Clarke-McNary Act of 1924. The legislation's stated purpose was ambitious indeed: "To provide for the protection of forest lands, for the reforestation of denuded areas, for the extension of national forests, and for other purposes, in order to promote the continuous production of timber on lands chiefly suitable therefor."[33] Yet the actual legislation was quite liberal, avoiding direct regulation of forest practices, while extending federal absorption of cutover private lands and creating federal programs and matching grants in the areas of forestry research and fire control.[34] All the while, the accumulating area of federally retained forests (see below) remained essentially off-limits to industrial logging. But with sustained-yield regulation, this would change dramatically.

Managing Extraction: Sustained-Yield Regulation

Precursors

If the pattern of social regulation in the forestry arena prior to the New Deal was more liberal in character, the fictions within this model were also apparent, laying the foundations for a more interventionist role for the state. These fictions concerned the social control and reproduction of the nation's forests (i.e., nature) on the one hand and the maintenance of an economic base for forest-dependent communities and regions (i.e., social labor) on the other. These twin fictions were made evident by an industry that was reluctant to undertake forestry until the accumulated wealth of "found" forests was exhausted and that had a deserved reputation for causing boom and bust development cycles in communities. These issues underpinned the politics of regulating forest practices on private lands and an increased profile for the federal state in providing the nation's timber supply (particularly in the western states) from public lands managed according to the principles of maximum sustained-yield forestry.

Concern was mounting through the 1910s and 1920s regarding the accumulation of cutover lands and the implications of this accumulation for future timber supply in the nation. Future Forest Service chief William Greeley, for example, published a paper in 1925 concerning an impending shortage of timber in America.[35] Denuded timberlands, abandoned logging camps, and moribund lumber mills littered the northeastern and southeastern states.[36] Estimates of the area of land converted from forest to agriculture and settlements totaled on the order of two hundred million acres by 1920, with the original standing volume of timber reduced to half its presettlement total by the same year, cut down by land conversion and, increasingly, by industrial logging and lumbering.[37] Although extensive inventories of old-growth timberland remained along the Pacific Slope, the region did not escape arguments that forestry needed to be reformed. Faith that clear-cut Douglas-fir forests would regenerate on their own, for instance, began to fade even within the industry's ranks as evidence accumulated to the contrary. These doubts were exacerbated by a prominent 1927 survey revealing that 40 percent of logged-over forestland in the Douglas-fir region was not growing back.[38]

At the same time, land consolidation was a major concern, particularly under the auspices of the railroad land grants and their abuses. The U.S. Bureau of Corporations Report of 1913 on concentrated timberland ownership in the United States, for instance, found that three companies in the Pacific Northwest had been able to accumulate 238 billion board feet of timber, an estimated 11 percent of the nation's total, relying centrally on use and abuse of railroad land grants. The most prominent of these companies was the Weyerhaeuser Timber Company (see chapter 5).

Finally, the social character of logging and lumbering became an additional basis of arguments for reform. The industry had become infamous for its transient character, in the Pacific Slope and elsewhere, undermining visions of a liberal, landed social order for the western states. As Vernon Jensen noted in his study of labor in the lumber sector, in 1900 about two-thirds of the workers in the lumber industry in California, Oregon, and Washington were not married. Moreover, he estimated this figure was closer to 90 percent in the logging camps as late as 1910.[39] Forest examiner William Gibbons remarked,

> The industry has to depend on a woods force composed in large part of restless, dissatisfied bachelors—old and young—largely foreign born, a large portion of whom shift from camp to camp via the larger centers of population.[40]

The fetid conditions in these camps have been extensively noted, and they were seen as havens for the footloose and ripe for radical organizing, not least by the Wobblies.[41] Industry-wide strikes in the Douglas-fir region paralyzed logging and lumber production in 1917 and again in 1935, and a general climate of antagonism prevailed between unionized workers and firms. The migratory and radical character of the workforce was perceived by many as directly related to the industry's cut-and-run forest practices, and momentum for reform brought together those worried about the power of the radical labor movement, those wishing for a more stable social order, and those rallying under the conservation banner.[42]

In this context, interest in more active forest management, in research and applied settings, and in reform of forest land tenure and its regulation was widely evident, and in numerous arenas. Most initiatives were quite moderate and posed little threat to the liberal pattern of regulating forestry. They emphasized cooperation between state and capital, most prominently in the emerging science of American forestry. Drawing on agricultural plant breeding, for example, foresters working under the auspices of the American Breeders Association had begun to express interest in forest genetics; this interest would, in time, lead to industrial tree improvement, including in the Pacific Slope (see chapter 5).[43] Forestry schools, the first established at Cornell in 1898, had begun to crop up around the country, training the first generation of professional foresters, many of whom became members of the Society of American Foresters, established in 1900.[44] And various federal and state initiatives in forestry research had been established, including the first of the Forest Service's Regional Forest Experiment Stations, established in 1908 at Fort Valley, Arizona, and, two years later, the Forest Products Laboratory established at Madison, Wisconsin. Overall, the approach taken by federal and state governments was to support the establishment of more active forest management in areas such as fire control, forest inventory, and, most significant, reforestation by means of the establishment of funding and extensive cooperation.

There were those, however, who argued for a more active role for the state, and particularly for the federal state, to ensure reforestation and to control the management of the nation's forest resources more generally. Gifford Pinchot was central among these figures. Pinchot was among the first and most prominent figures to argue that the federal government should assume jurisdiction over the regulation of private forest practices and that the federal government should also establish a program of sustained yield in the newly created system of federal forests. Pinchot's crusade for aggressive, state-centered forestry based on German and Scandinavian models resonated with widespread dissatisfaction over the management and consolidation of private forest lands, giving momentum to his "practical forestry" agenda.[45] Yet although legislation—including the key Clarke-McNary Act—did establish federal matching funds for state nursery programs and initiated the federal government's role in forestry education, the federal state was ultimately stopped short of regulating private forest practices in large part because of aggressive industry lobbying opposed to it.[46] Eventually, industry lobbying helped to deflect federal regulation on private lands by means of the establishment of local state jurisdiction, cemented by Oregon's passage of the nation's first forest-practices legislation in 1941.[47]

Although the federal state was impeded from directly regulating private forest lands (even if the threat of such an outcome propelled individual local state legislation), the politics of reform in American forestry nevertheless initiated a rethinking of disposal policies vis-à-vis federal forest lands. The resulting shift to land retention would have dramatic consequences for land tenure and forest management along the Pacific Slope, particularly in Oregon, as these lands subsequently became the basis of federal sustained-yield forest management programs.[48] As early as the 1870s, and influenced by George Perkins Marsh's *Man and Nature,* strong advocates of retaining forested lands in public ownership had emerged, including secretary of the Interior Department Carl Schurz (1877-1881).[49] By 1891, swept up in a growing tide of Progressive Era reform, Congress passed the General Revisions Act, which included a provision in Section 24 authorizing the president to create federal forest reserves. The origins and actual intent of the now (in)famous Section 24 have been hotly contested by historians, but conserving natural resources and countering abuse of homesteading land grants were certainly among the intentions.[50] By 1893, 17.5 million acres of federal land had been reserved, first by President Benjamin Harrison and then by President Grover Cleveland.[51] Most of these lands were in the West, where, because of the overall east to west movement of homesteading and railroad construction, most of the remaining unclaimed federal domain was concentrated.[52] Because of its extensive forest resources and its comparatively late inclusion in the homesteading and development processes that swept the West during the nineteenth century, Oregon in particular became a leading target for the creation of federal forest reserves. This is largely why the state ranks fifth behind California, Idaho, Alaska, and Montana with its 15.6 million acres of National Forest timberlands, representing about one-quarter of Oregon's total area.[53] Moreover, Oregon's 12 million acres of National Forest timberlands ranks second just behind Idaho among all states, and comprises more than half of the state's total timberland area.[54]

Retention in Oregon came not only in the form of the National Forests under control of the Forest Service but also in the guise of lands revested from the Oregon and California railroad grant in 1916. Here again, concern about the failure of the land disposal process and the social organization of timber extraction prompted congressional action. The fate of the O&C grant lands drew increasing scrutiny from the Oregon legislature and Congress after 1890 in response to the Southern Pacific's reluctance to transfer the lands to settlers as per the terms of the original land grant. The railroad administrators had made a practice of selling parcels at rates higher than allowed by the grant and in larger parcels than was prescribed, while also frequently logging the land in advance of offering it for sale. Yet logging it first eliminated the land's chief source of value and made it far less appealing to homesteaders.[55] Perhaps the final straw came in 1902, when the Southern Pacific disposed of four hundred thousand acres of grant lands to lumber industry figures, not homesteaders.[56] In 1908 the U.S. government filed suit, charging the railroad with violating the land grant of 1869. Although the 1915 Supreme Court decision went against the government,[57] it opened the door for congressional action; in 1916 Congress formally revested 2.3 million acres of O&C lands to the federal government, placing them under the jurisdiction of the Department of the Interior.

The legacy of this decision comprises a second portfolio of federal forest-land in Oregon, administered by the BLM.[58] These lands—still referred to in Oregon as the O&C lands—include more than two million acres of timberlands, representing about 10 percent of the state's total.[59] Because they are concentrated in western Oregon, where they make up about 14 percent of the timberland area, and because they tend to be at lower elevations than the National Forest lands, the O&C forests are among the most productive Douglas-fir timber-lands in the state. And they are of particular significance in southwestern Oregon, including Josephine County and the Illinois Valley (see Table 6.1).

Sustained-Yield Regulation

With this suite of federal lands retained as a reaction against the political and ecological contradictions of the disposal era, the emergence of the federal state as a major supplier of timber to the nation's forest products industry needed only a regulatory foundation.[60] The collapse of New Deal efforts to regulate industrial overproduction and cut-and-run forestry only exacerbated the climate of political and ecological crisis in American forestry, opening the door to this foundation in the guise of sustained-yield forest management on federal lands, including those of southwestern Oregon. Outspoken advocates of aggressive federal intervention in the form of an ambitious federal timber supply program governed by the principles of sustained-yield forestry predictably included fig-ures such as the ubiquitous Pinchot and, later, Chief Forester Ferdinand Silcox.[61] But advocates of sustained-yield forest management on federal lands and the parallel albeit voluntary adoption of these principles for managing private lands also included outspoken industry figures such as David Mason, who saw sustained-yield forest management as a panacea for the industry's problems of overproduction, boom and bust cycles, and the threat of a future timber famine.[62]

The idea was deceptively simple. Sustained-yield management and regulation would provide a more stable economic climate in the industry—and thus a less uncertain future in forest dependent communities rocked by industry booms and busts—by enabling more accurate predictions of future timber harvest volumes based on annual yields prescribed by scientific management, not price signals. At the same time, rational, science-based prescriptions would restrict cutting volumes to rates at which forests could regenerate. The federal government's now vast system of public forests could help lead the way.

Drawing on European forestry (a tradition in which many of America's first scientific foresters were trained, including such figures as Pinchot, Henry S. Graves, and Bernard Fernow), sustained yield prescribes the conversion of natural forests to *Normalbaum,* or normal forests. A normal forest has been defined as "an *ideally constituted forest* with such volumes of trees of various ages so distributed and growing in such a way that they produce equal annual volumes of produce which can be removed continually without detrimental impacts to future production" (emphasis added).[63] The idea is to organize a given forest into a set of even-aged blocks, each of which is harvested at a prescribed age—the rotation age—such that the annual harvest volume taken from the forest is constant over time and is preferably equal to the maximum amount possible for the area of forest in question.[64] This maximum corresponds to the "culmination of the mean annual increment," or the point at which the average annual rate of growth of the trees in the forest across their life span is maximized; this is the maximum sustainable yield. Critically, the model is typically envisioned and, to the extent possible, put into practice using monocultures of commercial timber species. At any rate, the accounting system underlying the management of forests according to this maxim only recognized as relevant the volume of commercially appealing timber growing within the management area.[65] Other trees, much less other types of organisms, were, according to the model, assumed away as irrelevant.

Although sustained-yield management was developed in the European forestry tradition, it offered several appealing facets in New Deal–era America. For an industry struggling to achieve some predictability in the context of rampant overproduction and price fluctuations, in part due to low-scale economies and low barriers to entry in the lumber sector (see chapter 4), sustained yield offered the promise of some restraint and more rational investment conditions. For communities faced with boom and bust cycles, sustained-yield forest management carried the promise of renewable forestry and constant harvest volumes to sustain industrial operations in place. For those worried about resource conservation, and the imminent threat of a timber famine, sustained yield offered a way to use science to regulate and renew forests, retaining their productivity for wise uses ad infinitum. And for progressives and New Dealers, sustained yield, particularly on federal lands, was consistent with the much broader advent of state-centered, science-based natural resource management[66] and, ultimately, with the onset of Keynesianism. As the American economy deteriorated in the early 1930s, reform of ecologically destructive forest management practices and the political economic problems of the lumber industry became a critical part of the emerging New Deal agenda. Although local states became the site of regulating private forest practices, advocates of a dramatically expanded federal timber sale program governed by sustained-yield management principles got their way.

Significantly, the explicit goals of sustained-yield regulation were to address the twin fictions of nature and labor as commodities in the forest sector. That is, if conserving and reproducing forests was obviously a key objective, so too was the need to address social tensions arising from a migratory, shifting industry and its effects on forest-dependent communities. Thus, sustained-yield regulation had a dual purpose. This is reflected by the language of the Oregon and California Act,[67] the first piece of legislation enabling sustained-yield doctrine in America, enacted to provide a regulatory framework for managing the lands revested from the O&C land grant. The act stated that O&C timberlands,

> Shall be managed . . . for permanent forest production, and the timber thereon shall be sold, cut, and removed in conformity with the principal of sustained yield for the purpose of providing a permanent source of timber supply, protecting watersheds, regulating stream flow, and *contributing to the economic stability of local communities and industries.* (emphasis added)

Similar verbiage is included in the Sustained Yield Management Act of 1944, enabling legislation for the introduction of sustained-yield management to the National Forest system.[68] Together, the Sustained Yield Management Act and the O&C Act established the regulatory basis for a greatly expanded federal timber sale program, which committed federal timber to capital in nondeclining annual volumes from each management area and offered it through auction-style bidding on individual timber sales. These auctions, as Walter Mead's work indicated, were frequently of questionable competitive character, bearing the imprint of substantial market control by dominant firms within particular timbersheds or timber supply areas.[69] Nevertheless, the fundamental division of labor was established; the state would manage the forests and offer the timber for sale, and private capital would process it into commodities. Communities and workers would benefit from the economic base thereby generated and could ostensibly be ensured a stable economic future and renewed forest resources by the promise of constancy that sustained-yield management seemed to bring.[70] Commitment to communities was further institutionalized by the distribution of resource rents to local communities in the form of remitted timber sale revenues guaranteed to counties in which federal timber was sold. Thus, the BLM would return 50 percent of gross receipts from O&C timber sales and contributions in lieu of property taxes to county general funds, whereas the Forest Service would commit 25 percent of gross receipts to counties, three-quarters of which was allocated to road construction and maintenance and the remainder to schools.[71]

The structured coherence comprised by this institutional configuration of industrial commodity production, state forest management and timber sales, and everyday life in forest dependent communities became the basis of a distinct postwar configuration of capital accumulation and social regulation in the Pacific Slope forest industry, particularly in those areas dominated by federal forestlands. In Oregon, where National Forest and BLM land comprise such a significant proportion of the state's timberland area, the expanded federal timber sale program under sustained-yield regulation offset declining harvests from private lands, becoming the mainstay of the state's postwar forest industry (refer to Figure 1.2).

Sustained Yield and the Illinois Valley

In Josephine County, harvest levels that had risen rapidly during the years immediately prior to World War II quickly exhausted local supplies on private lands. With the onset of sustained yield, pressure shifted to the county's more plentiful federal forests in the early postwar. These forests came to sustain fiber supply in the county (see Figure 6.5), which in turn became among the most dependent of all Oregon counties on receipts from federal timber sales. Thus, by fiscal years 1982-83 and 1985-86, for example, Josephine County drew an average of 42 percent of its total fiscal resources from BLM timber sale remissions.

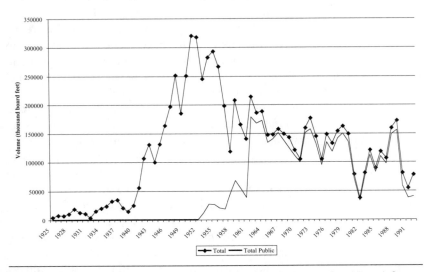

Fig. 6.5 Timber Harvest in Josephine County, 1925–1992. Source: Bourhill and Oregon Department of Forestry, 1994.

If Forest Service payments are included, the county's reliance on federal timber rents averaged more than half of the Josephine County revenue.[72] In the Illinois Valley, under the administration of the Medford district of the BLM and the Illinois Valley Ranger District of the Siskiyou National Forest, federal timber sales sustained the local forest industry almost entirely after 1950.

Yet despite the promise of stability in the structured coherence of maximum sustained-yield forest management, the experience of the Illinois Valley, and other similar, rural forest-dependent communities, indicates that this stability never materialized. This is apparent in several respects, including perhaps most glaringly in the failure of sustained-yield policies to result in the most obvious and basic measure of their own efficacy—sustained yields! Yet it was also manifest in significant industrial restructuring, including consolidation and industry attrition, raising serious questions about the equation of guaranteed federal timber—in constant volumes or not—with industrial stability. In turn, economic and social conditions in the valley, if anything, worsened during the period of sustained yield. Indeed, despite the reactions of industry figures blaming environmentalists and spotted owls for the woes of forest-dependent communities in the late 1980s and early 1990s, it is fully apparent that signs of trouble were manifest virtually throughout the sustained-yield era.

First, consider fluctuating timber volumes. As Figure 6.6 indicates, although planned harvest levels may have been constant, this hardly translated into constant volumes of either sales or harvests.[73] Instead, pronounced annual fluctuations are apparent in the amount of timber actually sold by the Siskiyou National Forest and in the amount of timber actually harvested through the heart of the sustained-yield regulatory era from 1960 through the late 1980s—before the listing of the spotted owl led to more dramatic harvest and sale reductions in the early 1990s. Similar fluctuations are also apparent in the amount cut and sold each year in the Illinois Valley Ranger District of the Siskiyou National Forest. Although such fluctuations might plausibly be apparent at more local scales without serious implications for the amounts of timber actually processed locally, as firms "shop around" for fiber, in fact, similar annual fluctuations are apparent in the total annual sales of timber in Oregon National Forests overall (refer to Figure 1.2). Moreover, as shown in Figure 6.5, it is apparent that aggregate harvests of public and private timber considered together, at least at the county level, were hardly constant. At the most basic level, then, this throws into question the translation of sustained-yield management policies into actually stable annual timber harvest levels at the local level.

Second, the industry hardly stayed constant during the sustained-yield era. In fact, the overall trend in the forest industry apparent in Josephine County and the Illinois Valley is one of contraction and consolidation virtually throughout the sustained-yield period. Josephine County's boom in timber harvesting and lumber milling actually peaked during the early 1950s (see Figures 6.3 and

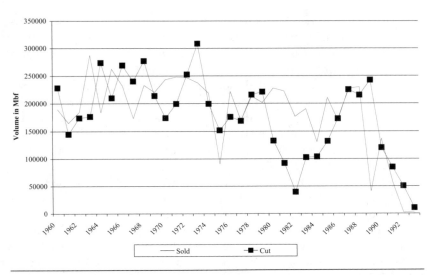

Fig. 6.6 Timber Cut and Sold in the Siskiyou National Forest, 1960–1993. Source: Siskiyou National Forest, Annual Reports, various years.

6.4), but the county remained a substantial producer of logs and lumber throughout the sustained-yield period, nominally ending in the early 1990s with the spotted owl listing. Yet the number of operating lumber mills in Josephine County dropped precipitously after 1950. Although the West Coast Lumbermen's Association reported sixty-four Josephine County lumber mills in 1951, ten years later there were only fourteen.[74] Direct data of this kind for the Illinois Valley are not available. However, according to one longtime local resident and industry participant, the number of mills operating in the Illinois Valley dropped from more than fifty in the early 1950s to eight by 1955.[75] These data are corroborated by anecdotal evidence available in the local archives. The dynamic here is one of consolidation in the lumber sector; despite its comparatively low-scale economies (see chapter 4), lumber production nevertheless became dominated by increasingly fixed, factory-like facilities in the late 1940s and early 1950s, propelled in part by a shift to electric power in rural areas.[76]

The impact of this consolidation on the social basis of the local economy was significant. From a network of small producers working alongside larger factory-style facilities emerged a handful of factory mills, reducing the ratio of owners to workers and increasing the size of the industrial working class in the area. Between 1950 and 1960, although the number of mills fell drastically, employment in wood-products manufacture in Josephine County fell only slightly, to just more than twenty-three hundred. Yet by 1960 most workers had been drawn out of the woods and into the mills. Workers had been pulled away from the smaller mobile mills to be employed by the larger, permanent, and centralized mills operated on a year-round basis. In turn, these few mills assumed primary responsibility for the development of the local economy, accounting for about one-quarter of the total workforce, and in excess of 80 percent of manufacturing employment.

Consolidation and associated mill closures, often triggered by economic downturns, have continued to reduce the number of mills in the area and, increasingly since the 1960s, the number of high-paying jobs in lumber production. In fact, although four or five larger mills survived the burst of consolidation in the 1950s and 1960s, only one mill remained following the turbulent economic conditions of the 1970s and the deep recession that hit Oregon and the nation hard from 1980 to 1982.[77] This phenomenon is reflected in part by the fact that between 1969 and 1990, the proportion of total income in Josephine County from lumber and wood products fell from more than 15 percent to less than 7 percent. In the Illinois Valley, the Rough 'n' Ready Lumber mill, operated by the Krauss family continuously since 1922, became the sole surviving large mill in the 1980s. The mill's 150 employees have the best-paying jobs in the valley, and they are joined by a fluctuating number of loggers and truck drivers who work to supply the mill with raw material, and who also are paid well by local standards. But these jobs—perhaps 250 in total—have come to occupy an increasingly large proportion of scarce sources of living wages in a community home to perhaps twelve thousand people.[78]

And as the number of local mills and associated jobs has declined, so too have local economic and social conditions. In fact, during the golden age of sustained-yield federal forest management stretching from 1960 to 1990, the Illinois Valley slowly slid into a morass of social and economic problems, earning a place among the most troubled communities in the state. For example, in 1960 median family income in Josephine County was 88 percent of the state median. Yet by1990 the county's median had fallen to 65 percent of the state figure, whereas median family income in the valley stood at just 54 percent of the state level.[79] Also, according to 1990 data, 23 percent of families in the valley were living in poverty, and the poverty rate among children was closer to 40 percent.[80] Overall, Illinois Valley household income distribution became heavily skewed toward the lower income categories in comparison to the state (see Figure 6.7). And as lumber and wood-products employment declined locally, the disparity between the more isolated Illinois Valley and portions of the county closer to Interstate 5 (primarily in the Grants Pass area) grew. Thus, 1990 census data showed unemployment figures for the valley at 16 percent, while the county overall had a 10 percent unemployment rate.

A range of other social indicators also point to the Illinois Valley as a troubled community. The area has been plagued by low levels of educational attainment and elevated high school drop-out rates, with Josephine County lagging behind the state in the percentage of adults ages twenty-five years and older with

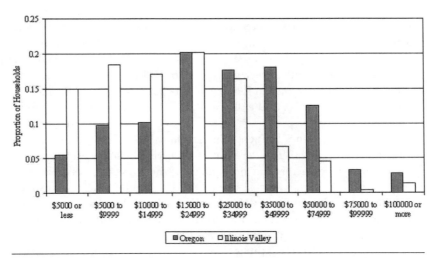

Fig. 6.7 Household Income Distribution, Oregon and Illinois Valley, 1990. Source: U.S. Department of Commerce, Bureau of the Census, Census of Population 1990.

high school diplomas (75 percent versus 82 percent) and college degrees (12 percent versus 21 percent). In part, this is a result of the historically low premium placed on educational attainment in the wood-products sector, which tradition-ally did not require a high school diploma of its employees or high rates of literacy and technical skills. Many residents with little formal education went to work in the mills when times were more buoyant. Yet this leaves many former and prospective workers with few options for employment in other activities or for the increasingly scarce jobs available in the mills. Low levels of educational attainment in the valley, including low high school graduation rates and low levels of adult literacy, are among the factors that reinforce perceptions by po-tential investing firms that the local labor market is backward, serving as an obstacle to diversification in the area widely cited by local observers, including economic planners and civic administrators.[81]

People in the valley are quite willing to discuss the chronically depressed local economic and social conditions. When pressed, most concede that these conditions largely predate the cyclical downturn of the early 1990s and the listing of the northern spotted owl as a threatened species. Glib attribution of all the area's problems to environmentalists and the federal government typically gives way in longer conversations to reflections on the dead-end character of extractive industries. Many residents—both friends and critics of the industry—recognize the dramatic transformation that has taken place in the age and size structure of local forests and the overall pattern of consolidation in the industry, and they understand that these are directly implicated in the gradual decline of the valley. Many in particular recall the 1980s, and the fallout from the reces-sion, as perhaps the most difficult period for the community. Indeed, there is considerable evidence that this recession marked a watershed of sorts. Among Josephine County residents, the incidence of poverty increased from 15.6 percent to 18.3 percent over the course of the 1980s. Out-migration from the area during the 1980s, combined with slower in-migration, also helped limit population growth in Josephine County to its slowest ten-year rate of increase since the decade from 1910 to 1920. In the Illinois Valley, population actually decreased during the 1980s.[82]

The troubles experienced by people in the Illinois Valley during the sustained-yield era, particularly in the latter years, are clearly not solely attributable to forest-products manufacture and forest policies. What these troubles do indicate, however, is the evident failure of sustained-yield regulation to achieve any semblance of community economic stability. In fact, prior to the listing of the northern spotted owl in June 1990, there is very little about the experience of Josephine County and the Illinois Valley to indicate the goals of sustained-yield forest management—stable levels of timber harvest from federal land, and stable economic and social conditions in the community—were being met by any reasonable interpretation of the meaning of stability. Why would this be so?

Internal Contradictions to the Sustained-Yield Model: Ecological Complexity, Industrial Geography, and Community Stability

As the Illinois Valley case indicates, although the crisis precipitated by the listing of the northern spotted owl forced a reregulation of federal forests, the tensions and contradictions running through the structured coherence of sustained-yield regulation were nevertheless apparent throughout. In particular, the twin fictions of nature and labor as commodities never were eliminated by sustained-yield regulation but were instead offset and reworked. They were to reemerge as the source of open political opposition to the sustained-yield model of regulation beginning in the 1970s and continuing into the 1990s, and they were propelled by the ecological contradictions of the Normalbaum together with the false promise of economic stability in forest-dependent communities.

Consider first that achieving the prescribed vision of a Normalbaum forest populated by even-aged stands of commercially important species—most significant, Douglas-fir—required a dramatic ecological conversion. Old-growth forests, rich in species diversity and complex in age structure, had to be remade and drastically simplified. That normal forest statistical accounting—or "statistical picturings" as they were described by David Demeritt[83]—did not recognize this conversion as significant now appears as a fatal flaw. In actual fact, the conversion from old-growth to young-growth was ecologically dramatic, whether measured simply in changes in the age and species composition of forests (see chapter 3) or by broader criteria such as changes in ecosystem composition and complexity more generally as the result of habitat transformations induced by logging—changes for which the northern spotted owl became the key indicator.[84] Moreover, predictions of future forest yields under the imposition of sustained-yield regulation were always best guesses, based as they were on little empirical data. These predictions were too often wildly optimistic as to what harvesting pressure could be sustained by future forests, undermined by ample evidence of reforestation failures and overly optimistic predictions of growth rates.[85]

At the same time, the conversion to Normalbaum would necessarily entail a "fall-down" effect in timber volumes as old-growth stands were liquidated and replaced with lower volume young-growth on a planned harvest rotation too short to regenerate old-growth.[86] As old-growth became increasingly scarce, it became harder and harder to sustain aggregate harvest volumes from young-growth forest stands with lower timber densities (see chapter 3, Figure 3.2).[87] Although these problems are implicated in the spotted owl crisis, and became politicized in the 1980s and into the 1990s (see epilogue), it is critical to note that they emerged out of the very logic of the sustained-yield model of forest regulation. In fact, the Normalbaum vision prescribed the ecological transformations, however implicitly, that were to become the subject of political contestation and litigation.

As for the fictions of labor, the naive equation of constant timber volumes with a stable economic future belied a flawed understanding of the basic organizing principles of industrial capitalism. To understand this, one must appreciate first that neither the Sustained Yield Management Act nor the O&C Act nor any subsequent legislation or regulation actually refined the notion of community stability; that is, how it could be defined, monitored, and achieved. The very notion of a community, its spatial boundaries, its membership, and so forth, is necessarily elusive and problematic, a tension that runs through much of the uncritical embrace of the concept in contemporary notions of community natural resource management.[88] Yet, despite such problems, the notion was seldom if ever refined within federal sustained-yield policy, and the relationship between forest management practices and timber supply schedules on the one hand and community economic development and stability on the other hand were never specified beyond the simple equation of constant raw material supply with local industrial, economic, and social stability. This implicit equation is thus the essence of how the community stability provision has been interpreted; that is, federal management agencies did not take any other steps to ensure community stability.[89]

Yet this equation reflects a fundamental misunderstanding of industrial capitalism, particularly in primarily rural, natural resource-based industries. This misunderstanding leads to three crucial contradictions that all point to the fictitiousness of labor as a commodity, as the needs of markets and industry on the one hand and the needs of workers, their families, and their communities on the other hand increasingly diverge. These contradictions stem from (1) the tendency of capitalist production, including the lumber industry and other facets of the forest sector, to shed employees under the influence of competition, innovation, and consolidation; (2) the periodic disruptions of capitalist market economies due to their inherent crisis-prone character; and (3) the inexorable pressures of restructuring nature-society relations in natural resource sectors due to changing political ecological circumstances, in this case propelled specifically by diminishing wood quality, increased competition for high-quality fiber and increased political opposition to capitalist resource access. Similar notions (with the exception of the third) are what propelled Freudenberg and Gramling to make a convincing case that long-term attrition in employment levels may be the critical link between extractive industries and rural poverty.[90]

That capitalist industries are inherently unstable and, more specifically, that they have a tendency to shed workers as a manifestation of competitive struggles between capitalist firms may be news in neoclassical circles, but it has become a truism in economic geography and among students of the uneven geography of capitalism. As Marx noted,

> The working population . . . produces both the accumulation of capital and the means by which it is itself made relatively superfluous; and it does this to an extent which is always increasing.[91]

The logic here is one in which capitalist firms, constantly competing with one another to achieve greater profits, seek ways to make production more efficient, involving, among other things, trying to produce more commodities with less labor. This tends to generate constant changes in industrial structure and organization, as firms adjust to new technologies, ways of doing things, and new market conditions. In an industry with insufficient growth in demand to offset the substitution of capital for labor, it also means total employment in the industry may be expected to go down. Even when total employment does not drop, however, considerable disruption in the space economy occurs by industry consolidation and rationalization. Petty commodity production gives way to generalized commodity production, as independent yeoman producers are put out of business by more and more modern and efficient facilities built to out-compete older ones. These larger and more efficient facilities in turn compete against one another to secure greater profits from expanded market share and lower costs, leading to shifts of employment from one place to another and to further rounds of substituting capital for labor.

Celebrated turns to more flexible production, and the apparent resurgence of small capital during the past few decades notwithstanding, the generally destabilizing tendencies of capitalist competition remain.[92] In fact, constant change is so pervasive in the eyes of some observers that it is seen as a signature feature of capitalist modernity: for example, Marshal Berman's classic *All That Is Solid Melts into Air* is a reflection on the wider implications of constant change and disruption whose title borrows from Marx's famous aphorism.[93] This is also the essential basis of most theories of uneven development and the inconstancy of capitalist space.[94] Critically, this understanding of the geography of capitalism directly contradicts notions of stability endorsed by sustained-yield regulation.

Upheaval related to the more or less steady or constant pressure to displace workers through productivity improvements is compounded by the macro-economic instability of capitalist economies. Theories of the business cycles and the crisis tendencies of capitalism are numerous and are a central feature of neoliberal and Marxist political economy.[95] No one, however, disputes that capitalist economies have been historically prone to recessions and depressions. These periodic and widespread events comprise an additional source of upheaval and can be particularly devastating to industries such as wood products, tied as they are to downstream sectors that fluctuate with business cycles; most directly, the housing industry. Periodic recessions and downturns are the source of changes in employment and income levels, as workers have their hours reduced, are temporarily laid off, or lose their jobs permanently as facilities close. Often, cyclical downturns and full-blown crises provide the impetus for firms to re-structure under changing competitive conditions, sometimes closing plants, sometimes investing in new equipment and adjusting employment levels accord-ingly, sometimes moving, sometimes consolidating, and sometimes combining all of these.

Added to the more general imperatives of capitalist production are a set of specific issues that characterize natural resource sectors. As Gavin Bridge noted in his discussion of the mining sector, notions of crisis tendencies in nature-based capital accumulation must account for the effects, both ecological and political, associated with local or generalized depletion as a result of industrial activities (see chapter 1).[96] Although this might seem unique to inherently non-renewable sectors like mining, it also applies to ostensibly renewable resource sectors such as forestry and fisheries if there is local depletion or if, as in the case of old-growth, renewable means converting the forests from one type (old-growth) to another (Normalbaum). These processes will also factor in to the long-term dynamics of the industry and may be implicated in some of the cyclical downturns and the direction that innovation takes (e.g., technological innovation and investment to offset resource depletion). I discussed aspects of such shifts in the Pacific Slope forest industry associated with the conversion from old-growth to young-growth in chapter 3.

At the same time, the remote rural geography typical of many land-based industries may exacerbate the destabilizing social effects of commodity pro-duction. Many raw materials industries have relatively low linkage densities;

that is, they have low spin-off effects in other industries. The reason for this is that natural resource production often involves simple conversion rather than more complex manufacturing processes. This creates a certain enclave character to raw material industries that is only compounded by their typically extensive geographies driven by the geographic availability of resources.[97] Communities located near raw materials and home to resource-based production may offer few geographic advantages as hosts to other types of production, and many of the linkages that do exist are based on inputs of goods and services from distant locations.[98]

All of these factors are implicated in the failure of sustained-yield regulation to overcome the fictions of labor as a commodity and to achieve any workable notion of community economic and social stability in locations such as Oregon's Illinois Valley. Although the particular experience of the Illinois Valley is unique, the more structural underpinnings of persistent poverty and declining activity in the local forest industry are apparent at much broader spatial scales. Industrial consolidation in the lumber sector, for instance, pronounced in the early years of sustained-yield regulation in southwestern Oregon, swept through Oregon's lumber industry during the late 1940s and into the 1950s (see chapter 4), demonstrating some of the tendencies of the dissolution of petty commodity production in the face of more industrial forms of production. As a result, the number of operating sawmills in western Oregon's Douglas-fir region dropped from a peak of almost thirteen hundred mills in 1947 to approximately three hundred mills in 1964, whereas aggregate lumber production dipped only slightly over the same period. Moreover, as noted in chapter 4, the total number of sawmills in Oregon overall dropped from three hundred to eighty-nine between 1968 and 1994, with very little change in aggregate lumber output. This consolidation has meant not only fewer mills but fewer jobs; employment in Oregon sawmills fell from a high of almost twenty-nine thousand jobs in 1966 to fewer than fifteen thousand by the mid-1990s, with parallel trends in the plywood and veneer sector (see Figure 6.8).[99]

Although there has been long-term consolidation and rationalization, there has also been periodic disruption from economic downturns. During the sustained-yield era, nothing demonstrated this more clearly than the devastating recession of 1980-82. Faced with persistent inflation, a weakening U.S. dollar, and a loss of American control over international capital markets, the Reagan administration forced interest rates up in early 1980s while sharply curtailing the money supply. Investment slowed dramatically, consumer confidence plummeted, industries laid off workers by the thousands, and the national economy was plunged into a recession. As housing starts fell by half, the solid wood-products industry was hit extremely hard.[100] In Oregon, aggregate employment in the wood-products sector fell by more than twenty-five thousand jobs

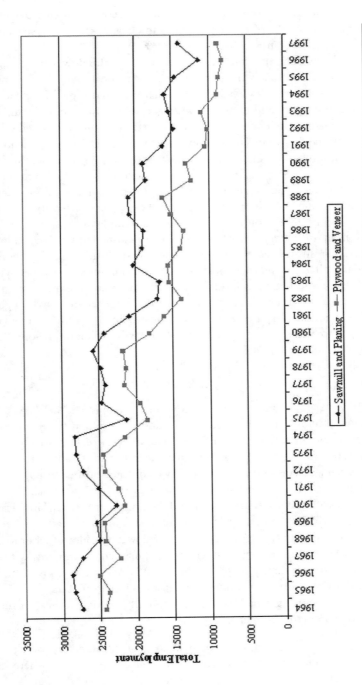

Fig. 6.8 Employment in Lumber and Plywood Manufacture, Oregon 1964–1997. Source: U.S. Bureau of Cenus (1958–1997) County Business Patterns "Oregon" Online: census.gov/qfd/index.html; Warren D. (2000).

between 1979 and 1982, representing one-third of prerecession employment in the sector; also by 1982 of 210 lumber mills in Oregon, 48 were closed, and 74 were operating substantially below capacity.[101]

Moreover, despite the apparently cyclical character of this downturn, its significance is marked in part by the fact that many of the lost jobs did not come back. Faced with the declining availability of high-value old-growth, and faced with eroding economic conditions, many Oregon solid wood producers made significant changes in their operations whose net effect was an accelerated substitution of capital for labor. Thus, for example, between 1979 and 1989, jobs in the Oregon lumber sector per million board feet of logs harvested in the state dropped from 3.35 to 2.2.[102] In short, although the sector is based on a resource that is at least nominally renewable, the Oregon solid wood-products sector has been buffeted by "normal" processes of capitalist restructuring compounded by the effects of resource extraction on rural employment and development.[103] None of these developments were foreseen by sustained-yield regulation of the federal forests in Oregon, or elsewhere, yet these tendencies are endemic to industrial capitalism, and they explained in significant measure by Polanyi's analysis of labor and nature as fundamentally fictitious commodities that pit market demands against broader social imperatives. The contradictions brought on by these fictions underpinned the collapse of sustained-yield social regulation, bringing to a close the long boom of old-growth exploitation along the Pacific Slope in the early 1990s.

Conclusion

Commenting on extractive economies, economist Thomas Michael Power[104] noted almost gleefully that the average duration of employment in wood-products manufacture has been no different than in other sectors, attempting to give the lie to notions of exceptionalism characterizing the literature on rural, resource-dependent communities. What Power did not bother to tell us is whether to laugh or cry at his results. He gave no apparent heed to the nakedness of the processes of capitalist restructuring and the compounded social implications of productivity or recession-induced job losses in communities where alternate sources of family wages are scarce. For Power, the idea of labor as a commodity, subject to the whimsical fluctuations of market demand irrespective of attachment to community or place is apparently natural, god-given, and hardly fictitious. In short, he just doesn't get it. Yet on the brutal transparency of capitalist restructuring processes in rural economies, E. P. Thompson's observations remain highly germane:

> In the mills and in many mining areas are . . . the large scale enterprise[s], the factory system with its new discipline, the mill communities—where the manufacturer not only made the riches out of the labour of the "hands"

but could be *seen* to make riches in one generation—all contributed to
the transparency of the process of exploitation.[105]

In communities throughout the Pacific Slope, and in academic and policy
circles, there was a broad consensus during the 1990s—spanning broad ideological divisions—that the decline of the region's forest-driven economy was
driven by growing political conflict over forest resources and, in particular, the
crisis of the spotted owl. Rare, sober calculations of the remarkably small direct
impact that the spotted owl listing had on levels of employment in the industry
were often ignored in the face of an almost hegemonic discourse counterposing
jobs and environment and, in particular, jobs and continued timber harvesting;
communities, it seemed, could have one or the other.[106] Few questioned the
veracity of this trade-off, or how even if true, it could have arisen historically to
be the choice faced by forest-dependent communities. Instead, academics, industry analysts, labor activists, and even environmentalists united in a "discourse
coalition"[107] that stubbornly refused to question the opposition of economy
and environment or the origins or circumstances surrounding this opposition.
Historian Richard White reflected the mood of the times, posing in 1995 the
rhetorical question, "Are You an Environmentalist, or Do You Work for a Living?"
in reference to the spotted owl controversy.[108] Yet even as owls were hung in
effigy, and as some owls were actually slaughtered by angry locals, careful historical reflection leads in multiple directions.

Such reflection first suggests that contradictory tendencies having to do with
the fictions of nature and labor as commodities ran through the sustained-
yield era of social regulation governing access to and control over timber on
the extensive federal lands of the Pacific Slope. In this sense, the spotted owl
controversy, about which I say more in the epilogue, reflected just as much the
underlying, structural crisis tendencies of industrial commodity production
along the Pacific Slope and the ultimate failure of sustained-yield regulation to
address these contradictions, as it did a sudden shift in cultural values associated
with old-growth, undermining forest capital and forest workers. As I argue,
new meanings were unquestionably conferred on "ancient forests" and translated
into social regulation by a combination of citizen action and litigation on the
one hand and new science on the other hand. But this must be seen as arising
in relation to the material changes wrought by commodity production and the
discursive prescriptions of sustained-yield regulation, including the
Normalbaum. If there were indeed conflicts between jobs and environment, as
most commentators and participants in the old-growth crisis seemed to accept,
then this conflict emerged not only from new attitudes about and demands on
old-growth forests per se but from the politicization of a scarcity endorsed and
prescribed by normal forest science and regulated under the guise of sustained-
yield policy.

Furthermore, the salience of shifting cultural attitudes toward forests in the region must also be linked to a precipitous decline in the number of workers employed by the industry during the sustained-yield era and prior to the listing of the owl. Far more workers lost their jobs from industrial restructuring than from the reregulation of the region's federal forestlands. And with each lost job, industry had one less political ally to count on. In this sense, although industry groups such as the Northwest Forestry Association and the Oregon Forest Industries Council blamed environmentalists for the loss of jobs and revenues during the 1990s, this reflects as much a cynical refusal to acknowledge the vastly greater number of jobs eliminated as the result of industrial reorganization, twinned with a regulatory system that failed to anticipate and address these job losses and their effects on forest communities. Acknowledging the direct linkages between industrial restructuring, ecological transformation, sustained-yield regulation, and the politics of old-growth in the 1990s was more than could be expected of most of the public discourse. Yet as I have tried to argue in this chapter, and as I discuss further in the epilogue, these connections are there to be made.[109]

Epilogue
Owls, Ecosystems, and the New Forestry

From Industrial Forestry to Ecoforestry

I opened this book with a look back at the spotted owl crisis, perhaps the most significant controversy in American environmental politics since Hetch-Hetchy. I argued that the crisis over the northern spotted owl and old-growth forests in the Douglas-fir region was a key moment not only in the construction of new meanings vis-à-vis environmental change, but also in the politics of ecological crisis tendencies arising from the production of commodities from nature. Having examined various elements of the problems surrounding capitalist nature in the preceding chapters, I would like to close by returning to and reexamining in greater depth the crisis over the northern spotted owl. Specifically, this chapter extends the previous one by exploring how the construction of new meanings associated with old-growth forests informed environmental politics in the region during the late 1980s and early 1990s and led to significant changes in the social regulation of industrial forestry and environmental change.

The sustained-yield model of forest regulation addressed the twin fictions of capitalist nature and labor in specific and ultimately problematic ways. In particular, the ideas of the *Normalbaum* and the hope of providing a stable economic foundation for community development through constant harvests of timber volumes from federal lands in the Douglas-fir region began to unravel under the weight of social, economic, and ecological contradictions that became increasingly politicized during the 1980s. This culminated in the listing of the northern spotted owl under the Endangered Species Act (ESA) in 1990, and the ensuing attempt by federal agencies, the courts, Congress, and the White House to respond. The ensuing crisis precipitated a reregulation of industrial forestry, at least on federal lands in the Douglas-fir region. This reregulation featured a bifurcation both toward and away from reenlisting the region's forests (specifically its public forests) to the circuits of capital. On the one hand, environmental nongovernmental organizations (ENGOs) led a successful campaign toward placing substantial remaining areas of old-growth off limits to commodity circulation, using both science and the courts. On the other hand, the lawsuits also precipitated a significant reregulation of industrial forestry, as science and the state embraced a new discourse of forest regulation that responded to the crises of the Normalbaum. This response entailed increasing the level of ecological detail visible to science and regulation, while also integrating formerly disparate elements of the forested landscape across space.

The project of reregulating and reenabling commodity production on federal lands in the Douglas-fir region proceeded along twin avenues: (1) the development of the New Forestry as a form of ecosystem management, providing an alternative discourse of scientific forest regulation to the Normalbaum, and (2) the campaign of civic opposition to sustained-yield regulation, particularly as expressed as lawsuits filed by ENGOs against the federal government. Together, these twin avenues constitute an example of the ways in which the politics of ecological crisis tendencies translate into new forms of social regulation through ecosocialization, as the politics of nature reacting to the environmental and social effects of capitalist nature rework social and cultural relations to nature in the regulatory arena. The twin avenues of ecosocialization also reinforce a theme discussed in the introduction; namely, that the politics of nature, including the construction of new meanings conferred on nature's transformations, are ultimately contingent. In this case, neither the construction of new meanings or understandings of old-growth forests nor the processes of ecosocialization and reregulation relied on discourses or political movements that were anticapitalist per se. Instead, this ecological crisis of capital accumulation, though arising in part as a response to capitalist nature, was largely addressed and resolved in the realm of social regulation in ways that ultimately adapt to and accommodate continued commodity production.

The New Forestry

A forest is not a mere collection of trees, just like a city is not a mere collection of unrelated men and women, or a Nation . . . merely a number of independent racial groups. A forest, like a city, is a complex community with a life of its own.[1]

The seeds of a new scientific basis for forest regulation in the Douglas-fir region actually predate the crisis over the northern spotted owl, and in some ways were germinated (literally and figuratively) by the eruption of Mount St. Helens, in southwestern Washington, on May 18, 1980. Here, in the midst of the devastation area around the volcano, critical research on recovery and succession in ecosystems of the Douglas-fir region would help to consolidate and extend new ways of thinking about the dynamics of old-growth and ecological disturbance, and contribute to the rise of the New Forestry. New Forestry is a variant of ecosystem management specifically developed in and adopted for the Douglas-fir region, one that draws on the concept of adaptive management and thus embraces a sort of experimental, management-based learning by doing.[2] It arose out of research conducted largely on old-growth forests during the 1970s and 1980s, and it must be understood as a challenge to conventional sustained yield forestry and a measure of the eroding legitimacy of sustained-yield regulation.

It bears noting that New Forestry is not the most significant shift in scientific forestry and forest regulation to emerge from the old-growth crisis. This is because New Forestry calls for active, adaptive management (i.e., logging), and more public land in the Douglas-fir region was set aside for preservation in the wake of the spotted owl crisis than remained available for commodity production (see following text).[3] By contrast, the New Forestry is hardly preservationist. Nevertheless, the significance of the New Forestry is that it provides a discursive foundation for a new form of science-based forest management that recasts and reenlists Douglas-fir region forests to commodity production. Moreover, it does so in ways that depart from the High Modernist forestry of the Normalbaum in two main respects: (1) New Forestry rescales forest science up to a broader landscape scale, linking the old fundamental unit of spatial analysis in forestry (the ambiguously defined forest "stand") to other elements of a forested (and deforested) landscape (e.g., streams, nearby forest fragments, wildlife corridors, etc.), resulting in a more integrated, ecosystem-like conception of spatial organization; and (2) New Forestry also scales down in the sense that it targets the management of a broader range of forest species, moving beyond the simple regeneration of commercially important timber trees and thereby rendering a greater range of biological diversity and ecological complexity visible to forest management.[4] The decidedly interventionist, transformative management orientation of New Forestry was conveyed unequivocally by the universally (and self-) acclaimed guru of the paradigm, Jerry Franklin, who in an article designed as much for political as for scientific purposes, derided in 1989 the "unhealthy" division of the Douglas-fir region landscape between commodity production and preservation. Franklin then posed and answered his own question, prescribing a role in management for the New Forestry: "Are there alternatives to the stark choice between tree farms and legal preservation? I believe ecological research is providing us with the basis for such alternatives."[5]

The historical lineage of the New Forestry is linked to much broader shifts in science, including the rise of ecology as a science in the context of postwar environmentalism, and to subsequent intellectual changes within ecology.[6] It also cannot be divorced from broader ideological and discursive shifts in the cultural politics of nature, including the rise of a new wilderness and outdoor recreation movement in the 1960s and its effects on the management of American forests, particularly on public lands.[7] At the same time, the institutional translation of these broader influences into New Forestry is remarkably straightforward and direct, based on publicly funded forest science conducted in old-growth forests of the Douglas-fir region beginning in the late 1960s. In fact much of the direct research basis for the New Forestry was conducted at a single location, in the H. J. Andrews Experimental Forest, a sixty-four-hundred-hectare tract of largely old-growth forest located about halfway between Eugene and Bend, Oregon, and set aside for research purposes in 1948. Though formally independent from any federal agency or university, the Andrews forest is jointly managed by the Forest Service's Pacific Northwest Research Station, the Willamette National Forest, and Oregon State University.

Through the 1950s and into the early 1960s, Andrews was the site of relatively prosaic logging studies pertaining to the conversion of old-growth to young-growth forests. However, Andrews research on geomorphology and hydrology—attempting among other things to link forest practices with flooding—began to point to the importance of downed woody debris and forest litter in nutrient cycling in the forest. This in turn helped establish a foundation for much more wide-ranging reexamination of the integrated ecological dynamics of old-growth forests.[8] This research direction was reinforced when the Andrews forest was named and funded as a site for research by the National Science Foundation as part of its commitment to the International Biological Programme (IBP) in 1970. Andrews was further reinforced as a focal point for new views on old-growth ecology under the auspices of the National Science Foundation's Long-Term Ecological Research Program (established in 1979), as scientists began to track the interrelations between the unique structural features of old-growth forests, complex material and energy flows, and biological diversity.[9] Significantly, though ostensibly geared to accumulating basic ecological knowledge, Andrews research had a decidedly instrumental or management orientation from the outset, key to securing funding from the U.S. Congress and dating in part to the IBP's original theme, "The Biological Basis of Human Productivity and Welfare."[10] As with the IBP, so too would it be with Andrews; although basic research would be conducted on forest ecology, it was ever with an eye to tying the science of forest ecosystems to forest regulation, seeking a balance between sustaining environmental systems on the one hand and allowing commodity production to continue essentially unchallenged on the other hand. With federal funding secured, and with scientific staff contributions from Oregon State University, the University of Washington, and the Forest Service, Andrews became *the* site in the Northwest for research on old-growth.[11]

Gradually, this research would help to confer new meaning and significance on old-growth, reinventing it not as the dead, dying, and decadent forest as characterized by conventional forestry but rather as a biologically productive, species-rich, and highly interconnected ecosystem.[12] Research at Andrews indicated that conventional forestry, by being focused on the regeneration of a narrow range of species, was eliminating a whole host of nontarget species, some of which played critical roles in sustaining forest ecosystems even from the narrowest of commodity orientations (e.g., nitrogen-fixing organisms).[13] In particular, at the level of individual forest stands, a view of old-growth emerged that contrasts sharply with perspectives typical of conventional, sustained-yield forestry in five key ways.[14] First, standing and downed woody materials provide critical habitat for certain species and are thus elements of the forest that must be retained to conserve biodiversity. By contrast, the prescriptions of the Normalbaum are typically to remove all standing biomass in the form of clear-cutting and slash burning prior to replanting. Second, ecological complexity (often vaguely defined, but including species diversity) tends to be highest when forests are young and when they are old. Because the Normalbaum is by definition even-aged and at most a middle-aged forest (when the average annual growth rate is maximized), complexity tends to be compromised in conventional sustained yield forestry in favor of maximizing timber volumes. Third, the input of nitrogen to the forest ecosystem by nitrogen-fixing organisms is most significant in early and late stages of growth. While some vegetation suppression in plantation forestry undermines nitrogen fixation, cutting all the trees before they become fully mature eliminates important sources of nitrogen fixation associated with the old-growth phase. Fourth, certain canopy dwelling organisms are important insect predators, and their existence can be compromised by the elimination of all canopy habitat. Management can counteract this by maintaining large, live trees in forest stands and maintaining multiple-layer forest canopies. Fifth, course, woody debris is important in stream ecology for retaining nutrients, controlling temperatures, and providing stream habitat for various organisms, including fish (notably juvenile salmonids). Logging right to the edge of streams under conventional forestry, and the removal or burning of slash, compromises this function.

Although research at Andrews helped recast old-growth stands, it was the eruption of Mount St. Helens in 1980 that reinforced new ways of understanding the ecology of disturbance, the interactions of landscape features across space (e.g., individual forest stands, streams, etc.), and the more direct integration of new insights on old-growth ecology with forest management.[15] In particular, Douglas-fir and other species characteristic of mature forests in areas devastated by the eruption returned much more quickly than researchers thought possible, indicating that the build up of ecological communities need not proceed through such strict sequences as had been believed. The reason for this, mirroring findings at Andrews, was that so-called biological legacies left over from the eruption became the basis for species and ecosystem reestablishment. In short, succession did not have to start from a clean slate. Rather, nature was more resilient and ecologically path dependent than had been thought.[16] Yet features of preexisting disturbance forests could only reestablish themselves quickly if, as at Mount St. Helens, remnants or legacies of the predisturbance ecosystems were left intact. Under conventional Normalbaum forestry, so-called biological legacies—particularly in the form of large old trees (dead or alive) and downed woody debris— were typically removed by the scorched earth policy of clear-cutting and slash burning. Moreover, management of individual stands in normal forestry was typically not integrated across space and did not take into consideration other features of the landscape. Yet research at Mount St. Helens indicated that greater degrees of connectivity existed, affecting the propagation of disturbance effects, and the dynamics of species and ecosystem reestablishment.[17]

The findings at Mount. St. Helens, combined with the accumulated research from Andrews, became the scientific basis of forestry's version of a "Third Way" for forest management, steering between the extremes of the Normalbaum on the one hand and complete preservation on the other hand. Specifically, if forest practices could better mimic so-called natural disturbances, then commodity production could coexist with the retention and more rapid recovery of some of the structural features of old-growth forests and associated ecological functions and species diversity. How to do this, it bears noting, is not completely clear, because of the adaptive character of New Forestry, because of local scale ecological complexity and differentiation in forests, and because of a lack of scientific and management experience with the application of New Forestry techniques. However, in general terms, New Forestry prescribes a "messier" kind of approach to harvesting stands that leaves biological legacies behind, including live and dead trees and downed woody debris in the form of logs and slash. It also prescribes the retention of a broader range of trees in terms of age and species structure than is typical of the even-aged, monoculture plantations characteristic of the Normalbaum.

At the landscape scale, New Forestry pushes for integrated planning to incorporate landscape features other than simply forest stands (e.g., streams) and to tie the management of distinct stands and harvested areas together, facilitated in large part by the use of geographical information systems. Thus, for example, harvesting near streams is to reflect the significance of woody debris in hydrology and fish habitat and in ways that are specific to particular streams. Moreover, New Forestry challenges not only large-scale clear-cutting (because it does not leave sufficient biological legacies) but also the method of widely dispersed, smaller (i.e., fifteen hectares) cut blocks typical of public forest management during recent decades. This is because the cumulative effects of these widely dispersed, small clear-cuts over time results in high degrees of habitat fragmentation at the landscape scale, with isolated patches and greater total perimeters between patches and harvested areas. This in turn leads to greater impacts on species in need of interior old-growth habitat and on the propagation of disturbances along patch edges; for example, from trees blown down in windstorms. New Forestry suggests instead mimicking natural disturbances with the retention of clusters of dead and live trees and snags, the maintenance of habitat corridors to connect patches of forest, and the arrangement of cut blocks to maximize internal forest spaces and to thereby minimize edges and the fragmentation of habitat.[18]

The end result is what Jerry Franklin referred to as "a kinder, gentler forestry that focuses equally on commodities and ecological values."[19] Yet translating this new science into the reregulation of forest commodity production would require growing civic opposition to the sustained-yield regulatory model, highlighted by a campaign of environmental litigation opposed to continued logging in the remaining publicly held old-growth forests of the Douglas-fir region. As the campaign gathered steam, federal agencies increasingly turned to public science to provide input to the regulatory response. No one factored into this mix more than New Forestry's principal champion Jerry Franklin, a product of the Douglas-fir region through and through. Raised in Camas, Washington, Franklin was trained as a forester in the region and in 1975 was hired on to the faculty in the College of Forestry at Oregon State University, where he remained until moving to the University of Washington in 1992. As an academic, with a research affiliation to the Andrews Experimental Forest and involvement in the research at Mount St. Helens, Franklin became the chief spokesperson for the New Forestry and its prescriptions for an intermediate path between preservation and Normalbaum. By the late 1980s, with the eroding legitimacy of sustained-yield forestry in the Douglas-fir region, and thanks in no small measure to Franklin's advocacy, the New Forestry was an established buzz phrase in the forest science community and was the subject of considerable attention in management circles. And as the spotted owl crisis boiled over during 1989 to 1993, Franklin and other advocates of the New Forestry became caught in the search for a policy response. Franklin was appointed to several key advisory panels, including the so-called Gang of Four committee appointed in 1991 to examine options for managing old-growth, and the Forest Ecosystem Management Advisory Team charged with compiling alternatives and recommending options for integrated management of old-growth following the Forest Summit in Portland, Oregon, in 1993.[20] Through such channels, New Forestry became twinned with the campaign of litigation around the northern spotted owl, bringing two avenues of ecosocialization together.

Canaries and Coal Mines

In the explosion of open political conflict over the disposition of remaining old-growth forests in the Douglas-fir region, the northern spotted owl (*Strix occidentalis caurina*) occupies a unique position. This is in part because the owl became a lightning rod for dissent regarding the management of federal forests and became the public's symbol for nature and the environmental costs of logging and forest commodity production. New meanings associated with old-growth—fragile, fleeting, beautiful, rare—are largely those that came to be attached to this stunning and fascinating bird. But the owl also came to occupy a unique scientific, legal, and administrative niche. A focal point for the concerted legal campaign to halt logging in the last remaining public old-growth forests, the owl was also the object of scientific studies that helped create new ways of understanding and representing old-growth and of measuring its decline. Linking science and litigation, the owl thus became a bridge between two avenues of ecosocialization, whereby growing opposition to the appropriation and transformation of nature under sustained-yield regulation led to new articulations between society, science, capital, and nature.[21]

One key reason for this critical link is that the spotted owl acted as an indicator species of broad ecological tendencies. There was, in this sense, considerable confusion about the significance of the owl even among those embroiled in the fight. That is, sometimes it seemed like the owl was the thing. Often, public and even scientific and policy discourse fixated on the owl, and conflict took on the character of one between people and owls. Stuffed owls were hung in effigy. Owls were shot and their corpses put proudly on display. Angry workers, community members, and those sympathetic to the plight of the logger, the sawmill worker, and the truck driver could not understand how an owl could come first. Some environmental activists added fuel to the flames by declaring that they cared more about owls than about people. And at times the scientific and legal struggle over old-growth did fixate narrowly on the owl, as though saving the owl could somehow reconcile the contradictions—ecological and political—of capitalist wood-commodity production and scientific, sustained-yield forestry.[22]

But the fight was never really about the bird per se (as important as it was and is). Rather, the true scientific, legal, and discursive significance of the northern spotted owl lay in what owl populations said about cumulative ecological changes in the forests of the region. Like canaries in coal mines, fewer owls increasingly became indicators that a fundamental remaking of nature had taken place. If forest management practices under the sustained-yield mode of forest regulation could not result in the retention of viable spotted owl populations, then this served as an indictment—legal but also ethical and moral to many— of these regulations and forest practices.[23] Thus, it is the owl's status as a legal, cultural, and ecological symbol of the crisis of nature's commodification, and the potency of this symbol for a social campaign organized against further liquidations of old-growth (i.e., as the locus of a Polanyian dual movement) that makes the fight over the northern spotted owl so significant in the annals of environmental politics.[24]

The northern spotted owl is one of three subspecies of spotted owl, the others being the California spotted owl (*Strix occidentalis occidentalis*) and the Mexican spotted owl (*Strix occidentalis lucida*), although the northern and California varieties may only be separate populations, not species. Yet of these three, the northern spotted owl is almost entirely unique to the Douglas-fir region.[25] Studies of northern spotted owl populations began in the late 1960s and early 1970s and quickly linked the owl to old-growth forests. And by the early 1970s, considerable evidence pointed to owl populations in decline as the area of remaining old-growth shrank; so much so in fact that the owl was considered for inclusion on the inaugural list of threatened and endangered species pursuant to the ESA of 1973.[26] Although the owl did not make the final list, scientific evidence of declining owl populations continued to mount.

By the mid-1980s, it had become evident that northern spotted owls were the best available barometer or indicator species uniquely associated with old-growth forests. Populations of owls were declining in the Douglas-fir region as remaining stands of old-growth became more and more scarce, more and more fragmented, and more and more isolated across the region.[27] According to one paper, the area of suitable owl habitat on privately owned forests in Oregon and Washington was reduced from approximately 4.5 million hectares in 1961 to 88,000 hectares in 1984, while the corresponding reduction on national forest lands was from 5.5 million hectares to 1.5 million hectares over the same period.[28] A second estimate, issued by the Forest Service and the Bureau of Land Management (BLM) in 1994, suggested roughly double this amount of remaining suitable owl habitat, but also it forecast that all foreseeable management alternatives would lead to further reductions by at best half of the then-remaining total—in short, the difference in estimates was only a matter of time.[29] With the population of northern spotted owls estimated at perhaps between four thousand and six thousand individuals in the 1980s, research suggested that each pair of nesting owls required perhaps eight hundred hectares of old-growth habitat to sustain them. Thus, it was becoming apparent that there was or would soon be just enough habitat on national forest lands to sustain then prevailing populations. And as more and more evidence accumulated through the 1980s and into the 1990s, results reinforced earlier findings that owls were in gradual decline throughout most of the region.[30] Forest management practices on federal lands, it seemed, if left unaltered would result in the elimination of the owl. The evidence was strong enough in fact for the Interagency Scientific

Committee assembled to evaluate evidence on the fate of the northern spotted owl in the region to state that Forest Service and BLM management plans prevailing in the late 1980s constituted a prescription for the extinction of the owl.[31]

Meanwhile, a suite of relatively new statutory provisions dubbed collectively an "Eco-Forest Statutory Regime" by Benjamin Cashore had created new federal obligations in the regulation of environmental change (including managing for biological diversity) and had also enabled direct citizen involvement in administering and enforcing these statutes by means of administrative reviews and court challenges.[32] This new statutory regime created the institutional framework in which the northern spotted owl was to take on formal status an indicator species, a bellwether for the status of old-growth forests. And at the same time, this statutory regime provided for citizen-originated environmental litigation targeting regulatory enforcement vis-à-vis owl management and for the enforcement of environmental regulations more broadly. The ensuing campaign of litigation marked a watershed in the use of court action by citizen groups to not only enforce but actively shape environmental policy, as the owl became a critical conduit for the ecosocialization of forest regulation.[33]

The precise sequence of litigation, appeals, injunctions, and administrative and legislative responses is a complex and compelling drama, full of twists and turns that have been recounted in full detail elsewhere.[34] However, it is important to note here some of the significant moments in the campaign in order to appreciate the ways in which the spotted owl crises not only catalyzed a new approach to forest regulation but at the same time threw into relief the tenuous, politically constituted, and highly contingent relationships among the commodification of nature, environmental change, ecological politics, and environmental regulation. Highlighting the stakes involved in this expanding ecological crisis, the courts, ENGOs, the public, the press, forest capital, Congress, and the White House all became embroiled in a protracted struggle ostensibly about the technical aspects of forest ecology but that was deeply infused with economic, political, and cultural meaning and implications. Town hall meetings, public protests for and against greater protection for old-growth and old-growth dependent species, newspaper articles and editorials, and the sheer number and sweeping character of the lawsuits all pointed to a conflict of great significance.[35] In the end, despite attempts by Congress and the first Bush White House to derail ENGO legal campaigns and to reassert the relative autonomy and discretion of federal agencies (which had until well into the 1980s been generally favorable to a commodity orientation), lawsuits drawing on the Eco-Forestry Statutory Regime helped usher in the management of federal forests in the Douglas-fir region under the auspices of the New Forestry and ecosystem management.[36]

Within the context of the Eco-Forest Statutory Regime, two specific institutional avenues were key to conferring indicator status on the northern spotted owl and for allowing citizen suits seeking enforcement based on this status. The first was the naming of the owl as a management species to be retained in the Forest Service's Draft Land and Resource Management Plan for the Pacific Northwest Region in 1981. As required by the National Forest Management Act of 1976 (NFMA), Land and Resource Management Plans must make provisions for managing a range of biodiversity, and specifically for planning to retain unique existing or desirable species.[37] Yet, critically, although the NFMA stipulates that the Forest Service is not allowed to let indicator species go extinct once designated the Forest Service proposed in its 1981 Draft Pacific Regional Plan and again in its 1984 Final Regional Guide and Final Environmental Impact Statement for the Pacific Region to reduce northern spotted owl populations in Oregon and Washington National Forests to 375 breeding pairs, with habitat protections set at four hundred hectares per pair.[38] Reflecting the deeply ingrained orientation of the Forest Service to maximizing timber volumes from federal forests in the region, the Forest Service's Preferred Alternative in the plan would, according to the agency's own scientific experts, result in a low probability at best that the owl would persist after 150 years.[39] It bears repeating that this was despite the agency having named the owl as an indicator and target species for management and retention in the same Land and Resource Management Plan![40] This inconsistency set the stage for litigation and conflict over business-as-usual, sustained-yield forestry, pivoting on the agency's contradictory conduct vis-à-vis the northern spotted owl; environmental groups sued the Forest Service for failing to comply with the NFMA, drawing also on the National Environmental Policy Act of 1969 (NEPA) and its stipulation that federal agencies complete environmental impact statements for proposed actions, plans, or policies and that they consult the public in doing so.[41]

Although the NFMA and the NEPA provided a first avenue by which the science of the spotted owl would become a focus of litigation and of ecosocialization in federal forest regulation, the second avenue was the ESA (also in combination with the NEPA). In 1987, Greenworld USA, a relatively obscure environmental nongovernmental organization petitioned the Fish and Wildlife Service to list the northern spotted owl as threatened under the terms of the ESA. When the agency declined, a coalition of ENGOs sued and won. Finding that the agency had acted in an arbitrary fashion, Judge Thomas Zilly of the federal district court for western Washington ordered the agency to reconsider. After also coming under fire by the General Accounting Office, the Fish and Wildlife Service reversed course in April 1989 and recommended listing the owl. On June 25, 1990, the owl was officially designated as threatened. The listing would in turn facilitate suits not only against the Forest Service regarding environmental impacts and public consultation processes of its management plans (drawing on the ESA, the NFMA, and the NEPA) but also against the BLM, and the provisions of this agency's timber management plans for lands in the Douglas-fir region.[42]

The first major breakthroughs in the courts for ENGOs came in the lawsuit against the Fish and Wildlife Service over whether to list the owl as a threatened species. This accomplished, the focus of litigation based first on the spotted owl, and later on other old-growth–dependent species, shifted to the BLM and the Forest Service. In the first of his dramatic rulings, Judge William H. Dwyer of the federal district court for western Washington ruled in 1989 in favor of a coalition of twenty-nine environmental groups led by the Sierra Club Legal Defense Fund, issuing an injunction against 140 planned Forest Service timber sales pending compliance with the ESA in the form of a northern spotted owl management plan.[43]

The first Bush White House and also the Congress initially responded to the emerging crisis by attempting to contain and deflect litigation and the ecological contradictions highlighted by the suits. Thus, Congress, for its part, on two occasions inserted language into fiscal appropriations bills for the Departments of the Interior and Agriculture that blocked pending lawsuits and injunctions against the Forest Service and BLM plans to allow logging in owl habitat. These congressional "riders" at the same time overrode (and some would say micromanaged) the agencies by requiring them to meet annual timber sale volume targets higher than could be warranted by any reasonable attempt to conserve owl habitat or remaining old-growth, and irrespective of the provisions of the NEPA, the ESA, the NFMA, or any lawsuit.[44] In addition, the White House specifically attempted to bolster the autonomy of the BLM in 1991 by convening the Endangered Species Committee—the so-called God-Squad—for just the third time since passage of the ESA in 1973. The God-Squad, following hearings in Portland, attempted to trump the Fish and Wildlife Service efforts at restricting BLM timber sales in owl habitat, acting to exempt thirteen enjoined BLM timber sales from review.[45]

Ultimately, however, attempts by the legislative and executive branches to contain the crisis failed. When the second congressional appropriations rider (the infamous Section 318) expired, the Seattle Audubon Society renewed its litigation against the Forest Service claiming violations of the NEPA and the NFMA. In a landmark decision in *Seattle Audubon Society v. Robertson*, on March 7, 1991, Judge Dwyer ruled again in favor of the plaintiffs, issuing a temporary injunction against logging on national forest land in spotted owl habitat, and this time noting in his decision that the Forest Service was under an obligation to plan "*for the entire biological community*—not one species alone" (emphasis added).[46] Following additional evidentiary hearings, on May 23, 1991, Dwyer made the injunction permanent pending the completion of a revised environmental impact statement by the Forest Service, one that would deal with the impacts of the agency's plans on the northern spotted owl in light of current scientific information. As the God-Squad met to try to free up BLM timber sales in the Douglas-fir region, the Forest Service hurriedly attempted to comply with Dwyer's ruling by issuing a revised, final environmental impact statement in January 1992 with an associated record of decision in March. Environmental groups again filed suit, and Dwyer again found in their favor. On July 2, 1992, Dwyer issued what was essentially a mirror image of the injunction he had granted a little more than one year prior, halting Forest Service timber sales in northern spotted owl habitat and reaffirming the Court's interpretation that the NFMA required planning for whole ecological communities or ecosystems, not individual components.[47]

On appeal, Dwyer's decision went to San Francisco's Ninth Circuit Court. In parallel, congressional protections shielding the BLM from litigation expired, precipitating additional litigation filed by the Portland Audubon Society against the agency's timber management plans, and specifically alleging that the BLM failed to conduct a supplementary environmental impact statement for logging in spotted owl habitat.[48] Here again, the plaintiffs won. On June 8, 1992, Oregon federal district court Judge Helen Frye ruled that the BLM had violated the NEPA. Frye also issued an injunction against BLM timber sales in spotted owl habitat pending compliance. The Ninth Circuit Court then held hearings dealing with both injunctions, and on July 8, 1993, it upheld that the Forest Service and the BLM had violated the NEPA by failing to complete supplemental environmental impact statement documents in light of changing scientific information regarding northern spotted owls, and planned actions by the agencies in light of the emerging science.[49]

With the forest-planning process on federal lands in Oregon and Washington essentially paralyzed, the crisis of sustained-yield regulation and industrial forestry in the Douglas-fir region as it had been (at least on federal lands) reached its apogee in 1993-94. Although ostensibly process-related rulings pertaining to completion of environmental impact statements and the specific requirement that federal agencies consult the public in compiling these statements, in fact the district court rulings, to be upheld by the Ninth Circuit Court, were essential moments punctuating the adoption of ecosystem management on federal lands in the northwest. Although the Normalbaum idea had allowed for and even prescribed dramatic ecological changes in the conversion of unruly, wild forests into tame, ordered ones, the sustained campaign of litigation directed at the Forest Service and the BLM affirmed, two decades after passage of the ESA and seventeen years after passage of the NFMA, that species in the forest other than commercially relevant timber trees would have to factor into the planning process on federal lands. It also affirmed that ENGOs were sufficiently empowered and committed to enforcing a different set of values in the regulation of federal forests. To meet the new imperatives, that is, to minimize the risk of allowing species to go extinct and to reduce the impasse created by litigation, as the spotted owl made all too clear, the Forest Service and the BLM would be forced to embrace planning at the scale of landscape and habitat. In the process, federal forest management came to render novel aspects of nature visible to the planning process, aspects hitherto erased by the discourse of rational, scientific forestry.

Reflecting this new reality, following the Dwyer injunction against the Forest Service in 1992, the agency commissioned a report to examine the range of species, like the northern spotted owl, dependent on old-growth habitat. The result was a list, drawing on the scientific legacy created by the New Forestry, that included 667 distinct species closely associated with old-growth.[50] It then took a Bush White House scrambling for cover on environmental issues in the lead-up to the 1992 presidential election to hurriedly and rather vacuously embrace an ill-defined concept of ecosystem management.[51] And it took the Forest Summit of 1993 as among the first major policy initiatives of the new administration in Washington, D.C. to formally commission a plan based on the principles of ecosystem management, aimed specifically at regulating forests at a broader spatial scale through integrated ecological planning and at a finer degree of resolution by making visible to the regulatory process a wider range of organisms. Following on a campaign promise to resolve the crisis of forestry in the Douglas-fir region, and in advance of the impending July Ninth Circuit Court deadline regarding district court injunctions, the Clinton administration presided over a one-day Forest Summit in Portland, Oregon, on April 2, 1993. Amid great fanfare, including televised debate, the summit resulted in establishment of the Forest Ecosystem Management Team, including Jerry Franklin as one of its members. The Forest Ecosystem Management Team in turn was charged with creating a set of alternatives and recommendations for managing old-growth on Forest Service and BLM lands, to be used as the basis for a new environmental impact statement to comply with the NEPA (and the Dwyer and Frye injunctions) and to provide a blueprint for implementing ecosystem management.

The result was so-called Option 9, a preferred alternative plan affirmed by a joint agency final supplemental environmental impact statement, published in February 1994.[52] Option 9 resulted in a reduction in the annual planned sale quantity of timber from federal lands in California, Oregon, and Washington by about 75 percent. Moreover, as part of the rescaling and reconstitution of forest management, the plan encompassed an unprecedented degree of integrated conservation planning across nineteen different National Forests and seven BLM districts. It also reduced the areas available for industrial timber harvesting to about one-quarter of the total area of Forest Service and BLM land in the region.[53] These lands, dubbed "Matrix," "Managed Late-Successional Areas," and "Adaptive Management Areas," would be the focus of implementing ecosystem management and the New Forestry, uniting a new paradigm of science and citizen involvement in the ecosocialization of forest regulation. Sustained-yield forestry, as it had been known, was dead.

Conclusion

The controversy over the northern spotted owl and the politics of old-growth in the Douglas-fir region continues in certain respects. Moreover, the implications of this ecological crisis for environmental regulation and forest management in particular are still unfolding. Subsequent struggles over other old-growth–dependent species, such as the marbled murrelet, and concerns about the viability of Pacific salmon populations suffering from the cumulative effects of dam construction and logging in the region have reinforced the inherently problematic relationship between capitalist commodity production and environmental change. These issues continue to be the source of political conflict, with implications for environmental policy and regulation on private and public lands in the Douglas-fir region.

At the same time, looking back on the spotted owl crisis, and particularly on the ways in which two avenues of ecosocialization—New Forestry and ENGO litigation—came together to install ecosystem management as the new approach to forest regulation in the region, several insights may be gained on the relationship between ecological crises, environmental politics, and environmental regulation. First, the politics of old-growth emerged from the ecological contradictions of more than 150 years of producing commodities from Douglas-fir forests, and approximately 50 years of regulating industrial forestry under the auspices of sustained-yield forest management. That is, this crisis had everything to do with the capitalist production of nature, in Neil Smith's terms, and with what I have called capitalist nature here. Second, however, the actual political construction of new meanings arising in the midst of this crisis (i.e., the politics of interpreting and describing environmental change) and the influence of these meanings on changes in the regulatory arena has been a contingent project that has rarely challenged commodity production per se. Rather, ecosocialization has acted to divide the landscape of public forests. Large areas of remaining old-growth were set aside, reinforcing the division between nature as commodity and nature as wilderness and aesthetic or spiritual refuge. If these two solitudes cannot be understood as equally capitalist natures, they are at the same time, as Neil Smith and Raymond Williams both suggested, each in their own way bourgeois natures.[54] The first is a nature reduced and demystified as a set of exchange values for capitalist commodity production. The other is reified and placed ostensibly outside the social realm, particularly removed from the realm in which social labor transforms nature for use values. It is nature as an artifact or remnant nature reified and fetishized, abstracted from the circumstances of its own production.[55]

Thus, the origins of the spotted owl crisis point to fundamental contradictions or to what Polanyi called the fictions of nature as commodity. The emergence of dissent—in science and civil society—to the wholesale subordination of nature's material transformation at the hands of the capitalist market indicate the enduring salience of Polanyi's fictitious commodity and dual movement ideas, along with O'Connor's elaborations on these ideas. But it is equally true that the politics of environmental meaning and their implications for environmental regulation may be a mix of the radical and prosaic, acting as much to accommodate as to oppose commodity circulation.

Consider, for instance, the New Forestry. The Third Way comparison is apt. New Forestry provides a key scientific and discursive basis for a challenge to the Normalbaum, while also prescribing elements of a new model of forest regulation. But it does so in ways that ultimately reenlist forests to commodity production at a finer scale of resolution by means of the embrace of biodiversity conservation and across broader spaces by means of landscape ecology. At the same time, New Forestry maintains the rigid separation between nature and society in forest science, never tackling the issue of nature as a set of use values versus a set of exchange values. That is, although New Forestry very explicitly responds to changing social, cultural, and economic demands on forests, as noted by Franklin and others, it does so in ways that take these demands as given, proceeding then to ask how forest management might respond. New Forestry never interrogates commodity production per se or examines the links between particular kinds of commodity production, labor processes, and social relations and use values on the one hand and forest ecology on the other hand. New Forestry instead seeks only to accommodate demands on the land that are treated as exogenous.

Thus, although New Forestry provides a mechanism for linking changing perspectives in scientific forestry with the broader ecosocialization of forest management in the Douglas-fir region, it does so in ways that are highly specific and that have ultimately prosaic political economic implications. Hardly a radical affront to capitalist nature, New Forestry is reformist. It in some ways strengthens state-centered regulation of nature's transformation, as O'Connor's theory of ecological crisis would suggest, yet also facilitates reenlisting the landscape to commodity production without ever questioning the social bases of environmental change that give rise to the need for new forms of regulation—commodity production. In fact, social relations to nature and the nature of capitalist nature are considered seemingly irrelevant to understanding ecological change in the discourse of New Forestry, conspicuous by their absence as "management" concerns. Yet, considered from another vantage point, this is clearly absurd. What, if not capitalist social relations and commodity production, is responsible for the destruction of old-growth that New Forestry, in its Third Way manner, is meant to conserve and reproduce?

Similarly, ENGO campaigns against the federal government are in one sense an inspiring episodes of citizen involvement in and influence on environmental regulation. ENGO litigation in particular represents the fruition of efforts to construct more participatory, democratic forms of environmental regulation. More specifically, the suits were integral to realizing some of the promise of an emergent Eco-Forestry Statutory Regime, as described by Benjamin Cashore. Undoubtedly, dedicated activists have made great sacrifices for little apparent personal gain to enforce their beliefs about right and wrong in the transformation of nature but equally, as Judge Dwyer was right to point out, to hold the federal government and its land management agencies accountable for obeying existing laws such as the NFMA, the ESA, and the NEPA. In the furor of the debate over whether jobs or nature should come first, it is easy to lose sight of who was breaking the law when it came to federal forest management or why these laws were being systematically flouted. Yet, that said, the spotted owl crisis and the efficacy of the environmental citizens campaign still leaves Richard White's haunting question, "Are You an Environmentalist or Do You Work for a Living?" unanswered.[56] This is because the ENGO campaign of litigation also ignored the issue of commodity production, and thus how nature might comprise a different set of use values—that is, how the crisis afforded an opportunity to examine fundamental social relations to nature in ways that would avoid a strict dichotomy between wilderness preservation and capitalist commodity production. Subsequent forays into community forest tenure and the like in the region are a start, but they emerged largely after the spotted owl crisis peaked, in part as a response to the lack of a compelling alternative vision of social nature in the new model of ecosystem management.[57]

One way we might begin to gain some perspective on why the spotted owl crisis did not more directly challenge capitalist nature is by comparing this crisis with that precipitating the emergence of sustained-yield regulation (reviewed in chapter 6). Specifically, it seems significant that the fictions of nature and labor as commodities shaped the emergence of sustained-yield regulation, whereas the fictions of labor factored little in the public, scientific, and regulatory discourses surrounding the crisis of old-growth and the spotted owl in the 1980s and 1990s. Why this would be so is a question beyond the scope of this book. But it does in my view suggest that the politics of nature must ultimately be cast against a broader backdrop of a liberal, capitalist order consolidated during the post–World War II period. That is, perhaps the changing dynamics of forest politics, and the politics of nature and labor as commodities in the realm of forest regulation, cannot be wholly divorced from the so-called Fordist compromise in American social regulation more broadly.[58] At any rate, we might expect that the politics of environmental regulation will continue to be addressed in ways that reflect as much as they challenge this social order.[59] In one sense, this serves as a challenge to the Polanyi-O'Connor thesis to the extent that the politics of opposition to capitalist crises and fictitious commodities are assumed to be, inherently, anticapitalist. On the other hand, as Michael Burawoy recently argued, Polanyi's recourse to a social realm (i.e., outside of production relations, the market, or the state) in which these politics are constituted also opens up important avenues of inquiry regarding the politics of capitalist societies and, as I have tried to argue, the politics of capitalist nature. Much remains to be done, politically and analytically. Ideally, they will be done together.

Final Words

In this book I have tried to use the production of forest commodities (and some limited forays into the commodification of trees) in the Douglas-fir region as an extended case study of the inherently problematic nature of capitalist nature. I have drawn on Polanyi's theory of the fictitious commodity, augmented by O'Connor's thesis of capitalism's second contradiction, and a range of more recent scholarship on the political economy and ecology of environmental change. I have done so to try to emphasize the structural disjuncture between capitalist and ecological production and the significance of this disjuncture as it translates into unique social and ecological outcomes that are products of intertwined accumulation strategies, political contestation, and biophysical nature as land, time, and form. At the same time, I have emphasized the inherently contingent, politically and discursively constructed character of manifest environmental problems such as the crisis over the northern spotted owl, and the implications of environmental politics for social regulation over capitalist nature.

So what? Why does any of this matter? After all, I have offered few explicit solutions. Perhaps this is an abdication. On the other hand, perhaps for someone in my position, solutions would be presumptuous. I neither study nor practice forestry specifically nor do I work in a mill. Neither do I care particularly deeply about spotted owls per se. However, it is my hope that this narrative and analysis makes a compelling case that to understand and address the character and implications of contemporary environmental problems, it is imperative to come to terms with the organization and character of contemporary society. One implication, surely, is that the origins of environmental problems are located in the very character of capitalist society, as are the politics by which these problems in whatever piecemeal fashion are addressed. To ignore this is to be condemned to the examination of environmental problems one at a time, in piecemeal fashion, as emergent and particular phenomena only, rather than as contradictions that are inherent to capitalist nature.

Notes

Chapter 1

1. My apologies to the memory of Earl Pomeroy who originally coined this term to refer more broadly to the cluster of northwestern and Pacific states he felt deserved their own distinct regional identity—Idaho, Washington, Utah, Oregon, California, and Nevada; E. S. Pomeroy, *The Pacific Slope: A History of California, Oregon, Washington, Idaho, Utah, and Nevada* (Lincoln: University of Nebraska Press, 1991). His notion was more expansive than the area to the west of the Cascades and denotes the states to the west of the continental divide. Yet, reading Pomeroy, there is no question that the distinct natural resource–based activities that helped forge this region's distinctness within the union were also integral to his idea of the Pacific Slope, and it is in this spirit that I use this term to refer to a region west of the Cascades, forged in significant measure (economically, ecologically, culturally, and politically) by its staple tree and the industrial complex it has in large part sustained for more than 150 years.

2. Walter A. Mead, in his definitive study of the organization of the Douglas fir lumber industry, used as his definition for the Douglas fir region that of the West Coast Lumbermen's Association, "the 19 counties in Western Oregon and the 19 counties in Western Washington which lie between the crest of the Cascade Mountains and the Pacific Ocean"; W. J. Mead, *Competition and Oligopsony in the Douglas Fir Lumber Industry* (Berkeley: University of California Press, 1966). However, this definition is largely for statistical convenience; within the industry itself, portions of northern California are generally included, and as a contiguous region defined by forest cover, large sections of northern California clearly belong.

3. Scottish botanist David Douglas gave the Douglas fir its name when he conducted a biological survey of the coastal Northwest in 1827 and identified the tree as unique to western science. He classified it as a true fir. However, it is not a fir but rather part of a distinct genus (*Pseudotsuga*), within which there are eight known species. The Latin name reflects the tree's similarity to fir (*Pseudotsuga*) and also honors Archibald Menzies, a naturalist who first correctly identified the species as distinct from the fir family. There are two varieties of Douglas fir, the coastal (var. *menziesii*) and the interior (var.*glauca*). B. Goldfarb and J. B. Zaerr, *Douglas-Fir. Biotechnology in Agriculture and Forestry,* Y. P. S. Bajaj (Berlin: Springer-Verlag, 1989), 5: 526–48.

4. W. B. Smith et al., *Forest Resources of the United States, 1997* (St. Paul, MN: U.S. Department of Agriculture, Forest Service, 2001).

5. U.S. Forest Service, *Forest Resources of the United States 2002* (draft, 2003). According to the 2002 draft inventory, Table 5, of 4,554,000 acres of Douglas fir forestlands in the Pacific Northwest, 3,130,000 are in the highest rated productivity class.

6. Smith et al., *Forest Resources of the United States, 1997.*

7. T. R. Cox, *Mills and Markets: A History of the Pacific Coast Lumber Industry to 1900* (Seattle: University of Washington Press, 1974).

8. As recently as 1992, Oregon and Washington ranked first and second, respectively, among all states in total lumber production, first and third in sawmill and planing mill employment, and first and fourth in plywood mill employment. Oregon in particular has been among the most dominant states in the U.S. forest sector; for example, leading the nation in 1991 in softwood timber removals, most of which was Douglas fir harvested on the Pacific Slope, with the western part of the state and its Douglas fir forests accounting for about 80 percent of the total volume. Data from United States Bureau of the Census, 1995; United States Bureau of the Census, 1995; and Western Wood Products Association, *Statistical Yearbook of the Western Lumber Industry* (Portland, OR: Author, 1998).

9. J. Hannum, *Oregon Lumber and Wood Products Employment as a Percentage of Total Manufacturing Employment* (personal communication, 1995).

10. On the listing, its lineage, and the court cases involving the spotted owl, see A. L. Hungerford, "Changing the Management of Public Land Forests: The Role of the Spotted Owl Injunctions (1993 Ninth Circuit Environmental Review)," *Environmental Law* 24, no. 3 (1994): 1395–1434; B. G. Marcot and J. W. Thomas, *Of Spotted Owls, Old Growth, and New Policies: A History since the Interagency Scientific Committee Report* (Portland, OR: U.S. Department of Agriculture, Forest Service, Pacific Northwest Forest Research Station, 1997); and V. M. Sher, "Travels with Strix: The Spotted Owl's Journey through the Federal Courts," *Public Land Law Review* 14 (1993): 41–79.

11. Despite a lot of confusion in the public discourse on owl conservation, the point was never really to conserve the owl per se, at least not according to scientists and forest managers. The owl had been named as a management indicator species by the Forest Service as per the requirements of the National Forest Management Act of 1976, a canary in the coal mine whose population dynamics were viewed as a barometer of ecological changes associated with the declining amount of old-growth forests in the Pacific Northwest region; see K. R. Dixon and T. C. Juelson, "The Political Economy of the Spotted Owl," *Ecology* 68, no. 4 (1987): 772–76; Marcot and Thomas, *Of Spotted Owls;* and D. Simberloff, "The Spotted Owl Fracas: Mixing Academic, Applied, and Political Ecology," *Ecology* 68, no. 4 (1987): 766–72.

12. This term was used in the joint Bureau of Land Management and Forest Service final environmental impacts statement to refer to the planned amount of annual allowable cut or sustained yield from federal lands, allowing for some uncertainty. PSQ is comparable to the term Annual Sales Quantity (ASQ) used in earlier publications, except for the inclusion of uncertainty.

13. A board foot is a piece of wood one-foot square, one-inch thick. This measure is commonly used to refer to finished lumber and to volumes of standing timber and logs. However, various log rules exist for estimating how many board feet will be produced by a round log of given dimensions. The two most commonly used standards are the Scribner log rule and the ¼-inch diameter international rule. The horrendous and needless confusion created by this archaic standard, combined with its technologically sensitive character, is leading to its gradual replacement in common usage by units of wood volume, either cubic feet or cubic meters. However, in this book, I typically report data as I found them, except where conversions were required for clarity or consistency. Typically, timber volumes are given in thousand board feet (Scribner). Short forms include mbf (thousand board feet), mmbf (million board feet), and bbf (billion board feet).

14. Although the PSQ in Oregon, Washington, and California in the 1980s averaged 4.5 bbf per annum, Option 9 called for a reduction of the combined PSQ for these states on federal lands to 1.1 bbf. See U.S. Forest Service and U.S. Bureau of Land Management, *Final Supplemental Environmental Impact Statement on Management of Habitat for Late-Successional and Old-Growth Forest Related Species within the Range of the Northern Spotted Owl* (Washington, DC: U.S. Department of Agriculture Forest Service; U.S. Department of the Interior Bureau of Land Management, 1994).

15. Data from Oregon Department of Forestry, *1995 Annual Reports* (Salem, OR: Author, 1997).

16. For commentary leading up to and beyond the crisis, see variously V. Alaric and D. C. Le Master, "Economic Effects of Northern Spotted Owl Protection," *Journal of Forestry* 90 (1992): 31–35; W. Dietrich, *The Final Forest: The Battle for the Last Great Trees of the*

Pacific Northwest (New York: Simon & Schuster, 1992); Dixon and Juelson, "The Political Economy of the Spotted Owl"; D. Doak, "Spotted Owls and Old Growth Logging in the Pacific Northwest," *Conservation Biology* 3, no. 4 (1989): 389–96; K. E. Franzsreb, "Perspectives on the Landmark Decision Designating the Northern Spotted Owl (*Strix Occidentalis Caurina*) as a Threatened Subspecies," *Environmental Management* 17, no. 4 (1993): 445–52; Marcot and Thomas, *Of Spotted Owls;* E. Roe, "Why Ecosystem Management Can't Work without Social Science: An Example from the California Northern Spotted Owl Controversy," *Environmental Management* 20, no. 5 (1996): 667–74; T. Satterfield, *Anatomy of a Conflict: Identity, Knowledge, and Emotion in Old-Growth Forests* (Vancouver and Toronto: University of British Columbia Press, 2002); Simberloff, "The Spotted Owl Fracas"; and R. White, "Are You an Environmentalist or Do You Work for a Living? Work and Nature," in *Uncommon Ground: Toward Reinventing Nature,* ed. W. Cronon (New York: W. W. Norton, 1995), 171–85.

17. See, for example, S. G. Boyce, "The Good Old New Forestry," *Bioscience* 41, no. 2 (1991): 67; D. S. Debell and R. O. Curtis, "Silviculture and New Forestry in the Pacific-Northwest," *Journal of Forestry* 91, no. 12 (1993): 25–30; J. F. Franklin, "Toward a New Forestry," *American Forests* 95, nos. 11, 12 (1989): 37–44; A. M. Gillis, "The New Forestry—An Ecosystem Approach to Land Management," *Bioscience* 40, no. 8 (1990): 558–62; and F. J. Swanson and J. F. Franklin, "New Forestry Principles from Ecosystem Analysis of Pacific-Northwest Forests," *Ecological Applications* 2, no. 3 (1992): 262–74.

18. D. Demeritt, "Scientific Forest Conservation and the Statistical Picturing of Nature's Limits in the Progressive-Era United States," *Environment and Planning D-Society and Space* 19, no. 4 (2001): 431–59.

19. Other species thought to depend on old growth include the northern goshawk, Vaux's swift, the silver-haired bat, the red tree vole, the marbled murrelet, and the northern flying squirrel; see Simberloff, "The Spotted Owl Fracas."

20. Dietrich, *The Final Forest.*

21. Dixon and Juelson, "The Political Economy of the Spotted Owl"; J. P. McCarthy, *The Political and Moral Economy of Wise Use. Geography* (Ph.D. dissertation, Dept. of Geography, University of California–Berkeley, 1999), 523; Satterfield, *Anatomy of a Conflict.*

22. Simberloff, "The Spotted Owl Fracas."

23. S. C. Peterson, *The Modern Ark: A History of the Endangered Species Act* (Ph.D. dissertation, Dept. of Hisotyr, University of Wisconsin–Madison, 2000), 293.

24. See White, "Are You an Environmentalist or Do You Work for a Living?" Also, see W. Cronon, ed., *The Trouble with Wilderness. Uncommon Ground: Toward Reinventing Nature* (New York: W. W. Norton, 1995), 69–90, and other essays in the volume for a sample of the increasingly orthodox view, at least in the humanities and social sciences, that meaning conferred on biophysical nature and environmental change is necessarily socially constructed. See also J. D. Proctor, "The Social Construction of Nature: Relativist Accusations, Pragmatist and Critical Realist Responses," *Annals of the Association of American Geographers* 88, no. 3 (1998): 352–76, for a discussion of this issue and its philosophical and political implications.

25. See, for example, B. Braun, *The Intemperate Rainforest: Nature, Culture, and Power on Canada's West Coast* (Minneapolis: University of Minnesota Press, 2002); Demeritt, "Scientific Forest Conservation"; Proctor, "The Social Construction of Nature"; and J. D. Proctor, "Whose Nature? The Contested Moral Terrain of Ancient Forests," in *Uncommon Ground: Toward Reinventing Nature,* ed. W. Cronon (New York: W. W. Norton, 1995), 269–97.

26. See most directly Satterfield, *Anatomy of a Conflict.*

27. See Braun, *The Intemperate Rainforest;* B. Willems-Braun, "Buried Epistemologies: The Politics of Nature in (Post)Colonial British Columbia," *Annals of the Association of American Geographers* 87, no. 1 (1997): 3–31, for useful and careful discussions of the ways that who we are shapes, often uncritically, the way we see nature, and some of the political implications of this. For interesting and elegant empirical work on this, see P. Robbins, "The Practical Politics of Knowing: State Environmental Knowledge and Local Political Economy," *Economic Geography* 76, no. 2 (2000): 126–44.

28. On the inherent need for capitalist production and markets to grow, see D. Harvey, *The Condition of Postmodernity: An Enquiry into the Origins of Cultural Change* (Oxford and Cambridge, MA: Blackwell, 1989); D. Harvey, *The Limits to Capital* (Oxford: Blackwell,

1982); D. Harvey, *Spaces of Hope* (Edinburgh: Edinburgh University Press, 2000); Neil Smith, *Uneven Development: Nature, Capital, and the Production of Space* (Oxford: Blackwell, 1984); R. A. Walker, "California's Golden Road to Riches: Natural Resources and Regional Capitalism, 1848-1940," *Annals of the Association of American Geographers* 91, no. 1 (2001): 167–99; R. A. Walker, "Regulation and Flexible Specialization as Theories of Capitalist Development," in *Spatial Practices: Markets, Politics, and Community Life,* ed. H. Liggett and D. Perry (Thousand Oaks, CA: Sage, 1995), 167–208. Economic geographers often have used this as the foundation for the view that the capitalist space economy is in a constant state of disruption and restructuring, thereby producing patterns of uneven development.

29. K. Polanyi, *The Great Transformation* (Boston: Beacon Press, 1944).

30. Smith et al., *Forest Resources of the United States, 1997.*

31. A. B. Carey, "A Summary of the Scientific Basis for Spotted Owl Management," in *Ecology and Management of the Spotted Owl in the Pacific Northwest,* ed. R. J. Gutierrez and A. B. Carey (Arcata, CA: U.S. Department of Agriculture Forest Service, 1985), PNW-185, 100–14. See also Simberloff, "The Spotted Owl Fracas."

32. S. L. Yaffee, *The Wisdom of the Spotted Owl: Policy Lessons for a New Century* (Covelo, CA: Island Press, 1994).

33. Also known as the neo-Cornucopian view, and the Pollyanna view.

34. For the classic case on the workings of the price mechanism (i.e., the market) to solve all ills, see H. J. Barnett and C. Morse, *Scarcity and Growth; the Economics of Natural Resource Availability* (Washington, DC, and Baltimore, MD: Resources for the Future by Johns Hopkins Press, 1963); J. L. Simon, *The Ultimate Resource* (Princeton, NJ: Princeton University Press, 1981); and J. L. Simon, *The Ultimate Resource 2* (Princeton, NJ: Princeton University Press, 1996). For critiques, see J. S. Dryzek, *The Politics of the Earth: Environmental Discourses* (Oxford and New York: Oxford University Press, 1997); and R. Norgaard, "Economic Indicators of Resource Scarcity: A Critical Essay," *Journal of Environmental Economics and Management* 18 (1990): 19–25.

35. See D. H. Meadows and Club of Rome, *The Limits to Growth: A Report for the Club of Rome's Project on the Predicament of Mankind* (New York: Universe Books, 1972); and P. Neurath, *From Malthus to the Club of Rome and Back: Problems of Limits to Growth, Population Control, and Migrations* (Armonk, NY: M. E. Sharpe, 1994). Again, for discussion, see Dryzek, *The Politics of the Earth.*

36. For more discussion on related points, see W. S. Prudham, "Regional Science, Political Economy, and the Environment," *Canadian Journal of Regional Science* 25 no. 2 (2003): 171–206.

37. See, for example, R. L. Bryant, "Political Ecology: An Emerging Research Agenda in Third-World Studies," *Political Geography* 11, no. 1 (1992): 12–36; S. B. Hecht and A. Cockburn, *The Fate of the Forest: Developers, Destroyers, and Defenders of the Amazon* (London and New York: Verso, 1989); N. L. Peluso, *Rich Forests, Poor People: Resource Control and Resistance in Java* (Berkeley: University of California Press, 1992); M. Watts, *Silent Violence: Food, Famine, and Peasantry in Northern Nigeria* (Berkeley: University of California Press, 1983).

38. P. M. Blaikie and H. C. Brookfield, *Land Degradation and Society* (London and New York: Methuen, 1987), 17. Note their insistence on a regional approach, based on recognition of important sources of variation across space in human and physical geographies.

39. M. Watts and R. Peet, "Conclusion: Toward a Theory of Liberation Ecology," *Liberation Ecologies: Environment, Development, Social Movements,* ed. R. Peet and M. Watts (New York: Routledge, 1996): 260–270.

40. I do not mean to suggest that taking the lessons of political ecology and applying them to the historical geography of the Pacific Slope is a straightforward or easy extension of what political ecology has come to mean. For one thing, the region is very different from the poorest and least industrialized regional and national economies where most political ecologists have trained their gazes; see Bryant, "Political Ecology." Very few scholars have successfully applied an explicitly political ecology approach to the rich and most heavily industrialized nations. One reason for this, although there are several, is the general absence of peasant, subsistence, or semisubsistence land users. Since such marginalized classes and their relationships to capitalist production networks comprise a central focus in much of the political ecology scholarship (particularly earlier political ecology research), adapting

lessons from political ecology to the rich countries, and to the United States in particular, requires careful thought; see J. McCarthy, "First World Political Ecology: Lessons from the Wise Use Movement," *Environment and Planning A* 34 (2002): 1281–302; P. Walker, "Reconsidering 'Regional' Political Ecologies: Toward a Political Ecology of the Rural American West," *Progress in Human Geography* 27, no. 1 (2003): 7–24. Moreover, it would be fair to say that political ecology has become highly fractured, to the point where it is barely a coherent field. In recent years, political ecology scholarship has shown great preoccupation with the micropolitics of local conflict and resistance. Fueling these developments in political ecology, there also has been a widespread poststructural turn informed particularly by neo-Foucauldian theory, with emphasis on the diverse ideological, cultural, and epistemological underpinnings of environmental discourses; for example, A. Escobar, *Encountering Development: The Making and Unmaking of the Third World* (Princeton, NJ: Princeton University Press, 1995); Robbins "The Practical Politics of Knowing"; J. C. Scott, *Seeing Like a State: How Certain Schemes to Improve the Human Condition Have Failed* (New Haven, CT: Yale University Press, 1998). It is fair to say I give less attention and focus to these dimensions here, though they do inform my attempt to combine a structural account of the origins of environmental crises with a poststructural understanding of how meaning is conferred on such crises. For review and discussion of related issues, see Bryant, "Political Ecology"; McCarthy, "First World Political Ecology"; R. Peet and M. Watts, "Introduction: Development Theory and Environment in an Age of Market Triumphalism," *Economic Geography* 69, no. 3 (1993): 227–53; R. Peet and M. Watts, *Liberation Ecologies: Environment, Development, Social Movements* (London and New York: Routledge, 1996).

41. Peet and Watts, "Introduction."
42. This rather jarring notion of nature as increasingly and meaningfully "produced" by capitalist processes draws on Smith, *Uneven Development*.
43. Polanyi, *The Great Transformation*.
44. Ibid. Polanyi wrote, "A market economy is an economic system controlled, regulated, and directed by markets alone; order in the production and distribution of goods is entrusted to this self-regulating mechanism," 68.
45. I disregard his argument about money here.
46. Although Engels deserves some mention here; see F. Engels, *The Condition of the Working Class in England* (Oxford and New York: Oxford University Press, 1993).
47. Polanyi discusses the idea of nature as a commodity in terms of the production of nature-based commodities for circulation in a capitalist market economy. However, in the age of industrial cultivation and proprietary, now genetically engineered varieties of organisms, it is necessary to distinguish a separate notion of the commodification of nature. I discuss such tendencies in chapter 5. I employ an expansive view of commodification, including the set of social relationships and material transformations along a commodity chain encompassing not only the conversion of logs into products but also logging, reforestation, and intensive forestry research, development, and technology transfer.
48. Again, see Smith, *Uneven Development,* on this notion of the production of nature.
49. J. O'Connor, "Capitalism, Nature, Socialism: A Theoretical Introduction," *Capitalism, Nature, Socialism* 1, no. 1 (1988): 11–38. See also O'Connor's revised version, which appears as chapter 8 in J. O'Connor, *Natural Causes: Essays in Ecological Marxism* (New York: Guilford Press, 1998).
50. O'Connor, "Capitalism, Nature, Socialism," 16.
51. Ibid., see especially p. 31. O'Connor's thesis has been quite significant in the rise of a more general ecological variant of Marxist scholarship. On Marx and natural limits and nature, see A. Schmidt, *The Concept of Nature in Marx* (London: NLB, 1971); and Smith, *Uneven Development*. The traditional Marxist antagonism to natural limits is based on suspicions, typically well founded, that invoking limits or generalized crises of an ecological character tends to obscure particular social and political roots of the problem and provides convenient rhetorical cover for profound inequalities that tend to act as underlying social causes of scarcity, in famines, for example; see T. Benton, "Marxism and Natural Limits: An Ecological Critique and Reconstruction," *New Left Review* 178 (1989): 51–86; N. Castree, "The Nature of Produced Nature: Materiality and Knowledge Construction in Marxism," *Antipode* 27, no. 1 (1995): 12–48; and D. Harvey, "Population, Resources and the Ideology of Science," *Economic Geography* 50 (1974): 256–77.

52. O'Connor, *Natural Causes,* 159.

53. D. Harvey, *Justice, Nature, and the Geography of Difference* (Cambridge, MA: Blackwell, 1996).

54. E. Leff, *Green Production: Toward an Environmental Rationality* (New York and London: Guilford Press, 1995), 14.

55. D. Harvey, *The Urban Experience* (Baltimore, MD: Johns Hopkins University Press, 1989), 83. See also Harvey, *The Limits to Capital.*

56. K. Kautsky, *The Agrarian Question* (London and Winchester, MA: Zwan Publications, 1988), 147.

57. Ibid.

58. Kautsky downplayed the significance of land as a constraint in forestry largely because he was interested more narrowly in whether capitalism was observed to develop at all. Forestry and pasturage, he noted, comprised the first form of large-scale, land-extensive capital accumulation in Europe and the United States. Hence, these spheres clearly did not resist the penetration of capital, at least not to the same degree as intensive crop agriculture. However, if the organization of industrial forestry and associated commodity production is the focus rather than the development of capitalism, per se land can indeed be seen as a significant structural constraint.

59. In fact, many lumber mills used to be built as mobile operations, small scale, and easily assembled and disassembled in order to move them to where the timber was, as I discuss in chapter 4 and in chapter 6.

60. Of course, as I discuss in chapter 5, over time this has become less the dominant strategy as the supply of natural or nonplantation fiber gradually gives way to cultivated forests, and as social regulation of cut-and-run strategies becomes more and more stringent. On the changing availability of global fiber, see R. Sedjo, *The Forest Sector: Important Innovations* (Washington, DC: Resources for the Future, 1997); R. A. Sedjo and K. S. Lyon, *The Long-Term Adequacy of World Timber Supply* (Washington, DC, and Baltimore, MD: Resources for the Future, 1990) (distributed worldwide by Johns Hopkins University Press). For a history of extraction, see M. Williams, *Americans and Their Forests: A Historical Geography* (Cambridge Cambridgeshire, and New York: Cambridge University Press, 1989). For a perspective on the shift from extraction to cultivation, see R. A. Clapp, "The Resource Cycle in Forestry and Fishing," *Canadian Geographer* 42, no. 2 (1998): 129-44; and M. P. Marchak, *Logging the Globe* (Montreal and Buffalo: McGill-Queen's University Press, 1995).

61. K. Marx, *Capital: A Critique of Political Economy. Volume 2. The Process of Circulation of Capital* (New York: International Publishers, 1967), 248.

62. B. Adam, "The Temporal Gaze: The Challenge for Social Theory in the Context of GM Food," *British Journal of Sociology* 51, no. 1 (2000): 125–42, 137. See also B. Adam, *Timescapes of Modernity: The Environment and Invisible Hazards* (London: Routledge, 1998).

63. This also draws on Kautsky's seminal observations. See S. A. Mann, *Agrarian Capitalism in Theory and Practice* (Chapel Hill: University of North Carolina Press, 1990); S. A. Mann and J. M. Dickinson, "Obstacles to the Development of a Capitalist Agriculture," *Journal of Peasant Studies* 5, no. 4 (1978): 466–81. Goodman, Sorj, and Wilkinson also invoke "biological time in plant growth and animal gestation" as a challenge that helps to shape the organization of industrial agriculture; D. Goodman, B. Sorj, and J. Wilkinson, *From Farming to Biotechnology: A Theory of Agro-Industrial Development* (Oxford and New York: Blackwell, 1987), 1. Also on biological time and agricultural political economy, see D. Goodman and M. R. Redclift, *Refashioning Nature: Food, Ecology, and Culture* (London and New York: Routledge, 1991); and G. Henderson, "Nature and Fictitious Capital: The Historical Geography of an Agrarian Question," *Antipode* 30, no. 2 (1998): 73–118.

64. Stephen Bunker (p. 590) noted, "Industrial production starts and ends at the same time as the labor that defines it. In agriculture, much of the production occurs after labor has modified the conditions of that production and before labor harvests that product, while in extraction, actual production is anterior to human labor, and in the case of minerals occurs within much longer time frames than humans normally use; S. G. Bunker, "Staples, Links, and Poles in the Construction of Regional Development Theory," *Sociological Forum* 4, no. 4 (1989): 589–610. As also noted by Bunker, limited or uneven availability of natural inputs is a temporal and spatial issue, the implications of which are highly specific to biophysical processes and the geography of particular raw material availability, as well as to the

institutional strategies that arise around these; see also B. Barham, S. G. Bunker, and D. O'Hearn, *States, Firms, and Raw Materials: The World Economy and Ecology of Aluminum* (Madison: University of Wisconsin Press, 1994).

65. R. A. Sayer, *Radical Political Economy: A Critique* (Oxford and Cambridge, MA: Blackwell, 1995), 64. Though Alfred Chandler is less likely to acknowledge the diversity of possibilities in organizational outcomes within particular industries, his classic study of the rise of the modern American corporation also makes room for differences across sectors, as a partial function of their specific material problems; A. D. Chandler, *The Visible Hand: The Managerial Revolution in American Business* (Cambridge, MA: Belknap Press, 1977). For example, writing about the U.S. petroleum industry in the late nineteenth century, he noted, "Because the pipeline could carry crude oil to processing facilities but not refined products to markets, the completion of long-distance lines called for relocation of refinery capacity at centers close to the market" (p. 323).

66. Goodman, Sorj, and Wilkinson used this notion of social production revolving around nature to great effect in their discussion of the land-based character of agriculture; Goodman, Sorj, and Wilkinson, *From Farming to Biotechnology.*

67. A brilliant example is provided by Cronon's examination of the ways in which continuous, natural variation in wheat had to be "made" amenable to market demands for clear quality standards prior to the emergence of a true commodities market in Chicago. Sadly, Cronon ignores wheat breeding as an ancillary aspect of this and also seems content to accept neoclassical notions of commodities as simply generic raw material products without further reflection on the commodity form and its implications for making and remaking nature; W. Cronon, *Nature's Metropolis: Chicago and the Great West* (New York: W. W. Norton, 1991).

68. Ted Benton refers to exactly this distinction in noting that what he terms "primary labor processes" (extractive industries) "are highly dependent on both naturally given contextual conditions"; that is, the land issue "and the properties of the subjects of labour" (emphasis added). In agriculture, as numerous observers have noted, such qualities as texture, weight, and perishability are major influences on the development of commodity-specific systems of provisioning or supply chains; see Benton, "Marxism and Natural Limits," 71. The importance of physical resource properties in shaping the social and spatial organization of natural resource industries more generally is also one of the insights of Harold Innis's staples theory of regional development. On commodity characteristics and commodity chains, see W. H. Friedland, A. E. Barton, and R. J. Thomas, *Manufacturing Green Gold: Capital, Labor, and Technology in the Lettuce Industry* (Cambridge Cambridgeshire and New York: Cambridge University Press, 1981); Mann, *Agrarian Capitalism in Theory and Practice;* M. J. Wells, *Strawberry Fields: Politics, Class, and Work in California Agriculture* (Ithaca, NY: Cornell University Press, 1996). For a geographical perspective on commodity chains, see E. Hartwick, "Geographies of Consumption: A Commodity-Chain Approach," *Environment and Planning D-Society & Space* 16, no. 4 (1998): 423–37; E. R. Hartwick, "Towards a Geographical Politics of Consumption," *Environment and Planning A* 32, no. 7 (2000): 1177–92. For Innis's staples perspective, see H. A. Innis, *Essays in Canadian Economic History* (Toronto: University of Toronto Press, 1956); M. Q. Innis, *An Economic History of Canada* (Toronto: Ryerson Press, 1954); and T. J. Barnes, R. Hayter, and E. Hay, "Stormy Weather: Cyclones, Harold Innis, and Port Alberni, BC," *Environment and Planning A* 33 (2001): 2127–47.

69. N. Rosenberg, *Exploring the Black Box: Technology, Economics, and History* (Cambridge and New York: Cambridge University Press, 1994), 244.

70. Harvey, *Justice, Nature, and the Geography of Difference,* 193. Also, as George Henderson rightly stated, "Natural processes are at one and the same time an invitation and barrier to capital"; Henderson, "Nature and Fictitious Capital," 76.

71. Leff, *Green Production,* 14.

72. O'Connor, *Natural Causes,* 164.

73. On biotechnology and firm strategies in American agriculture, see W. Boyd, "Wonderful Potencies: Deep Structure and the Problem of Monopoly in Agricultural Biotechnology," in *Engineering Trouble: Biotechnology and Its Discontents,* ed. R. Schurman and D. Takahashi (Berkeley: University of California Press, 2003), 24–62. More generally, see W. Boyd, S. Prudham, and R. Schurman, "Industrial Dynamics and the Problem of Nature," *Society and Natural Resources* 14, no. 7 (2001): 555–70.

74. In geography, Neil Smith's thesis is highly germane; Smith, *Uneven Development*. For Smith, overcoming the separation of nature and culture can be achieved only by understanding that nature is more and more socially produced at ideological and material levels. Yet, as important as this insight may be in rethinking nature as that which is not social, and in breaking down the unhelpful categories of "first" and "second" nature (i.e., original and transformed), there is more than hint of Prometheanism here, rendering any notion of a material nature external to human control insignificant or incidental to the dynamics of capital and to environmental politics. What Noel Castree called "nature's materiality," that is, the capacity of the natural world to operate according to its own logic in time and space, seems to retain little implication for structuring human activities and relationships; Benton, "Marxism and Natural Limits"; and Castree, "The Nature of Produced Nature." Harvey, harkening to his earlier essay on Malthus and Marx, rejects Benton's argument and clings steadfastly to his resistance to the notion of limits; see Harvey, *Justice, Nature, and the Geography of Difference*. However, a thoughtful if difficult reflection on the need to resurrect some idea that nature is not infinitely malleable comes from Demerritt, who argues in part based on Actor Network Theory that nature and society (or culture) are locked in an embrace of what he called "conjoined materiality"; see D. Demeritt, "The Construction of Global Warming and the Politics of Science," *Annals of the Association of American Geographers* 91, no. 2 (2001): 307–37. See also D. J. Haraway, *Modest-Witness, Second-Millennium: Femaleman Meets Oncomouse: Feminism and Technoscience* (New York: Routledge, 1997). Dick Walker's paper on California's emergence as an economic powerhouse harkens back to Harold Innis's work in trying to combine emphasis on the social relations and institutions underpinning regional capitalist development with the role of biophysical factors, including resource endowments, in dialectically shaping growth trajectories; Walker, "California's Golden Road to Riches."
75. Goodman, Sorj, and Wilkinson, *From Farming to Biotechnology*.
76. J. R. Kloppenburg, *First the Seed: The Political Economy of Plant Biotechnology, 1492–2000* (Cambridge Cambridgeshire and New York: Cambridge University Press, 1988).
77. For Kloppenburg, this has been accomplished with three interwoven strategies. The first strategy was to gain biological control over plant breeding, initially through the use of hybrid varieties whose productivity in subsequent generations declines rapidly. The second strategy was in the realm of property, pursing in the political realm the creation of excludable and enforceable property rights over plant varieties. The third strategy was in the transformation of public science, again through political intervention, to limit competition with private capital from the public sector.
78. I worked on a paper with William Boyd and Rachel Schurman in which we developed and extended these and related insights to a general approach for all nature-centered sectors. We argued that the dynamic tendencies of nature-based industries need to be understood in terms of natural obstacles and opportunities, contrasting the different possibilities for capital to take hold of and intensify natural processes under extraction and cultivation. We emphasized in particular the ways that biologically based industries are being reshaped by the industrialization of natural growth processes, with important implications for industry structures and regional economic growth, but are accompanied also by serious environmental risks and concerns. We argued for case studies that explore how natural obstacles and opportunities shape firm strategies, industrial organization, and regional development processes, including social and environmental implications based on production-centered analyses of regional industrial systems; Boyd, Prudham, and Schurman, "Industrial Dynamics and the Problem of Nature."
79. See, for example, Bryant, "Political Ecology"; A. Escobar, "Constructing Nature: Elements for a Poststructural Political Ecology," in *Liberation Ecologies: Environment, Development, Social Movements,* ed. R. Peet and M. Watts (New York: Routledge, 1996), 46–68; Escobar, *Encountering Development;* Peet and Watts, "Introduction"; Peet and Watts, *Liberation Ecologies;* and Robbins, "The Practical Politics of Knowing." This is also a theme in environmental history and environmental politics more broadly. See for example, the collection of essays in Cronon, *The Trouble with Wilderness*. This is a strong theme that also runs through Beck's notion of the risk society, and is clearly reflected in Maarten Hajer's influential book on the politics of acid rain regulation in Europe during the 1980s; U. Beck, *World Risk Society* (Malden, MA: Polity Press, 1999); U. Beck and M. Ritter, *Risk Society: Towards*

a *New Modernity* (London and Newbury Park, CA, and New Delhi: Sage, 1992); and M. Hajer, *The Politics of Environmental Discourse* (New York: Oxford University Press, 1995).

80. Braun, *The Intemperate Rainforest;* N. Castree and B. Braun, "The Construction of Nature and the Nature of Construction: Analytical and Political Tools for Building Survivable Futures," in *Remaking Reality: Nature at the Millennium,* ed. B. Braun and N. Castree (London and New York: Routledge, 1998), 3–42; D. Demeritt, "Ecology, Objectivity and Critique in Writings on Nature and Human Societies," *Journal of Historical Geography* 20, no. 1 (1994): 22–37; D. Demeritt, "Science, Social Constructivism and Nature," in *Remaking Reality: Nature at the Millenium,* ed. B. Braun and N. Castree (London and New York: Routledge, 1998), 173–93; and Willems-Braun, "Buried Epistemologies."

81. See Hajer, *The Politics of Environmental Discourse.*

82. R. C. Lewontin, *Biology as Ideology: The Doctrine of DNA* (Concord, Ontario: Anansi, 1991), 93.

83. Hajer, *The Politics of Environmental Discourse;* Harvey, *Justice, Nature, and the Geography of Difference;* N. Low, "Ecosocialisation and Environmental Planning: A Polanyian Approach," *Environment and Planning A* 34 (2002): 43–60.

84. Proctor, "The Social Construction of Nature." See also Demeritt, "The Construction of Global Warming and the Politics of Science." I am indebted to my good friend William Boyd for this basic aphorism when it comes to relativistic claims about the politics of knowledge.

85. Terre Satterfield's excellent book on the old-growth conflict is largely an examination of this struggle over meaning. Although she reduced the struggle to one over meaning in a way I find problematic, she recognized and documented the struggle all the same; Terre Satterfield, *Anatomy of a Conflict.*

86. See, for example, Proctor, "The Social Construction of Nature"; Proctor, *Whose Nature? The Contested Moral Terrain of Ancient Forests;* Satterfield, *Anatomy of a Conflict.*

87. Here, I am influenced by Adam Tickell and Jamie Peck; A. Tickell and J. Peck, "Accumulation, Regulation, and the Geographies of Post-Fordism: Missing Links in Regulation Research," *Progress in Human Geography* 16 (1992): 190–218. A regime of accumulation refers broadly to historically specific tendencies in the organization of production and the allocation of social product. Key issues include divisions of labor and the hierarchical organization of labor and management within the firm; production, information, and communication technologies; patterns of vertical and horizontal integration; the predominance of large versus small firms; investment time horizons for firms and their decisions about the deployment of fixed capital; the social distribution of income; markets; and relations between capitalist and noncapitalist production. A mode of social regulation refers largely to the institutional context of the economy. This includes wider patterns of capital-wage relations (e.g., industry-wide collective bargaining frameworks, labor laws), property relations, the credit system, the rules governing interfirm competition and market power, the relationship of the domestic and international economies, and various other dimensions of state and nonstate intervention in managing the economy and its social effects (e.g., affirmative action programs, state educational and health care provisioning, social movements and citizen groups). On the merits of adapting this approach to nature-based or rural production, see D. Goodman and M. Watts, "Reconfiguring the Rural or Fording the Divide? Capitalist Restructuring and the Global Agro-Food System," *Journal of Peasant Studies* 22, no. 1 (1994): 1–49. For critical commentary on the Regulation School, some more sympathetic than others, see inter alia A. Amin, "Post-Fordism: Models, Fantasies and Phantoms of Transition," in *Post-Fordism: A Reader,* ed. A. Amin (Oxford and Cambridge, MA: Blackwell, 1994), 1–39; R. Brenner and M. Glick, "The Regulation Approach: Theory and History," *New Left Review* 188 (1991): 45–120; M. Gertler, "The Limits to Flexibility: Comments on the Post-Fordist Vision of Production and Its Geography," *Transactions, Institute of British Geographers* 13 (1988): 419–32; Tickell and Peck, "Accumulation, Regulation, and the Geographies of Post-Fordism"; Walker, *Regulation and Flexible Specialization;* M. J. Webber and D. L. Rigby, *The Golden Age Illusion: Rethinking Postwar Capitalism* (New York: Guilford Press, 1996).

88. See M. Aglietta, *A Theory of Capitalist Regulation: The US Experience* (London: NLB, 1979); and R. Boyer, *The Regulation School: A Critical Introduction* (New York: Columbia University Press, 1990).

89. I should add here that there are intriguing parallels between the emergence of a specifically ecological variant of the Regulation approach and political ecology. One of the themes of

political ecology, particularly as developed during the 1980s by the likes of Blaikie and Brookefield, is that the articulation of specific regional combinations of biophysical nature and the institutional and political aspects of resource access and control with broader processes of capital accumulation and integration into the world market give rise to distinct regional development processes; S. Berry, "Social Institutions and Access to Resources," *Africa* 59, no. 1 (1989): 41–55; Blaikie and Brookefield, *Land Degradation and Society*. Much might be gained from exploring the tensions and parallels between political ecology and the Regulation approach to environmental politics and governance. Although political ecology could perhaps contribute acute sensitivity to biophysical variation, and the social constructions that confer meaning on this variation, the Regulation approach could reinforce attention to the central and increasing profile of nature-based capital accumulation and market integration in environmental change and environmental politics across the globe. For discussion of related themes, see Peet and Watts, "Introduction."

90. S. Corbridge, "Marxism, Post-Marxism, and the Geography of Development," in *New Models in Geography: The Political-Economy Perspective,* ed. R. Peet and N. J. Thrift (London and Winchester, MA: Unwin-Hyman, 1989), 224–54.

91. See, for example, K. J. Bakker, "Privatising Water, Producing Scarcity: The Yorkshire Drought of 1995," *Economic Geography* 76, no. 1 (2000): 4–27; G. Bridge, "The Social Regulation of Resource Access and Environmental Impact: Nature and Contradiction in the US Copper Industry," *Geoforum* 31, no. 2 (2000): 237–56; G. Bridge and P. McManus, "Sticks and Stones: Environmental Narratives and Discursive Regulation in the Forestry and Mining Sectors," *Antipode* 32, no. 1 (2000): 10–47; D. Gibbs, "Ecological Modernisation, Regional Economic Development, and Regional Development Agencies," *Geoforum* 31 (2000): 9–19; D. Gibbs, "Integrating Sustainable Development and Economic Restructuring: A Role for Regulation Theory?" *Geoforum* 27, no. 1 (1996): 1–10; and D. Gibbs and A. E. G. Jonas, "Governance and Regulation in Local Environmental Policy: The Utility of a Regime Approach," *Geoforum* 31, no. 3 (2000): 299–313. For related discussion, see Prudham, "Regional Science, Political Economy, and the Environment."

92. He also claims money as a third category of fictitious commodity, but I leave this aside for the current discussion. See Polanyi, *The Great Transformation.*

93. Ecosocialization, drawing on Polanyi, refers to the processes of resistance to market coordination of nature's transformation arising from a society seeking to protect itself from the negative consequences and to adapt capitalism to ecological limits; see Low, "Ecosocialisation and Environmental Planning," 51. The concept is critical in part because it forces consideration of how ecological crises are worked out in the realm of society, and thus that some understanding of society figure into explanations of crises and their regulation. On the significance of Polanyi for theorizing society, see M. Burawoy, "For a Sociological Marxism: The Complementary Convergence of Antonio Gramsci and Karl Polanyi," *Politics and Society* 31, no. 2 (2003): 193–261.

94. O'Connor extended Polanyi in positing that the politics of the dual movement and ecological crisis tendencies necessarily entail a movement toward greater and greater state coordination in the reproduction of nature. In short, the state responds to the politics of ecological crisis by undertaking that which capital alone will not do. For O'Connor, this provides an additional pathway to eventual socialist transformation, as coordination of environmental governance and management become increasingly socialized. But this may or may not prove to be an accurate prognosis. For instance, as I discuss, and as O'Connor suggested, it is also possible for the capitalization of nature to offset crisis tendencies, at least temporarily, including, for example, in the sphere of industrial forest tree cultivation; Boyd, Prudham, and Schurman, "Industrial Dynamics and the Problem of Nature." Moreover, nothing guarantees that the state will respond in this way, nor that environmental social movements will push for it. Indeed, in the age of neoliberal assaults on state capacities, it has become increasingly obvious that numerous institutional and political outcomes may arise from social opposition to the capitalist underproduction of nature, including, for example, consumer boycotts, ecocertification campaigns, and market-based schemes to address environmental problems. As Steven Bernstein pointed out, environmentalism has, if anything, become more compatible with liberalism in recent decades, not less. See S. Bernstein, "Ideas, Social Structure, and the Compromise of Liberal Environmentalism," *European Journal of International*

Relations 6, no. 4 (2000): 464–512; S. Bernstein, "Liberal Environmentalism and Global Environmental Governance," *Global Environmental Politics* 2, no. 3 (2002): 1–16.

95. O'Connor, *Natural Causes.*
96. Harvey, *The Limits to Capital.*
97. I drew some inspiration, or reinspiration perhaps, on the need to reconcile critiques and accounts of capitalism on one hand and environmental change and its politics on the other hand at a series of panel discussions at the Association of American Geographers annual meetings in Los Angeles in 2002, commemorating twenty years since the publication of the *Limits to Capital.* One of the common themes among panelists (including Eric Sheppard, Trevor Barnes, Richard Walker, Erica Schoenberger, Eric Swyngedouw, Neil Brenner, Bob Jessop, and others) was that, brilliant though Harvey's analysis was, it echoed Marx's own failure to provide an account of capitalist nature and its politics.
98. Benton, "Marxism and Natural Limits."

Chapter 2

1. T. Benton, "Marxism and Natural Limits: An Ecological Critique and Reconstruction," *New Left Review* 178 (1989): 51–86.
2. On structure and agency in explanation, see A. Giddens, *Central Problems in Social Theory: Action, Structure, and Contradiction in Social Analysis* (Berkeley: University of California Press, 1979); A. Giddens, *A Contemporary Critique of Historical Materialism* (Houndmills, England: Macmillan, 1995).
3. G. Norcliffe and J. Bates, "Implementing Lean Production in an Old Industrial Space: Restructuring at Corner Brook, Newfoundland, 1984-94," *Canadian Geographer* 41, no. 1 (1997): 41–60.
4. R. Coase, "The Nature of the Firm," *Economica* 4 (1937): 386–405.
5. A. J. Scott, "Industrial Organization and Location: Division of Labor, the Firm and Spatial Process," *Economic Geography* 63 (1987): 215–31. For a more recent example, see B. Ó hUallacháin and D. Wasserman, "Vertical Integration in a Lean Supply Chain: Brazilian Automobile Component Parts," *Economic Geography* 75, no. 1 (1999): 21–42. Economies of scope, which I discuss in more detail in the context of the Douglas-fir industry, in chapter 4, are essentially advantages that may be gained by producing multiple products from technologically or organizationally related or parallel processes. Examples would be petroleum products produced from multiple fractions of oil and all generated in parallel as by-products of the crude refining process.
6. See, for example, M. Castells, *The Rise of the Network Society* (Cambridge, MA: Blackwell, 1996); and M. J. Piore and C. F. Sabel, *The Second Industrial Divide: Possibilities for Prosperity* (New York: Basic Books, 1984).
7. M. Storper, *The Regional World: Territorial Development in a Global Economy* (New York: Guilford Press, 1997).
8. S. Bowles and H. Gintis, "The Revenge of Homo-Economicus: Contested Exchange and the Revival of Political Economy," *Journal of Economic Perspectives* 7, no. 1 (1993): 83–102.
9. Scott, "Industrial Organization and Location," 220.
10. A. Sayer, "Postfordism in Question," *International Journal of Urban and Regional Research* 13 (1989): 666–95; and A. Sayer, *Radical Political Economy: A Critique* (Oxford and Cambridge, MA: Blackwell, 1995).
11. See, for example, T. Barnes and R. Hayter, "The Little Town that Did: Flexible Accumulation and Community Response in Chemainus, British Columbia," *Regional Studies* 26 (1992): 647–63; T. J. Barnes and R. Hayter, "Economic Restructuring, Local Development and Resource Towns: Forest Communities in Coastal British Columbia," *Canadian Journal of Regional Science* 17, no. 3 (1994): 289–310; R. Hayter and T. Barnes, "Troubles in the Rainforest: British Columbia's Forest Economy in Transition," in *Troubles in the Rainforest: British Columbia's Forest Economy in Transition,* ed. T. Barnes and R. Hayter (Victoria, Canada: Western Geographical Press, 1997), 1–11; J. Holmes, "In Search of Competitive Efficiency: Labour Process Flexibility in Canadian Newsprint Mills," *Canadian Geographer* 41, no. 1 (1997): 7–25; and Norcliffe and Bates "Implementing Lean Production in an Old Industrial Space."

12. R. A. Walker, "California's Golden Road to Riches: Natural Resources and Regional Capitalism, 1848-1940," *Annals of the Association of American Geographers* 91, no. 1 (2001): 167–99.

13. S. A. Mann, *Agrarian Capitalism in Theory and Practice* (Chapel Hill: University of North Carolina Press, 1990), 3. See also S. A. Mann and J. M. Dickinson, "Obstacles to the Development of a Capitalist Agriculture," *Journal of Peasant Studies* 5, no. 4 (1978): 466–81; and D. Goodman and M. R. Redclift, *Refashioning Nature: Food, Ecology, and Culture* (London and New York: Routledge, 1991). For a somewhat critical discussion and reframing, see G. Henderson, "Nature and Fictitious Capital: The Historical Geography of an Agrarian Question," *Antipode* 30, no. 2 (1998): 73–118.

14. B. Adam, *Timescapes of Modernity: The Environment and Invisible Hazards* (London: Routledge, 1998).

15. Mann, *Agrarian Capitalism in Theory and Practice,* 39. On contracting in agriculture more broadly, see W. Boyd and M. Watts, "Agro-Industrial Just-in-Time," in *Globalising Food: Agrarian Questions and Global Restructuring,* ed. D. Goodman and M. Watts (London and New York: Routledge, 1997), 192–225; and M. J. Watts, "Life under Contract: Contract Farming, Agrarian Restructuring, and Flexible Accumulation," in *Living under Contract: Contract Farming and Agrarian Transformation in Sub-Saharan Africa,* ed. P. D. Little and M. J. Watts (Madison: University of Wisconsin Press, 1994), 21–77.

16. Benton, "Marxism and Natural Limits." See also Boyd, Prudham, and Schurman "Industrial Dynamics and the Problem of Nature.".

17. See, for example, W. Boyd, S. Prudham, and R. Schurman, "Industrial Dynamics and the Problem of Nature," *Society and Natural Resources* 14, no. 7 (2001): 555–70; Boyd and Watts, "Agro-Industrial Just-in-Time"; Goodman and Redclift, *Refashioning Nature;* D. Goodman, B. Sorj, and J. Wilkinson, *From Farming to Biotechnology: A Theory of Agro-Industrial Development* (Oxford and New York: Blackwell, 1987); Henderson, "Nature and Fictitious Capital"; G. L. Henderson, *California and the Fictions of Capital* (New York: Oxford University Press, 1999).

18. Boyd, Prudham, and Schurman, "Industrial Dynamics and the Problem of Nature"; Henderson, "Nature and Fictitious Capital."

19. Watts, "Life under Contract," 71.

20. U.S. Department of Commerce Bureau of the Census, *1997 Economic Census Manufacturing Industry Series: Logging* (Washington, DC: Bureau of the Census, 1999, 2000).

21. These differences vary considerably, because gyppo contractors use different pay scales. Data on average salaries between contract and company loggers are not available. However, according to the 1997 census of manufacturing, the average hourly production wage in the Oregon logging sector was $14.75; ibid. In the interviews I conducted with gyppo contractors, I found that typical gyppo employees could expect a salary more in the $10 to $12 range.

22. These aspects of production contracts in general are also identified by B. Harrison, *Lean and Mean : The Changing Landscape of Corporate Power in the Age of Flexibility* (New York: Basic Books, 1994); J. Holmes, "The Organization and Locational Structure of Production Subcontracting," in *Production, Work, Territory: The Geographical Anatomy of Industrial Capitalism,* ed. A. J. Scott and M. Storper (Boston: Allen & Unwin, 1986), 80–106; A. Sayer and R. Walker, *The New Social Economy: Reworking the Division of Labor* (Cambridge, MA: Blackwell, 1992). Watts argued that the specific salience of (agricultural) contracting from a political standpoint is that it represents a means by which firms subordinate contract workers and at the same time draw them into the production sequence; Watts, "Life under Contract."

23. This estimate is based on conversations with several people, including an interview conducted with a union representative on September 2, 1997, in Gladstone, Oregon, with a follow-up telephone call on October 28, 1998, and an interview conducted on September 25, 1997, also with a union representative, in Springfield, Oregon. It is also based on inferences drawn from the census of manufactures, and information provided by the AOL. According to the most recent AOL estimates, there are about 840 active independent logging contractors in Oregon. The 1997 census of manufactures reports 1,130 logging establishments in the state. This would indicate that on the order of one-quarter to one-third of the logging establishments are company logging establishments. However, the census allows for firms to be counted more than once if they have multiple establishments, and the AOL survey indicates that the average logging contractor maintains 1.8 logging sides.

24. Barney Warf stated that early logging operations along the coast of Washington and Oregon during what he called the "First Wave" (1850s and 1860s) were typically disintegrated from lumber mills and that integration between logging and lumber manufacture paralleled the expansion of the industry in the region and the establishment of large steam-powered mills; B. Warf, "Regional Transformation, Everyday Life, and Pacific Northwest Lumber Production," *Annals of the Association of American Geographers* 78, no. 2 (1988): 326–46. Prouty seemed to contradict this view, however, and suggested that the most common arrangement was integration between mill and logging camp, even during the 1870s; A. M. Prouty, *More Deadly Than War! Pacific Coast Logging, 1827-1981* (New York: Garland, 1988).

25. W. H. Gibbons, *Logging in the Douglas Fir Region,* Government Printing Office (Washington, DC, 1918), 256.

26. Although there are a few gyppos who buy timber sales directly and sell logs to firms, this practice was much more common historically than it is today; see ibid. As a result of actual and feared collusion among loggers earlier in the century, mills have actively opposed gyppo control of timber; see W. J. Mead, *Competition and Oligopsony in the Douglas Fir Lumber Industry* (Berkeley: University of California Press, 1966); U.S. Bureau of Corporations, *The Lumber Industry* (Washington, DC: Government Printing Office, 1913). Consequently, the typical arrangement is one in which a gyppo is retained on a production contract to log timber that is owned by the mill or has been purchased through a separate transaction. In very recent years (since 1992), some gyppos have started to get involved in scouting for timber and buying their own sales. This is largely a function of the turmoil in regional timber markets following cutbacks in sales of federal timber and an associated increase in the rate of cutting on small, privately owned tracts.

27. V. H. Jensen, *Lumber and Labor* (New York: Farrar and Rinehart, 1945).

28. Tellingly, between 1910 and 1929, the number of automobiles in Oregon increased from 2,500 to about 250,000, and by 1941, Oregon had established the basis of its modern state highway system with more than 7,000 miles of primary and secondary roads. Data from S. N. Dicken and E. F. Dicken, *The Making of Oregon: A Study in Historical Geography* (Portland: Oregon Historical Society, 1979). It is not at all certain that automobile ownership and travel was available to mill workers and loggers en masse. But the road network and the automobile certainly facilitated moving workers on a daily basis from towns to logging camps. Moreover, extension of the roads allowed log transport by truck as opposed to rail, making logging camps much more mobile by liberating them from rail lines. Also, consider that between 1930 and 1940, the proportion of Northwest logs hauled by truck increased from 6 percent of the total to half; see M. Williams, *Americans and Their Forests: A Historical Geography* (Cambridge Cambridgeshire, and New York: Cambridge University Press, 1989).

29. According to Lucia the first successful portable chain saw was developed by Charlie Wolf at the Peninsula Iron Works of Portland, Oregon, in 1920. However, the gasoline-powered portable chain saw was developed by Germany's Andreas Stihl in the late 1920's, with critical improvements to the chain design by Oregon logger Joe Cox in the 1940s; E. Lucia, *The Big Woods: Logging and Lumbering, from Bull Teams to Helicopters, in the Pacific Northwest* (Garden City, NY: Doubleday, 1975).

30. Warf, "Regional Transformation."

31. Williams, *Americans and Their Forests,* 319.

32. There is still a significant difference between logging technologies used in the Pacific Northwest and those used in other parts of the world. Specifically, advanced mechanized logging equipment (e.g., feller-buncher) is much more prevalent in other regions P. MacDonald and M. Clow, " 'Just One Damn Machine after Another'? Technological Innovation and the Industrialization of Tree Harvesting Systems," *Technology in Society* 21 (1999): 323–44. Such machines are not designed to handle the size of timber in Douglas fir forests of the Northwest or to negotiate the steep slopes characteristic of the Coast and Cascade ranges.

33. On these more extensive geographies, see also Mead, *Competition and Oligopsony;* Williams, *Americans and Their Forests.*

34. Susan Mann's analysis of biologically determined production times and labor time draws on Marx's distinction between production time and labor time more generally, and capital's imperative to close the gap; see Mann, *Agrarian Capitalism in Theory and Practice.* Again, however, on the "problems" posed by social time colliding with natural time, see B. Adam,

"The Temporal Gaze: The Challenge for Social Theory in the Context of GM Food," *British Journal of Sociology* 51, no. 1 (2000): 125–42""; Adam, *Timescapes of Modernity*.

35. My description and analysis of the labor process in logging was assisted by site visits and interviews with numerous loggers and industry officials, all conducted between June 1996 and December 1997.

36. J. C. Scott, *Seeing Like a State: How Certain Schemes to Improve the Human Condition Have Failed* (New Haven, CT: Yale University Press, 1998).

37. A "'crummy'" is a logger's term for the vehicle, usually a van or truck, used to transport workers to and from the job site. This quote is from an interview conducted on December 17, 1997, with a longtime logger in Cottage Grove, Oregon.

38. These comparisons are based on data taken from "1992 Census of Manufactures, Industry Series," (Washington, DC: U.S. Department of Commerce, Bureau of the Census; U.S. Government Printing Office); Reports MC92-I-24A "Logging Camps, Sawmills, and Planing Mills" and MC92-I-24B "Millwork, Plywood, and Structural Wood Members, Not Elsewhere Classified."

39. See H. M. Somers and A. Somers, *Workmen's Compensation: Prevention, Insurance, and Rehabilitation of Occupational Disability* (New York: Wiley, 1954). On accidents and logging, see also Jensen, *Lumber and Labor;* Prouty, *More Deadly Than War!* For recent data and discussion, see E. F. Sygnatur, "Logging Is Perilous Work," *Compensation and Working Conditions* (Winter 1998): 3–9.

40. Here I differ with the account offered by R. Rajala, *Clearcutting the Pacific Rain Forest: Production, Science, and Regulation* (Vancouver, Canada: UBC Press, 1998).

41. A. Egan and C. Alerich, " 'Danger Trees' in Central Appalachian Forests in the United States: An Assessment of Their Frequency of Occurrence," *Journal of Safety Research* 29, no. 2 (1998): 77–85; and C. Slappendel et al., "Factors Affecting Work-Related Injury among Forestry Workers: A Review," *Journal of Safety Research* 24, no. 1 (1993): 19–32.

42. Sygnatur, "Logging Is Perilous Work."

43. Injured workers have sued logging employers, alleging unsafe working conditions in violation of existing safety standards. See Egan and Alerich, " 'Danger Trees' in Central Appalachian Forests in the United States."

44. Scott, "Industrial Organization and Location."

45. A. J. Scott, *New Industrial Spaces: Flexible Production Organization and Regional Development in North America and Western Europe* (London: Pion, 1988); Storper, *The Regional World*.

46. Interview conducted on June 30, 1997, with a gyppo logger in Springfield, Oregon. This was echoed by an academic researcher whose expertise is Oregon logging and by several other loggers I spoke with.

47. Interviews conducted on July 10 and 11, 1997, with mill procurement managers in central western Oregon, and echoed by gyppos.

48. Interview with a gyppo logger, December 23, 1997, in Springfield, Oregon.

49. My observations are based on research done in the late 1990s prior to Weyerhaeuser's acquisition of Willamette Industries.

50. Interview with a gyppo logger, December 17, 1997 in Cottage Grove Oregon. This perspective was echoed in several interviews.

51. See M. Widenor, "Diverging Patterns: Labor in the Pacific Northwest Wood Products Industry," *Industrial Relations* 34, no. 3 (1995): 441–63; M. Widenor, "Pattern Bargaining in the Pacific Northwest Lumber and Sawmill Industry, 1980–1990," in *Labor in a Global Economy: Perspectives from the U.S. and Canada*, ed. S. Hecker and M. Hallock (Eugene: Labor Education Research Center, University of Oregon, 1991), 252–62.

52. I say that these profits might be even higher because, as noted earlier, wage and benefit rates for gyppo loggers are typically lower than they are for company loggers.

53. Interview with a Weyerhaeuser logger, December 18, 1997, in Springfield, Oregon.

54. Interview with a union representative, July 2, 1997, in Springfield, Oregon.

55. Weyerhaeuser owns approximately two million acres of land in the Douglas fir region of Oregon and Washington alone; Weyerhaeuser Company, *Form 10-K* (Washington, DC: Securities and Exchange Commission, 1999).

56. R. L. Heilman, *Overstory— Zero: Real Life in Timber Country* (Seattle, WA: Sasquatch Books, 1995), 21–22.

57. Personal communication, State of Oregon Employment Department, April 2, 1994.
58. Testimony of Rick Herson, president of Hoedads cooperative, a reforestation company, before the House Subcommittee on Forests; see U.S. Congress House Committee on Agriculture, S. o. F., *Reforestation Efforts in Western Oregon: Hearing before the Subcommittee on Forests of the Committee on Agriculture, House of Representatives,* 95th Cong., 1st sess., July 8, 1977, Roseburg, Oregon (Washington, DC: U.S. Government Printing Office), iv, 257.
59. Ibid.
60. G. Mackie, *The Rise and Fall of the Forest Workers' Cooperatives of the Pacific Northwest* (master's thesis, Eugene: University of Oregon, Department of Political Science, 1990), 154.
61. Rajala, *Clearcutting the Pacific Rain Forest;* Williams, *Americans and Their Forests.*
62. This figure is corroborated by a statement given by Richard Koven of the Northwest Forest Workers Association before the House Committee on Agriculture, Subcommittee on Forests, delivered May 15, 1980, in Eugene, Oregon. Koven estimated that there were seven hundred Oregon employees working for the twelve cooperatives that formed the Northwest Forest Workers Association out of a state total of three thousand reforestation workers; U.S. Congress House Committee on Agriculture, S. o. F. (1981), *Use of Illegal Aliens in Government Reforestation Contracts: Hearing before the Subcommittee on Forests of the Committee on Agriculture, House of Representatives,* 96th Cong., 2nd sess., May 15, 1980, Eugene, Oregon (Washington, DC: U.S. Government Printing Office). See also Mackie, *The Rise and Fall of the Forest Workers' Cooperatives of the Pacific Northwest.*
63. U.S. Congress House Committee on Agriculture, *Reforestation Efforts in Western Oregon.*
64. Mackie, *The Rise and Fall of the Forest Workers' Cooperatives of the Pacific Northwest.*
65. Ibid.
66. U.S. Congress House Committee on Agriculture, *Reforestation Efforts in Western Oregon.*
67. Mackie, *The Rise and Fall of the Forest Workers' Cooperatives of the Pacific Northwest.*
68. *United States v. Robert Felix Gonzalez,* 388 F. Supp. 892 (Oregon 1974).
69. Mackie, *The Rise and Fall of the Forest Workers' Cooperatives of the Pacific Northwest.*
70. Testimony by Martin Desmond, executive director of the Northwest Reforestation Contractors Association; U.S. Congress House Committee on Government Operations, I., Justice, Transportation, and Agriculture Subcommittee (1994), *Allegations of Contract Abuse in the U.S. Forest Service Reforestation Program: Hearing before the Information, Justice, Transportation, and Agriculture Subcommittee of the Committee on Government Operations, House of Representatives,* 103rd Cong., 1st sess., June 30, 1993 (Washington, DC: U.S. Government Printing Office). (For sale by the U.S. Government Printing Office Supt. of Docs. Congressional Sales Office.)
71. INS data and the possible connection between this order and the expanded use of undocumented workers was provided by Carl Houseman, deputy district director of the INS, in testimony before a congressional hearing on undocumented workers in Oregon reforestation; U.S. Congress House Committee on Agriculture, *Use of Illegal Aliens in Government Reforestation Contracts.*
72. E. Galarza, *Merchants of Labor: The Mexican Bracero Story: An Account of the Managed Migration of Mexican Farm Workers in California 1942–1960* (San Jose: Rosicrucian Press, 1964).
73. M. J. Wells, *Strawberry Fields: Politics, Class, and Work in California Agriculture* (Ithaca, NY: Cornell University Press, 1996).
74. Galarza, *Merchants of Labor.*
75. Oregon Bureau of Labor, *And Migrant Problems Demand Attention* (the final report of the 1958–59 migrant farm labor studies in Oregon including material from the preliminary report of the Bureau of Labor titled "We Talked to the Migrants," Salem, 1959); Oregon Bureau of Labor, *"Vamonos Pal Norte" (Let's Go North): A Social Profile of the Spanish Speaking Migratory Farm Laborer* (Salem, 1958); Oregon Interagency Committee on Migratory Labor, *1965 Report of the Interagency Committee on Migratory Labor* (Salem: Oregon Department of Agriculture, 1966), 30 leaves, 1 folder in pocket.
76. Oregon State University, *Seasonal Agricultural Labor in Oregon* (task force report, Salem, 1968).
77. Oregon Bureau of Labor, *And Migrant Problems Demand Attention.*
78. R. White, *"It's Your Misfortune and None of My Own": A History of the American West* (Norman: University of Oklahoma Press, 1991).

204 • Notes

79. Oregon Bureau of Labor *"Vamonos Pal Norte"*, Oregon State University, *Seasonal Agricultural Labor in Oregon.*
80. For example, the Oregon Farm Labor Contractors Act of 1959 and the federal Farm Labor Contractor Registration Act of 1964; see Oregon Interagency Committee on Migratory Labor, *1965 Report of the Interagency Committee on Migratory Labor.*
81. Interview conducted in Woodburn with a Pineros y Campesinos Unidos del Noroeste representative on September 4, 1997.
82. Evidence on abusive labor practices was collected from several interviews and published materials, including transcripts of a congressional hearing held in Eugene, Oregon, on May 15, 1980; U.S. Congress House Committee on Agriculture, *Use of Illegal Aliens in Government Reforestation Contracts;* transcripts of a congressional hearing held in Washington, D.C., on June 30, 1993: U.S. Congress House Committee on Government Operations, *Allegations of Contract Abuse in the U.S. Forest Service Reforestation Program;* and "U.S. Hires Undocumented Workers," *Sacramento Bee,* June 8, 1993, A14; "Peonage in the Pines," *The Progressive,* November 1987, 24–27.
83. U.S. Congress House Committee on Government Operations, *Allegations of Contract Abuse in the U.S. Forest Service Reforestation Program,* 3.
84. Ibid., 8–9.
85. Former Lane County commissioner Jerry Rust, in testimony before the U.S. Congress House Committee on Agriculture, Subcommittee on Forests; ibid., 3.
86. Until IRCA reforms of 1986, the INS did not care about employers who hired undocumented workers. INS only cared about deporting these workers when they were discovered. Although INS did ask employers to pay their undocumented workers for completed work, either on the spot or by forwarding money to the workers in their home countries, INS took no responsibility for ensuring that back wages were paid. As a result, numerous contractors used the INS to cheat their employees.
87. Mann, *Agrarian Capitalism in Theory and Practice.*

Chapter 3

1. D. N. Bengston and H. M. Gregerson, "Technical Change in the Forest-Based Sector," in *Emerging Issues in Forest Policy,* ed. P. N. Nemetz (Vancouver: UBC Press, 1992), 187–211, especially pp. 187–88.
2. W. Boyd, S. Prudham, and R. Schurman, "Industrial Dynamics and the Problem of Nature," *Society and Natural Resources* 14, no. 7 (2001): 555–70; G. Bridge, "Resource Triumphalism: Postindustrial Narratives of Primary Commodity Production," *Environment and Planning A* 33, no. 12 (2001): 2149–73.
3. G. Bridge, "The Social Regulation of Resource Access and Environmental Impact: Nature and Contradiction in the US Copper Industry," *Geoforum* 31, no. 2 (2000): 237–56.
4. For related discussion, see, for example, B. Barham and O. Coomes, "Reinterpreting the Amazon Rubber Boom: Investment, the State, and Dutch Disease," *Latin Amercian Research Review* 29, no. 2 (1994): 73–109; Boyd, Prudham, and Schurman, "Industrial Dynamics and the Problem of Nature"; Bridge, "The Social Regulation of Resource Access and Environmental Impact"; S. G. Bunker, "Staples, Links, and Poles in the Construction of Regional Development Theory," *Sociological Forum* 4, no. 4 (1989): 589–610.
5. The term "industrial ecology" is of course not mine, and my use of it here is somewhat at odds with what it has come to signify in environmental analysis and policy circles; that is, comprehensive analysis and reduction of the environmental burden associated with particular industrial products and commodities; see, for example, R. U. Ayres and L. Ayres, *Industrial Ecology: Towards Closing the Materials Cycle* (Cheltenham, UK and Brookfield, VT: Elgar, 1996); R. H. Socolow, *Industrial Ecology and Global Change* (Cambridge and New York: Cambridge University Press, 1994). My intent is not to displace but rather to incorporate this focus, adding to it a broader perspective on the relationships between industrial organizations and processes on one hand and raw material inputs on the other.
6. Goodman, Sorj, and Wilkinson developed the idea of appropriation as a way to describe how discrete elements of farm processes are carved off and made the basis of capital accumulation and circulation, resulting in industrially produced farm inputs that replace specific

antecedents once produced on the farm (see chapter 1 for additional discussion). The classic example is the production of chemical fertilizers displacing manure spreading in crop cultivation but otherwise leaving the farm-based crop production processes unchanged. The cumulative result is a kind of "hollowing out" of farm production, creating a farm that takes in specific industrially generated inputs and spits out specific outputs, a shell of what was once a complex assemblage of interlinked and more self-sufficient farm-based production processes. They also develop the idea of substitution, a stronger version of appropriation involving the outright displacement of farm-based commodity chains with new commodities using industrial processes and relying on wholly new bases of crop and animal production. The classic example here is the use of plant-based oils and industrial processing to displace butter with margarine. See D. Goodman, B. Sorj, and J. Wilkinson, *From Farming to Biotechnology: A Theory of Agro-Industrial Development* (Oxford and New York: Blackwell, 1987). Obviously these are very agriculture-centered ideas, not least because they focus on the displacement of labor processes and raw material inputs, all specific to traditionally farm-based production. However, insofar as particular ecological processes are displaced or augmented by these dynamics, there is the potential to draw analogies with other kinds of nature-based industries. See Boyd, Prudham, and Schurman "Industrial Dynamics and the Problem of Nature."

7. See also W. B. Greeley, "The Relation of Geography to Timber Supply," *Economic Geography* 1, no. 1 (1925): 1–14; W. G. Robbins, *Lumberjacks and Legislators: Political Economy of the U.S. Lumber Industry, 1890-1941* (College Station: Texas A&M University Press, 1982); U.S. Bureau of Corporations, *The Lumber Industry* (Washington, DC: Government Printing Office, 1913); M. Williams, *Americans and Their Forests: A Historical Geography* (Cambridge Cambridgeshire, and New York: Cambridge University Press, 1989).

8. Although there has been an ongoing feud in the Pacific Northwest of the United States and Canada regarding how much old-growth is left and how old-growth should be defined (see chapter 1 for definition), there are also revisionist narratives that question the degree to which what is generally accepted as old growth was ever characteristic of the region's forests; see, for example, A. Chase, *In a Dark Wood: The Fight over Forests and the Rising Tyranny of Ecology* (Boston: Houghton Mifflin, 1995). This is despite scientific and officially accepted estimates that the area of old-growth forests in the Douglas fir region are less than 5 percent of their historic levels prior to the era of industrial extraction; see J. F. Franklin et al., *Ecological Characteristics of Old-Growth Douglas-Fir Forests* (U.S. Department of Agriculture Forest Service, 1981). It also flies in the face of common sense and of reams of photographic and folk history, all clearly indicating that a massive transformation of the forest landscape of the region has taken place—nature remade by capital.

9. U.S. Bureau of Corporations, *The Lumber Industry,* vol. 1, 52. "Cruisers" are people who estimate standing volume (usually per acre) and the quality of timber, often in advance of a sale. This was traditionally done by walking sections of forest and taking sample measurements; contemporary techniques involve greater use of aerial photography and remote sensing. The resulting estimate is called a timber "cruise."

10. The Pacific Northwest is defined in the report as comprising California, Oregon, Washington, Montana, Idaho, and California—essentially Pomeroy's Pacific Slope minus Nevada.

11. Data from U.S. Bureau of Corporations, *The Lumber Industry.*

12. M. Hibbard and J. Elias, "The Failure of Sustained-Yield Forestry and the Decline of the Flannel-Shirt Frontier," in *Forgotten Places: Uneven Development in Rural America,* ed. T. A. Lyson and W. W. Falk (Lawrence: University Press of Kansas, 1993), 195–215; C. R. Howd and U.S. Bureau of Labor Statistics, *Industrial Relations in the West Coast Lumber Industry: December, 1923* (Washington, DC: U.S. Goverment Printing Office, 1924); and U.S. Bureau of Corporations, *The Lumber Industry.*

13. I do recognize that difference can and should be measured or evaluated in more ways than simply how much volume of timber each acre of forestland holds, and I understand the political and ecological significance of the fact that stands have been abstracted by capital and scientific forestry in exactly these terms for much of the past 100 to 150 years, including during the sustained-yield era; see, for example, D. Demeritt, "Scientific Forest Conservation and the Statistical Picturing of Nature's Limits in the Progressive-Era United States," *Environment and Planning D-Society and Space* 19, no. 4 (2001): 431–59, and also chapter 6. For now, my concern is to follow this notion of difference through by reflecting on the

dynamic relationship between regional differences in forest type and the industrial geography of commodity manufacture.

14. I stress that size and age of timber are not the only or even necessarily the best indicators of old growth; see Franklin et al., *Ecological Characteristics of Old-Growth Douglas-Fir Forests*. But from an industrial commodity production standpoint, these along with timber density are the prime indicators of the value of the resource. Moreover, age of the stand is a reasonable proxy for old growth, because most characteristics of old-growth Douglas fir forests begin to develop after 175 to 250 years.

15. All of these data are from W. B. Smith et al., *Forest Resources of the United States, 1997* (St. Paul, MN: U.S. Department of Agriculture, Forest Service, 2001).

16. Ibid., 33.

17. Ibid. Of this, about two-thirds is in the Douglas fir region. The rest of the Douglas fir is located in the Southwest and Rocky Mountain West.

18. For data, see chapter 1 and also D. R. Gedney, "The Private Timber Resource," in *Assessment of Oregon's Forests 1988*, ed. G. J. Lettman (Salem: Oregon State Department of Forestry, 1988), 53–57; and Smith et al., *Forest Resources of the United States, 1997*.

19. The culmination age (or more formally the culmination of the mean annual increment) is the most commonly cited prescribed cutting age in forestry, corresponding to the point at which a tree's annual growth, averaged over its life span, is maximized. This is also the age at which the annual increment, or annual growth, is maximized. Subsequent to this age, all other things being equal, the annual rate of growth of the tree is positive but declining. Management of forests to this age theoretically generates the most volume over time, not necessarily the most revenue (see chapter 6).

20. The dominant view in the Northwest is that Douglas fir should be grown over longer rotations and that it is best suited for solid wood-products manufacture. There are financial risks associated with these long delays, but most analysts believe that Douglas fir stumpage rates and the price of the timber will appreciate in real terms during the coming decades. Information on the solid wood-products connection with Douglas fir was drawn from various interviews, including one conducted on July 7, 1997, in Albany, Oregon, with the timberlands manager of a large firm; another conducted on August 13, 1997, in Portland with an acquisitions manager for a major timberland investor; and another conducted on July 9, 1997, in Cottage Grove, Oregon, with a silviculture expert employed by a major forest-products company. On Douglas fir stumpage appreciation, see C. J. Cleveland and D. I. Stern, "Productive and Exchange Scarcity: An Empirical Analysis of the U.S. Forest Products Industry," *Canadian Journal of Forest Research* 23 (1993): 1537–49.

21. B. Zobel and J. B. Jett, *Genetics of Wood Production* (Berlin and New York: Springer-Verlag, 1995).

22. This may seem counterintuitive in the case of lumber, but it is true. Consider that the largest straight-sided area that can be taken from a circular cross-section is a square. The area of this square rises directly in proportion to the area of the circle. But so does the residual. As bigger logs are used, larger and more valuable pieces can be salvaged from the squared off residuals, giving a higher total yield of merchantable lumber. See D. M. Smith et al., *The Practice of Silviculture: Applied Forest Ecology* (New York: Wiley, 1997); and J. Zaremba, *Economics of the American Lumber Industry* (New York: R. Speller, 1963). This dynamic has implications in the context of conversion from old growth to plantation, and thus smaller diameter timber. For example, in a 1988 assessment of Bureau of Land Management forest management practices in western Oregon, allowable annual cuts were projected at constant cubic volume, yet because of the gradual conversion of BLM forests from old-growth large-diameter trees to younger, smaller trees, "As the cubic volume is held constant and average stand diameters are reduced, board foot volume levels decline over time from approximately 1,181 million board feet in 1990 to 1,136 million board feet in 2080"; D. Preston and B. Alverts, "Oregon's BLM Timber Resources," in *Assessment of Oregon's Forests 1988*, ed. G. J. Lettman (Salem: Oregon State Department of Forestry, 1998), 33–41.

23. F. C. Zinkhan, *Timberland Investments : A Portfolio Perspective* (Portland, OR: Timber Press, 1992).

24. Some of these observations are based on an interview conducted with an official with the Oregon State Department of Forestry on March 23, 1995. See also B. Goldfarb and J. B. Zaerr, *Douglas-Fir. Biotechnology in Agriculture and Forestry*, Y. P. S. Bajaj (Berlin: Springer-

Verlag, 1989), 5: 526–48; R. H. Kunesh and J. W. Johnson, "Effect of Single Knots on Tensile Strength of 2- by 8-Inch Douglas-Fir Dimension Lumber," *Forest Products Journal* 22, no. (1) (1972): 32–36. Insight into these issues was aided by several other interviews, including one conducted on March 24, 1995, with a forester employed by the Oregon State Department of Forestry in Salem, Oregon, and one conducted with a nursery manager employed by a major wood-products firm on August 5, 1997, in Cottage Grove, Oregon.

25.   Douglas fir specific gravity is typically in the range of 0.45; A. Van Vliet, "Strength of Second-Growth Douglas-Fir in Tension Parallel to the Grain," *Forest Products Journal* 9, no. 4 (1959): 143–48; this expression measures the density of the wood as a ratio of the density of water. It can be calculated by dividing the mass of a sample by the mass of an equal volume of water.

26.   In 1997 softwood timber harvests in the three leading states of Georgia, Alabama, and Oregon measured in million cubic feet totaled 1,062, 891, and 824, respectively; see Smith et al., *Forest Resources of the United States, 1997,* Table 36.

27.   Oregon Department of Forestry, *Timber Harvest Report* (Salem: Author, 2002); F. R. Ward, *Oregon's Forest Products Industry, 1992* (Portland, OR: U.S. Department of Agriculture Forest Service Pacific Northwest Research Station, 1995).

28.   Data from F. R. Ward, *Oregon's Forest Products Industry, 1994; Resource bulletin PNW; RB-216* (Portland, OR: U.S. Department of Agriculture Forest Service Pacific Northwest Research Station, 1997), 70.

29.   Typically one of each is integrated on each industrial site. There is some ambiguity in the exact number of mills depending on how they are counted. Some facilities have two pulp mills on site, but they are part of the same facility. These data are as of 1995, taken from Miller Freeman, *1995 Lockwood-Post 'S Directory of the Pulp, Paper and Allied Trades* (San Francisco, CA: Author, 1994).

30.   Ward, *Oregon's Forest Products Industry, 1994.* For discussion of this phenomenon, see D. R. Gedney and S. E. Corder, " Residue Use Is Basis for Pulp Industry Expansion," *The Timberman* 57 (1956); R. M. Samuels, "Expanding the Use of Wood Residues for Pulp Production," *Forest Products Journal* 7, no. 8 (1957): 253–55.

31.   Western Wood Products Association, *Statistical Yearbook of the Western Lumber Industry* (Portland, OR: Author, 1998).

32.   T. D. Perry, *Modern Plywood* (New York: Pitman, 1948).

33.   The Portland Manufacturing Company, asked by the Lewis and Clark Exposition directors to put together something "unusual," sponsored an exhibition of Douglas fir plywood. They developed special, animal-based glues for the occasion, which apparently smelled so bad that operators of the plywood-making machinery had to seek frequent relief; R. F. Baldwin, *Plywood Manufacturing Practices* (San Francisco, CA: Miller Freeman, 1981). See also R. M. Cour, *The Plywood Age: A History of the Fir Plywood Industry's First Fifty Years* Portland, OR: Binfords and Mort for the Douglas Fir Plywood Association, 1955).

34.   So called because they are literally peeled or stripped in the veneer production process.

35.   Baldwin, *Plywood Manufacturing Practices.*

36.   Output of plywood is typically quoted in terms of the standard 3/8-inch thickness; such is the case with the figures used here unless otherwise specified.

37.   Source: Data from U.S. Bureau of the Census (1957) and Ruderman (1974–85).

38.   Baldwin, *Plywood Manufacturing Practices;* H. Montrey and J. M. Utterback, "Current Status and Future of Structural Panels in the Wood Products Industry," *Technological Forecasting and Social Change* 38 (1990): 15–35; N. Rosenberg, *Exploring the Black Box: Technology, Economics, and History* (Cambridge and New York: Cambridge University Press, 1994).

39.   Handling costs in veneer production include those that originate in mounting the log into position for veneer stripping, and these are relatively constant per log. As a result, the ratio of stripped wood to handling costs goes up with larger logs, all other things being equal. This is one of the main reasons why veneer mills want larger diameter logs.

40.   The size of the core left behind after the veneer is stripped is determined by the size of the blocks between which the log is spun against the blades.

41.   Baldwin, *Plywood Manufacturing Practices,* 27.

42.   Much of the information about Georgia-Pacific was gathered from a series of articles by Anthony Bianco profiling the company. These articles were published in the *Willamette*

Week on March 12, 19, 26, and April 2, 1979, and were obtained in the Knight Collection of the Oregon Archives at the University of Oregon in Eugene.

43. W. J. Mead, *Competition and Oligopsony in the Douglas Fir Lumber Industry* (Berkeley: University of California Press, 1966).

44. J. Graham and K. St. Martin, "Resources and Restructuring in the International Solid Wood Products Industry," *Geoforum* 20, no. 24 (1989): 11–24; Rosenberg, *Exploring the Black Box.*

45. Cleveland and Stern, "Productive and Exchange Scarcity."

46. D. Warren, *Harvest, Employment, Exports, and Prices in Pacific Northwest Forests, 1965–2000* (Portland, OR: U.S. Department of Agriculture, Forest Service, Pacific Northwest Research Station, 2002).

47. B. Adam, "The Temporal Gaze: The Challenge for Social Theory in the Context of GM Food," *British Journal of Sociology* 51, no. 1 (2000): 125–42, 137. See also B. Adam, *Timescapes of Modernity: The Environment and Invisible Hazards* (London: Routledge, 1998).

48. Franklin et al., *Ecological Characteristics of Old-Growth Douglas-Fir Forests.*

49. Baldwin, *Plywood Manufacturing Practices,* 28.

50. Ibid.

51. Georgia-Pacific Corporation, *Form 10-K405* (Washington, DC: Securities and Exchange Commission, 1999).

52. U.S. Department of Commerce, Bureau of the Census. *1992 Census of Manufactures. Industry Series. Millwork, Plywood, and Structural Wood Members, Not Elsewhere Classified* (Washington, DC: Author, 1995).

53. Graham and St. Martin, "Resources and Restructuring in the International Solid Wood Products Industry," 16.

54. Rosenberg, *Exploring the Black Box,* 244.

55. J. A. Youngquist, "Wood-Based Composites and Panel Products," in *Wood Handbook—Wood as an Engineering Material* (Madison, WI: U.S. Department of Agriculture Forest Service, Forest Products Laboratory, 1999), 463, especially pp. 10–11.

56. The real decline of Douglas fir stumpage prices during the 1980s interrupted a long-term trend. The deep recession in 1980 to 1982, and in particular the collapse of the U.S. housing market resulted in a dramatic decline in timber harvest, product output, and employment in the Northwest wood-products sector, and this led to a decline in stumpage prices. During the latter part of the decade, the Reagan administration forced the Forest Service and the Bureau of Land Management to increase timber harvest levels as an element of its supply side policies to boost the economy; P. W. Hirt, *A Conspiracy of Optimism: Management of the National Forests since World War Two* (Lincoln: University of Nebraska Press, 1994). The resulting escalation of federal harvest levels depressed stumpage prices and interrupted the long-term trend. This is, however, an aberration. Moreover, with the more recent restrictions related to the northern spotted owl, the escalation of stumpage prices appears to have been resumed.

57. Others have of course made this connection, including Bengston and Gregerson, "Technical Change in the Forest-Based Sector"; Rosenberg, *Exploring the Black Box;* and R. L. Youngs, "Reconstituted Wood Materials—New Opportunities and New Responsibilities" (Proceedings of the 1982 Waferboard Symposium, Special Publication Sp508e, Forintek Canada Corp, 1984). In specific reference to varieties of glued laminated veneer lumber, Moody, Hernandez, and Liu noted, "Structural composite lumber was developed in response to the increasing demand for high quality lumber at a time when it was becoming difficult to obtain this type of lumber from the forest resource." R. C. Moody, R. Hernandez, and J. Y. Liu, "Glued Structural Members, " in *Wood Handbook—Wood as an Engineering Material* (Madison, WI: U.S. Department of Agriculture Forest Service, Forest Products Laboratory, 1999), 463, especially p. 11–1.

58. Quoted from "The Timber Resource—Catalyst for Change" (Proceedings of Engineered Wood Products, Processing and Design, Atlanta, Georgia, March 26–27, 1991).

59. J. F. Weigand, *Composition, Volume, and Prices for Major Softwood Lumber Types in Western Oregon and Washington 1971–2020* (Portland, OR: U.S. Department of Agriculture, Forest Service, Pacific Northwest Research Station, 1998).

60. The importance of stumpage price increases is apparent in the timing of market penetration for some of the new products. Although the important technological innovations underlying

manufacture of OSB and waferboard were developed during the 1950s, it was not until the 1970s that serious investment in production capacity resulted in competition with plywood for the softwood structural panels market; Rosenberg, *Exploring the Black Box.*

61. There are different terms used as umbrellas for this range of commodities and different conceptions of which ones are similar and why. Youngquist, for example, uses composites and reconstituted wood products interchangeably and defines composites as those that are produced using adhesives, including in this plywood; see Youngquist, "Wood-Based Composites and Panel Products." Youngs uses the term "reconstituted wood products" expansively but does not include plywood; Youngs, "Reconstituted Wood Materials." "Engineered wood products" is also a term that is used, which may or may not include plywood. I prefer the term "reconstituted" because of its descriptive appeal; however, I restrict this discussion to structural wood products only. I therefore exclude products such as particleboard and medium density fiberboard. One could argue that these products reflect the same tendencies; that is, product innovation motivated by raw material constraints. But they are not typically used as structural wood and are therefore not really competitors with Douglas fir lumber and plywood.

62. Moody, Hernandez, and Liu, "Glued Structural Members."

63. In fact, testing has revealed that glulam can actually be superior in strength to comparable lumber. See R. H. Falk and F. Colling, "Laminating Effects in Glued-Laminated Timber Beams," *Journal of Structural Engineering* 121, no. 12 (1995): 1857–63; and E. S. Padjen, "Engineered Lumber's Strengths," *Architecture* 86, no. 2 (1997): 104–108.

64. Padjen, "Engineered Lumber's Strengths."

65. Montrey and Utterback, "Current Status and Future of Structural Panels in the Wood Products Industry."

66. K. E. Skog et al., "Wood Products Technology Trends: Changing the Face of Forestry," *Journal of Forestry* 93, no. 12 (1995): 30–33; and Spelter, *Capacity, Production, and Manufacture of Wood-Based Panels in the United States and Canada* (Madison, WI: U.S. Department of Agriculture, Forest Service, Forest Products Laboratory, 1996).

67. Skog et al., "Wood Products Technology Trends."

68. This spectrum is similar to that offered by O. Suchsland and G. E. Woodson, *Fibreboard Manufacturing Practices in the United States* (Washington, DC: U.S. Department of Agriculture, 1986); however, although their classification scheme includes all reconstituted wood products, I am primarily interested in those intended for structural applications. For additional discussion along these lines, see Rosenberg, *Exploring the Black Box.*

69. M. Smith, *The U.S. Paper Industry and Sustainable Production: An Argument for Restructuring* (Cambridge, MA: MIT Press, 1997).

70. See Bengston and Gregerson, "Technical Change in the Forest-Based Sector."

71. Skog et al., "Wood Products Technology Trends."

72. *Spelter Capacity, Production, and Manufacture of Wood-Based Panels in the United States and Canada.*

73. Falk and Colling, "Laminating Effects in Glued-Laminated Timber Beams."

74. Mill counts from the "1999 Directory of the Wood Products Industry."

75. U.S. Census Bureau, *1997 Economic Census, Manufacturing, Industry Series: Engineered Wood Member (except Trusses) Manufacturing* (Washington, DC: U.S. Department of Commerce, 1999); U.S. Census Bureau, *1997 Economic Census, Manufacturing, Industry Series: Reconstituted Wood Product Manufacturing* (Washington, DC: U.S. Department of Commerce, 1999).

76. Life cycle assessment is an important analytical approach used in industrial ecology and in the study of the environmental impacts of commodities and lifestyle choices more generally. It involves the assessment of cradle to grave environmental burdens of a commodity or activity, including all aspects of production, use, and disposal. To date, life cycle assessment has been applied primarily to the pulp and paper side of the forest industry and less on solid wood products; see S. L. LeVan, "Life Cycle Assessment: Measuring Environmental Impact" (Forest Products Society 49th Annual Meeting, Portland, Oregon, 1995). On industrial ecology more generally, see Ayres and Ayres, *Industrial Ecology;* Socolow, *Industrial Ecology and Global Change.*

77. S. H. Imam et al., "Environmentally Friendly Wood Adhesive from a Renewable Plant Polymer: Characteristics and Optimization," *Polymer Degradation and Stability* 73, no. 3 (2001):

529–33. Others classify the most commonly used adhesives as urea-formaldehyde, phenol-formaldehyde, or isocyanates. For additional discussion of adhesives in use, see Moody, Hernandez, and Liu, "Glued Structural Members"; Youngquist, "Wood-Based Composites and Panel Products."

78. Baldwin, *Plywood Manufacturing Practices*; Perry, *Modern Plywood*.

79. Imam et al., "Environmentally Friendly Wood Adhesive from a Renewable Plant Polymer."

80. On occupational exposures, and off-gassing from finished products, see variously T. Kauppinen, "Occupational Exposure to Chemical Agents in the Plywood Industry," *Annals of Occupational Hygiene* 30, no. 1 (1986): 19–29; T. Kauppinen et al., "Respiratory Cancer and Chemical Exposures in the Wood Industry: A Nested Case-Control Study," *British Journal of Industrial Medicine* 43, no. 2 (1986): 84–90; S. Simon, "Update on Formaldehyde Emission as It Relates to the Production of Panel Products" (proceedings of the 1982 Canadian Waferboard Symposium, Special Publication SP508E, Forintek Canada Corporation, 1984); C. B. Vick, "Adhesive Bonding of Wood Materials," in *Wood Handbook—Wood as an Engineering Material* (Madison, WI: U.S. Department of Agriculture Forest Service, Forest Products Laboratory, 1999), 463; Wood-Products-Sub-Council, "Principal Pollution Problems Facing the Solid Wood Product Industry," *Forest Products Journal* 21, no. 9 (1971): 33–36; T. W. Zinn, D. Cline, and W. F. Lehmann, "Long-Term Study of Formaldehyde Emission Decay from Particleboard," *Forest Products Journal* 40, no. 6 (1990): 15–18.

81. U.S. Consumer Product Safety Commission, *An Update on Formaldehyde: 1997 Revision* (Washington, DC: Author, 2003).

82. Kauppinen, "Occupational Exposure to Chemical Agents in the Plywood Industry," reported that exposures in Finnish plywood plants from 1965 to 1975 were typically in the one to two ppm range. Freeman and Grendon mention casually that in the early 1970s, the rule of thumb in U.S. plywood plants was to keep ambient formaldehyde levels below ten ppm, a scandalously high number by contemporary standards, more than one hundred times higher than levels now considered the threshold of safety; H. G. Freeman and W. C. Grendon, "Formaldehyde Detection and Control in the Wood Industry," *Forest Products Journal* 21, no. 9 (1971): 54–57.

83. See Kauppinen, "Occupational Exposure to Chemical Agents in the Plywood Industry"; Kauppinen et al., "Respiratory Cancer and Chemical Exposures in the Wood Industry"; T. Malaka and A. M. Kodama, "Respiratory Health of Plywood Workers Exposed to Formaldehyde," *Archives of Environmental Health* 45 (1993): 288–94.

84. Freeman and Grendon, "Formaldehyde Detection and Control in the Wood Industry."

85. See M. Makinen, P. Kalliokoski, and J. Kangas, "Assessment of Total Exposure to Phenol-Formaldehyde Resin Glue in Plywood Manufacturing," *International Archives of Occupational and Environmental Health* 72, no. 5 (1999): 309–14.

86. Zinn, Cline, and Lehmann, "Long-Term Study of Formaldehyde Emission Decay from Particleboard." See also R. Margosian, "Initial Formaldehyde Emission Levels for Particleboard Manufactured in the United States," *Forest Products Journal* 40, no. 6 (1990): 19–20.

87. For reference, see U.S. Environmental Protection Agency, *Technology Transfer Network, Clearinghouse for Inventories & Emission Factors* (Washington, DC: Author, 2003).

88. The full list of hazardous air pollutants identified under the EPA program and associated with the Plywood and Composite Wood Products sector includes phenol; formaldehyde; acetaldehyde; acrolein; benzene; bromomethane; chloroethane; chloroethene; cumene; 1,2-dichloroethane; methanol; methyl ethyl ketone; methyl isobutyl ketone; methylene chloride; propionaldehyde; styrene; toluene; 1,2,4-trichlorobenzene; o,m,p-xylene; methyl diphenyl diisocyanate; benzo-a-pyrene; o,m,p-cresol; naphthalene; chloromethane; ethyl benzene; styrene; xylenes; biphenyl; bis-_2-ethylhexyl phthalate; cumene; di-N-butyl phthalate; hydroquinone; acrolein; 1,1,1-trichloroethane; 4-methyl-2-pentanone; carbon disulfide; carbon tetrachloride; chloroform; n-hexane; methylene chloride. For summary and analysis of the EPA's regulatory campaign regarding Plywood and Composite Wood Products and the pollutants in question, see M. G. D. Baumann, "Air Quality and Composite Wood Products," *Women in Natural Resources* 20, no. 4 (1999): 4–6; and K. Bigbee, "Emissions Control: Background and Status of EPA's Maximum Achievable Control Technology Rule," *Engineered Wood Journal* 3, no. 1 (2000). Also, see U.S. Environmental Protection Agency, "Wood Products Industry," in *Compilation of Air Pollutant Emission Factors, AP-42, 5th Edition, Volume 1: Stationary Point and Area Sources* (Washington, DC: Author,

2002); RTI International, "Plywood and Composite Wood Products: Final Background Report," in *Emission Factor Documentation for AP-42* (Research Triangle Park, NC: U.S. Environmental Protection Agency Office of Air Quality Planning and Standards, Emission Factors and Inventory Group, 2003). See also U.S. Environmental Protection Agency, *Technology Transfer Network*.

89. Imam et al., "Environmentally Friendly Wood Adhesive from a Renewable Plant Polymer"; LeVan, "Life Cycle Assessment."

90. An example is provided by Ò hUallachàin and Matthews in their study of the Arizona copper industry; B. Ò hUallachàin and R. A. Matthews, "Restructuring of Primary Industries: Technology, Labor, and Corporate Strategy and Control in the Arizona Copper Industry," *Economic Geography* 72, no. 2 (1996): 196–215. Although they did not theorize raw material dependence as I do here, their analysis privileges geographic variation in raw material quality in combination with changing production technologies and the institutions governing resource access in an account of industry restructuring.

Chapter 4

1. J. Zaremba, *Economics of the American Lumber Industry* (New York: R. Speller, 1963), 84.

2. A. D. Chandler and T. Hikino, *Scale and Scope: The Dynamics of Industrial Capitalism* (Cambridge, MA: Belknap Press, 1990).

3. Debate over the historical and contemporary importance of big firms and small firms in the American economy is far from settled; see, for example, B. Harrison, *Lean and Mean: The Changing Landscape of Corporate Power in the Age of Flexibility* (New York: Basic Books, 1994); M. J. Piore and C. F. Sabel, *The Second Industrial Divide: Possibilities for Prosperity* (New York: Basic Books, 1984); and A. J. Scott, "Flexible Production Systems and Regional Development: The Rise of New Industrial Spaces in North America and Western Europe," *International Journal of Urban and Regional Research* 12 (1988): 171–85. However, the origins, prevalence, and power of corporate capital is an important topic when one considers that many of these companies grew into the globe spanning giants that are central to contemporary notions of economic globalization; R. J. Barnet and J. Cavanagh, *Global Dreams: Imperial Corporations and the New World Order* (New York: Simon & Schuster, 1994); P. Dicken, *Global Shift: Transforming the World Economy* (New York: Guilford Press, 1998); M. J. Webber and D. L. Rigby, *The Golden Age Illusion: Rethinking Postwar Capitalism* (New York: Guilford Press, 1996).

4. Chandler and Hikino, *Scale and Scope*, 22.

5. It is worth pointing out that one of the things Chandler ignored is the ruthlessness, and even questionable legality, of tactics used by some of the pioneering managerial firms in their rise to bigness. Price fixing and illegal land deals were hardly peripheral to the railroad business for instance; see, for example, A. Fishlow, *American Railroads and the Transformation of the Ante-Bellum Economy* (Cambridge, MA: Harvard University Press, 1965); D. Jensen and G. Draffan, *Railroads and Clearcuts: Legacy of Congress's 1864 Pacific Railroad Land Grant* (Spokane, WA: Inland Empire Public Lands Council, 1995); E. Richardson, U.S. Bureau of Land Management, Oregon State Office and Forest History Society, *BLM's Billion-Dollar Checkerboard: Managing the O and C Lands* (Santa Cruz, CA, and Washington, DC: Forest History Society, 1980) (For sale by the U.S. Government Printing Office Supt. of Docs.); R. White, *"It's Your Misfortune and None of My Own": A History of the American West* (Norman: University of Oklahoma Press, 1991). Rockefeller's use of pricing schemes to undermine competitors and his attempts to consolidate market power in the face of early antitrust regulation deserve some consideration as contributors to constructing the Standard Oil empire alongside the more benign notions of technology driven economies of scale; D. Yergin, *The Prize: The Epic Quest for Oil, Money, and Power* (New York: Simon & Schuster, 1991).

6. Chandler and Hikino, *Scale and Scope*, 8.

7. Ibid., 24.

8. See W. G. Robbins, *Lumberjacks and Legislators: Political Economy of the U.S. Lumber Industry, 1890–1941* (College Station: Texas A&M University Press, 1982).

9. U.S. Bureau of Corporations, *The Lumber Industry* (Washington, DC: Government Printing Office, 1913), 33.

10. Ibid., 35.

11. U.S. aggregate lumber production peaked at roughly forty-five billion board feet, after which time substitution by competing materials and commodities, including concrete and steel but also plywood, cut into lumber's importance to the construction trades; see A. J. Van Tassel, *Mechanization in the Lumber Industry: A Study of Technology in Relation to Resources and Employment Opportunity* (Philadelphia, PA: Work Projects Administration, National Research Project on Employment Opportunities and Recent Changes in Industrial Techniques, 1940).

12. U.S. Bureau of Corporations, *The Lumber Industry*. See also W. S. Prudham, "Timber and Town: Post-War Federal Forest Policy, Industrial Organization, and Rural Change in Oregon's Illinois Valley," *Antipode* 30, no. 2 (1998): 177–96. This information was also corroborated by interviews conducted by the author with several longtime residents of the Illinois Valley of southwestern Oregon, an area with a long history of lumbering. The accuracy of early mill counts must be viewed with some suspicion in part because of the very mobility of mills, particularly when they began to use diesel-powered headrigs. Because these mills were so small and mobile, the census takers very likely missed many of them.

13. V. H. Jensen, *Lumber and Labor* (New York: Farrar and Rinehart, 1945).

14. Ibid., 17.

15. This index is the proportion of aggregate sectoral output produced by the four largest firms.

16. Jensen, *Lumber and Labor,* 18.

17. Van Tassel, *Mechanization in the Lumber Industry,* 10.

18. Ibid., 23. The report noted that the proportion of aggregate horsepower supplied to sawmills by electricity rose from about 7 percent in 1914 to about 45 percent in 1929. However, aggregate industry horsepower was declining during this period, as production swung back to smaller mills.

19. Data from the Census of Manufacturing, U.S. Department of Commerce, Bureau of the Census, and from the West Coast Lumbermen's Association annual statistical report for 1958.

20. As the Work Projects Administration noted, energy delivery was the key to differentiating large and small mills. For example, in the Douglas fir region in 1936, although more than three-quarters of the mills in the largest size class (daily capacity in excess of one hundred thousand board feet) were powered by electricity or steam, only one-fifth of the small mills used either; Van Tassel, *Mechanization in the Lumber Industry.*

21. Data for 1947 taken from W. J. Mead, *Competition and Oligopsony in the Douglas Fir Lumber Industry* (Berkeley: University of California Press, 1966); whereas data for 1972 are from the 1972 Census of Manufactures, U.S. Department of Commerce, Bureau of the Census.

22. Ibid.

23. Data from 1994 are used here because they are the most recent available from a series of mill surveys conducted in Oregon by the Pacific Northwest Research Station of the U.S. Forest Service. See E. R. Manock, G. A. Choate, and D. R. Gedney, *Oregon Timber Industries: Wood Consumption and Mill Characteristics* (Salem: State of Oregon Department of Forestry, 1968); F. R. Ward, *Oregon's Forest Products Industry, 1994; Resource bulletin PNW; RB-216* (Portland, OR: U.S. Department of Agriculture Forest Service Pacific Northwest Research Station, 1997).

24. See Prudham, "Timber and Town."

25. U.S. Department of Commerce Bureau of the Census, *Concentration Ratios in Manufacturing, 1997 Census* (Washington, DC: Author, 2001). The data from 1997 are slightly different than previous years because of a shift in the classification of industries. This figure is for NAICS 321113 "Sawmills." For the more aggregate NAICS 3211 "Sawmills and Wood Preservation," the number is lower at 14.5 percent. See also P. V. Ellefson and R. N. Stone, *U.S. Wood-Based Industry: Industrial Organization and Performance* (New York: Praeger, 1984). The authors quote a four-firm concentration index of 38 percent in 1981, which would represent the peak; this number was likely inflated by recession-driven mill closures. However, the authors use total production data rather than value added or value of shipments, as the census does.

26. U.S. Department of Commerce Bureau of the Census, *Concentration Ratios in Manufacturing, 1997 Census.* The exact figure for NAICS 321212 "Softwood Veneer and Plywood Mfg" is 48.8 percent. For NAICS 32211 "Pulp Mills," it was 58.6 percent. NAICS 3221 "Pulp, Paper, and Paperboard Mills" stood at 28 percent.

27. Ibid.
28. Mill production and employment data were obtained from Miller Freeman, *Directory of the Wood Products Industry 1999* (San Francisco, CA: Author, 1999), and supplemented by interviews, including an interview conducted with a longtime independent industry observer on June 24, 1996, and again on June 18, 1997 in Eugene, Oregon, and an interview conducted with a mill manager on July 10, 1997, in Springfield, Oregon.
29. Information on specific mills was obtained from Miller Freeman, *Directory of the Wood Products Industry 1999,* and supplemented by several interviews, including an interview conducted with a mill manager on July 10, 1997, in Springfield, Oregon; an interview conducted with a longtime independent industry observer in Eugene, Oregon, on June 24, 1996 and on June 18, 1997; an interview conducted with a mill manager in Cottage Grove, Oregon, on July 11, 1997; and an interview with a mill manager in Cave Junction, Oregon, on June 6, 1996.
30. When I conducted my field research, Willamette Industries was second only to Weyerhaeuser as the major corporate player in the Douglas fir region, particularly in Oregon. However, on January 21, 2002, Willamette Industries management reached an agreement to be sold to Weyerhaeuser, and they are now one company.
31. Information on the type of ownership was obtained from Miller Freeman, *Directory of the Wood Products Industry 1999.*
32. Productivity is measured a number of ways (e.g., returns to all inputs—total factor productivity—or individually to labor or capital inputs). Productivity improvement measures the change in output per unit input over a given period, typically one year. No measure of productivity is unproblematic, but for the purposes of comparison, I assume here that there is no systematic bias of measures with respect to industry.
33. The measures are not the same. However, in as much as attempts to capture economies of scale involve investment in expanded output, this must be accompanied by improved productivity of capital investments and other inputs—otherwise, there would be no return to larger scale.
34. J. W. Kendrick, *Productivity Trends in the United States* (Princeton, NJ: Princeton University Press, 1961).
35. Recent data taken from Bureau of Labor Statistics, *Selected Industries: Employment and Annual Rates of Change in Output Per Hour, Selected Periods* (Washington, DC: Author, 1997). More rapid rates of improvement in labor productivity in the 1970s and 1980s are due in large measure to the introduction of computer-guided saws and laser scanners in lumber mills, resulting in the loss of many jobs per unit output and a corresponding reduction in labor requirements per unit of production. Also, many less efficient mills went out of business during the recession of the early 1980s; see B. J. Greber, "Impacts of Technological Change on Employment in the Timber Industries of the Pacific Northwest," *Western Journal of Applied Forestry* 8, no. 1 (1993): 34–37.
36. Van Tassel, *Mechanization in the Lumber Industry.*
37. Ibid., 2-3.
38. Average annual log consumption by sawmills was calculated based on data in Ward, *Oregon's Forest Products Industry, 1994.* Average timber density in the Douglas fir region was estimated based on data from W. B. Smith et al., *Forest Resources of the United States, 1997* (St. Paul, MN: U.S. Department of Agriculture, Forest Service, 2001); D. Preston and B. Alverts, "Oregon's BLM Timber Resources," in *Assessment of Oregon's Forests 1988,* ed. G. J. Lettman (Salem: Oregon State Department of Forestry, 1998), 33–41; and D. R. Gedney, "The Private Timber Resource," in *Assessment of Oregon's Forests 1988,* ed. G. J. Lettman (Salem: Oregon State Department of Forestry, 1988), 53–57. For private land, the figure used here is likely high, but this leads to a conservative estimate of required area of timberlands.
39. Interview conducted on June 11, 1996, with a sawmill owner. According to Ward, *Oregon's Forest Products Industry, 1994,* the aggregate rate of recovery in Oregon sawmills is about 40 percent by volume, but this is based on dated technical information and is probably low; M. Smith, *The U.S. Paper Industry and Sustainable Production: An Argument for Restructuring* (Cambridge, MA: MIT Press, 1997), estimates the figure is closer to 50 percent.
40. This practice has become more prevalent over time as a strategy to limit postsale distortions of lumber under natural or uncontrolled drying conditions and to particularly limit warping after the lumber has been installed.

41. Longer hauls are possible if they can be linked as two-way journeys.
42. The best sources on these points in the Douglas fir region are by Walter J. Mead; see Mead, *Competition and Oligopsony in the Douglas Fir Lumber Industry;* W. J. Mead et al., *Competitive Bidding for U.S. Forest Service Timber in the Pacific Northwest, 1963–83* (Washington, DC: U.S. Department of Agriculture Forest Service Timber Management, 1983).
43. Van Tassel, *Mechanization in the Lumber Industry,* xviii.
44. The quote is from W. B. Greeley, "The Relation of Geography to Timber Supply," *Economic Geography* 1, no. 1 (1925): 1–14, and appears in W. G. Robbins, "Lumber Production and Community Stability: A View from the Pacific Northwest," *Journal of Forestry* 31, no. 4 (1987): 187–96, especially p. 189.
45. See W. G. Robbins, *American Forestry: A History of National, State, and Private Cooperation* (Lincoln: University of Nebraska Press, 1985); W. G. Robbins, *Colony and Empire: The Capitalist Transformation of the American West* (Lawrence: University Press of Kansas, 1994); and M. Williams, *Americans and Their Forests: A Historical Geography* (Cambridge Cambridgeshire, and New York: Cambridge University Press, 1989).
46. Greeley, "The Relation of Geography to Timber Supply"; W. G. Robbins, *Hard Times in Paradise: Coos Bay, Oregon, 1850–1986* (Seattle: University of Washington Press, 1988); Robbins, "Lumber Production and Community Stability"; and Robbins, *Lumberjacks and Legislators.*
47. C. R. Howd and U.S. Bureau of Labor Statistics, *Industrial Relations in the West Coast Lumber Industry: December, 1923* (Washington, DC: U.S. Goverment Printing Office, 1924); Williams, *Americans and Their Forests.*
48. Williams, *Americans and Their Forests.*
49. W. Cronon, *Nature's Metropolis: Chicago and the Great West* (New York: W. W. Norton, 1991), 169.
50. Williams, *Americans and Their Forests.*
51. Howd and U.S. Bureau of Labor Statistics, *Industrial Relations in the West Coast Lumber Industry;* U.S. Bureau of Corporations, *The Lumber Industry.*
52. T. R. Cox, *Mills and Markets: A History of the Pacific Coast Lumber Industry to 1900* (Seattle: University of Washington Press, 1974).
53. E. S. Pomeroy, *The Pacific Slope: A History of California, Oregon, Washington, Idaho, Utah, and Nevada* (Lincoln: University of Nebraska Press, 1991); Robbins, *Hard Times in Paradise;* Van Tassel, *Mechanization in the Lumber Industry;* and Williams, *Americans and Their Forests.*
54. Note that, because of the way the data are reported, some of the counties have been aggregated. These are designated by hyphenated names. Thus, Clat-Till-Linc-Yam is the aggregation of Clatsop, Tillamook, Lincoln, and Yamhill counties; HoodR-Mult-Wash is the aggregation of Hood River, Multnomah, and Washington counties; Coos and Curry counties also are combined.
55. Robbins, *Hard Times in Paradise;* Williams, *Americans and Their Forests.*
56. Richard Rajala's discussion of the underlying economic imperatives reinforcing clear-cutting is excellent; see R. Rajala, *Clearcutting the Pacific Rain Forest: Production, Science, and Regulation* (Vancouver, Canada: UBC Press, 1998).
57. Van Tassel, *Mechanization in the Lumber Industry,* 7.
58. To develop a familiarity with these and related issues, I conducted numerous mill tours and interviews during the summers of 1996 and 1997, including an interview with a mill owner in John Day, Oregon, on July 31, 1996; an interview with a mill manager in White City, Oregon, on June 19, 1996; an interview with a mill union representative in Klamath Falls, Oregon, on June 20, 1996; an interview with a mill owner in Selma, Oregon, on July 17, 1996; an interview with a former lumber mill employee in Cave Junction, Oregon, on July 20, 1996; an interview with the operator of a stud mill in North Powder, Oregon, on July 30, 1996; an interview with the manager of a lumber mill in North Powder, Oregon, on July 30, 1996; an interview with a lumber mill manager in John Day, Oregon, on July 31, 1996; an interview with the owner of a number of lumber mills, conducted in John Day, Oregon, on July 31, 1996; an interview with a union representative in Springfield, Oregon, on July 2, 1997; an interview with a log buyer for a lumber mill, conducted in Springfield, Oregon, on July 10, 1997; and an interview with the manager of a lumber mill in Cottage Grove, Oregon, on July 11, 1997. For another useful description of lumber mills and their basic technolo-

gies, see E. M. Williston, *Lumber Manufacturing: The Design and Operation of Sawmills and Planer Mills* (San Francisco, CA: Miller Freeman, 1988).

59. Van Tassel, *Mechanization in the Lumber Industry,* 29.
60. B. Barham, "Strategic Capacity Investments and the Alcoa-Alcan Monopoly, 1888–1945," in *States, Firms, and Raw Materials: The World Economy and Ecology of Aluminum,* ed. B. Barham, S. G. Bunker, and D. O'Hearn (Madison: University of Wisconsin Press, 1994), 69–110.
61. U.S. Bureau of Corporations, *The Lumber Industry,* 37–38.
62. For example, the Supreme Court decided to break up the Standard Oil Trust in 1911, two years prior to publication of the lumber industry report.
63. On the acquisition and its questionable legality, see Jensen and Draffan, *Railroads and Clearcuts.* On the details of the transaction, see C. E. Twining, *George S. Long, Timber Statesman* (Seattle: University of Washington Press, 1994).
64. Twining, *George S. Long, Timber Statesman,* 29.
65. See, for example, Greeley, "The Relation of Geography to Timber Supply," for an example of the prevailing idea that timber was becoming scarce in the United States.
66. See, for example, J. A. Young and J. M. Newton, *Capitalism and Human Obsolescence: Corporate Control vs. Individual Survival in Rural America* (Montclair, NJ: LandMark Studies, 1980). Also, see P. W. Hirt, *A Conspiracy of Optimism: Management of the National Forests since World War Two* (Lincoln: University of Nebraska Press, 1994); Richardson et al., *BLM's Billion-Dollar Checkerboard;* and White, *"It's Your Misfortune and None of My Own,"* especially pp. 149–50.
67. U.S. Bureau of Corporations, *The Lumber Industry.*
68. One of the implications of increasing federal timber supply after World War II was a diminished potential for the exercise of oligopoly power by integrated firms. This is true even though there is evidence of irregularities in federal timber sales in timbersheds dominated by large firms in the Douglas fir region; Mead, *Competition and Oligopsony in the Douglas Fir Lumber Industry;* and Mead et al., *Competitive Bidding for U.S. Forest Service Timber in the Pacific Northwest, 1963–83.* On competitive conditions and the control of timberlands in the Douglas fir region, see also Jensen, *Lumber and Labor.*
69. See Ellefson and Stone, *U.S. Wood-Based Industry;* and Mead, *Competition and Oligopsony in the Douglas Fir Lumber Industry.* Weyerhaeuser produced slightly less than five billion board feet of softwood lumber in the United States and Canada combined in 1998; Weyerhaeuser Company, *Form 10-K* (Washington, DC: Securities and Exchange Commission, 1997). This compares to U.S. national output of about thirty-five billion board feet.
70. Historical data from Mead, *Competition and Oligopsony in the Douglas Fir Lumber Industry.* More recent data on Georgia-Pacific from Georgia-Pacific Corporation, *Form 10-K405.*
71. B. Adam, "The Temporal Gaze: The Challenge for Social Theory in the Context of GM Food," *British Journal of Sociology* 51, no. 1 (2000): 125–42; B. Adam, *Timescapes of Modernity: The Environment and Invisible Hazards* (London: Routledge, 1998); and G. Henderson, "Nature and Fictitious Capital: The Historical Geography of an Agrarian Question," *Antipode* 30, no. 2 (1998): 73–118.
72. D. C. Le Master, *Mergers among the Largest Forest Products Firms 1950–1970* (Pullman: Washington State University, College of Agriculture Research Center, 1977). The tax break given to integrated companies through capital gains treatment of timber was reduced, though not eliminated, by the Tax Reform Act of 1986.
73. Industrial ownerships are those retained by firms that operate at least one wood-products facility. Data from D. S. Powell et al., *Forest Resources of the United States, 1992* (Fort Collins, CO: U.S. Department of Agriculture, Forest Service, 1993).
74. Atterbury Consultants, *Major Forest Landownership: Western Oregon and Southwest Washington* (Portland, OR: Author, 1996).
75. WTD Industries, *Form 10-K405* (Washington, DC: Securities and Exchange Commission, 1998).
76. Willamette Industries, *Form 10-K* (Washington, DC: Securities and Exchange Commission, 1999).
77. A similar logic propelled Seneca Timberlands, the timberland arm of Eugene's Seneca Sawmill, to purchase 120,000 acres of timberland in coastal Oregon in June 1992.

78. Information on Willamette Industries was obtained from the company's 1999 10-K report to the Securities and Exchange Commission and from an interview conducted on July 7, 1997, with a company official, from "Buyers Divvy up Bohemia holdings," *Portland Oregonian*, August 15, 1991, A1, and from "Willamette Buys Huge Chunks of Timberland," *Portland Oregonian*, March 13, 1996, A1.

79. Also noted by Mead, *Competition and Oligopsony in the Douglas Fir Lumber Industry.*

80. Information on firm strategies and relationships between different facilities was obtained through a series of corporate interviews conducted in the summers of 1996 and 1997, including an interview with the president of a large, independent sawmill conducted on June 11, 1996, in Cave Junction, Oregon; an interview with a timber sales manager for facilities owned by a major wood-products firm conducted on June 20, 1996, in Klamath Falls, Oregon; interviews conducted with a mill manager and forester conducted on July 30, 1996, in North Powder, Oregon; an interview with the president of a chain of lumber mills conducted on July 31, 1996, in John Day, Oregon; an interview with a Northwest region raw material procurement manager for a major wood-products company conducted on August 4, 1997, in St. Helens, Oregon; and an interview with a fiber supply manager for a paper mill conducted on September 5, 1997, in Camas, Washington. This information was supplemented with data from Miller Freeman, *Directory of the Wood Products Industry 1999* and from Boise Cascade Corporation, *Form 10-K* (Washington, DC: Securities and Exchange Commission, 1997); Weyerhaeuser Company, *Form 10-K* (Washington, DC: Securities and Exchange Commission, 1997); and Willamette Industries, *Form 10-K* (Washington, DC: Securities and Exchange Commission, 1997).

81. Information on the raw material procurement strategies of mills operated by these firms is based on several interviews, including one conducted on June 5, 1997, with a mill fiber procurement manager in Longview, Washington; one conducted on July 8, 1997, with a mill fiber procurement manager in Springfield, Oregon; one conducted on July 16, 1997, with a mill fiber procurement manager in Albany, Oregon; one conducted on August 4, 1997, with a wood supply manager in St. Helens, Oregon; and one conducted on August 5, 1997, with a fiber procurement manager in Toledo, Oregon

82. One notable exception to this is the recent development of laminated veneer lumber and the close relationship between the production of this product and plywood.

83. This phrase belongs to David Demeritt, who has struggled to retain a space for the ways in which material nature shape and constrain social outcomes without positing a rigid determinism, and without denying that the meaning associated with any environmental phenomenon is inescapably socially and culturally based; D. Demeritt, "The Construction of Global Warming and the Politics of Science," *Annals of the Association of American Geographers* 91, no. 2 (2001): 307–37. I use it here to suggest a coevolutionary relationship between material nature as raw material on the one hand and firm strategies (geographical, technological, etc.) on the other.

Chapter 5

1. K. Marx, *Capital: A Critique of Political Economy. Volume 2. The Process of Circulation of Capital* (New York: International Publishers, 1967), 248.

2. A. Escobar, "Constructing Nature: Elements for a Poststructural Political Ecology," in *Liberation Ecologies: Environment, Development, Social Movements*, ed. R. Peet and M. Watts (New York: Routledge, 1996), 46–68; and O'Connor, *Natural Causes: Essays in Ecological Marxism* (New York: Guilford Press, 1998).

3. Neil Smith, *Uneven Development: Nature, Capital, and the Production of Space* (Oxford: Blackwell, 1984).

4. W. Boyd, S. Prudham, and R. Schurman, "Industrial Dynamics and the Problem of Nature," *Society and Natural Resources* 14, no. 7 (2001): 555–70; and J. R. Kloppenburg, *First the Seed: The Political Economy of Plant Biotechnology, 1492–2000* (Cambridge Cambridgeshire and New York: Cambridge University Press, 1988).

5. Boyd, Prudham, and Schurman, "Industrial Dynamics and the Problem of Nature."

6. See D. Harvey, *Justice, Nature, and the Geography of Difference* (Cambridge, MA: Blackwell, 1996), for eloquent and often quoted passages about the need to understand the complex interactions between social and environmental change as an essential element of any historical geography of social change. Advocating historical materialism as an approach is well suited to examining these interactions, Harvey argued for "a way of depicting the funda-

mental physical and biological conditions and processes that work their way through all social, cultural [and] economic projects to create a tangible historical geography, and to do it in such a way as not to render those physical and biological elements as a banal and passive background to human historical geography" (p. 192). Enrique Leff made a closely related argument, stating, "The availability of nonbiotic resources and the conditions for biological reproduction of different ecosystems affect the form and appropriation of natural resources. These factors also establish potentials and set certain limits to the expansion, reproduction, and sustainability of capital. These then, are the reasons to insist on thinking about *how ecological processes are inscribed in the dynamics of capital*" (emphasis added; see chapter 1 for discussion); E. Leff, *Green Production: Toward an Environmental Rationality* (New York and London: Guilford Press, 1995), 14.

7. M. A. Bordelon, "Genetic Improvement Opportunities," in *Assessment of Oregon's Forests 1988*, ed. G. J. Lettman (Salem: Oregon State Department of Forestry, 1988), 231–39; and T. T. Munger and W. G. Morris, *Growth of Douglas-Fir Trees of Known Seed Source* (Portland, OR: U.S. Department of Agriculture, Pacific Northwest Forest Experiment Station, 1936), 40. Munger went on to become the first director of the U.S. Forest Service's Pacific Northwest Forest Experiment Station in Portland, Oregon, beginning in 1924.

8. Munger's experiments in many ways reflected and reinforced important contemporaneous developments in American forestry, including the shift of U.S. forest capital to the Pacific Slope, the increasing profile of the U.S. federal government in the political economy of American forestry, and the rise of scientific forest management alongside the westward march of lumbering. See D. Demeritt, "Scientific Forest Conservation and the Statistical Picturing of Nature's Limits in the Progressive-Era United States," *Environment and Planning D-Society and Space* 19, no. 4 (2001): 431–59; W. B. Greeley, "The Relation of Geography to Timber Supply," *Economic Geography* 1, no. 1 (1925): 1–14; and W. G. Robbins, *Lumberjacks and Legislators: Political Economy of the U.S. Lumber Industry, 1890–1941* (College Station: Texas A&M University Press, 1982). It also bears noting that American foresters, by expressing increased interest in forest genetics, were following the lead of agricultural plant breeders inspired by the rediscovery of Mendel's work in plant breeding and its applications to American agriculture by the likes of Liberty Hyde Bailey and Rowland Harry Bifkin. See J. H. Perkins, *Geopolitics and the Green Revolution: Wheat, Genes, and the Cold War* (New York and Oxford: Oxford University Press, 1997).

9. W. Boyd and S. Prudham, "Manufacturing Green Gold: Industrial Tree Improvement and the Power of Heredity in the Post-War United States," in *Industrializing Organisms: Introducing Evolutionary History,* ed. S. Schrepfer and P. Scranton (New York: Routledge, 2003), 107–42; E. K. Morgenstern, *Geographic Variation in Forest Trees: Genetic Basis and Application of Knowledge in Silviculture* (Vancouver, Canada: UBC Press, 1996); Munger and Morris, *Growth of Douglas-Fir Trees of Known Seed Source;* and B. Zobel and J. B. Jett, *Genetics of Wood Production* (Berlin and New York: Springer-Verlag, 1995).

10. D. Harvey, *The Limits to Capital* (Oxford: Blackwell, 1982).

11. M. Williams, *Americans and Their Forests: A Historical Geography* (Cambridge Cambridgeshire, and New York: Cambridge University Press, 1989).

12. Robbins, *Lumberjacks and Legislators.*

13. W. D. Hagenstein, *Growing 40,000 Homes a Year* (Berkeley: School of Forestry and Conservation, University of California, Berkeley, 1973).

14. Morgenstern, *Geographic Variation in Forest Trees;* and B. J. Zobel, "Forest Tree Improvement—Past and Present," in *Advances in Forest Genetics,* ed. P. K. Khosla (New Delhi, India: Ambika, 1981), 11–22.

15. Boyd, Prudham, and Schurman, "Industrial Dynamics and the Problem of Nature."

16. B. Adam, "The Temporal Gaze: The Challenge for Social Theory in the Context of GM Food," *British Journal of Sociology* 51, no. 1 (2000): 125–42. On timescapes more generally, see B. Adam, *Timescapes of Modernity: The Environment and Invisible Hazards* (London: Routledge, 1998).

17. D. Goodman and M. R. Redclift, *Refashioning Nature: Food, Ecology, and Culture* (London and New York: Routledge, 1991); and S. A. Mann, *Agrarian Capitalism in Theory and Practice* (Chapel Hill: University of North Carolina Press, 1990).

18. See variously W. Boyd, "Making Meat: Science, Technology, and American Poultry Production," *Technology and Culture* 42, no. 4 (2001): 631-64; Boyd, Prudham, and Schurman, "Industrial Dynamics and the Problem of Nature"; N. Castree, "Commodifying

What Nature?" *Progress in Human Geography* 27, no. 3 (2003): 273–97; Kloppenburg, *First the Seed.* Also, George Henderson's argument about the creation of distinct circuits of capital accumulation by means of agricultural credit systems underwriting investment in cultivation projects with built-in delays and risks is also pertinent; see G. Henderson, "Nature and Fictitious Capital: The Historical Geography of an Agrarian Question," *Antipode* 30, no. 2 (1998): 73–118; and G. L. Henderson, *California and the Fictions of Capital* (New York: Oxford University Press, 1999).

19. G. Namkoong, R. A. Usanis, and R. R. Silen, "Age-Related Variation in Genetic Control of Height Growth in Douglas-Fir," *Theoretical and Applied Genetics* 42 (1972): 151–59. The same may be said of research on the genetic determination of drought resistance; W. K. Ferrell and S. E. Woodward, "Effects of Seed Origin on Drought Resistance of Douglas-Fir (Pseudotsuga Menziesii) (Mirb.) Franco," *Ecology* 47, no. 3 (1966): 499–503.

20. Volume returns have been estimated at 20 percent for Douglas fir in industrial settings; K. Jayawickrama, *The Northwest Tree Improvement Cooperative* (Wilsonville, OR: 2001).

21. D. Copes, F. Sorensen, and R. Silen, "Douglas-Fir Seedling Grows 8 Feet Tall in Two Seasons," *Journal of Forestry* 67, no. 3 (1969): 174–75; Pacific Northwest Tree Improvement Research Co-operative, *Annual Report 1995–1996* (Corvallis: Forest Research Laboratory, Oregon State University, 1996); K. H. Ritters, "Early Genetic Selection in Douglas-Fir: Interactions with Shade, Drought, and Stand Density" (Ph.D. dissertation, College of Forestry, Oregon State University, Corvallis, 1986); and J. B. St. Clair, "Evaluating Realized Genetic Gains from Tree Improvement" (IUFRO S4.01 Conference, College of Forestry, Virginia Polytechnic Institute, Blacksburg, Virginia, 1993).

22. K. K. Ching, T. M. Ching, and D. P. Lavender, *Flower Induction in Douglas-Fir (Pseudotsuga Menziezii (Mirb.) Franco) by Fertilizer and Water Stress. Part I: Changes in Free Amino Acid Composition in Needles* (Siberian Branch, Nauka, USSR: Sexual Reproduction of Conifers, Academy of Science, 1973); and Hagenstein, *Growing 40,000 Homes a Year.*

23. R. R. Silen and D. L. Copes, "Douglas-Fir Seed Orchard Problems—A Progress Report," *Journal of Forestry* 70, no. 3 (1972): 145–47; R. R. Silen and J. G. Wheat, "Progressive Tree Improvement Program in Coastal Douglas-Fir," *Journal of Forestry* 77 (1979): 78–83.

24. K. K. Ching, *Controlled Pollination of Douglas-Fir* (Corvallis: Oregon State University, Forest Research Center, 1960). The problems stem largely from the fact that flowering trees may be fifty to sixty feet tall, making it difficult to reach the flowers. Trees tend also to be widely spaced, meaning that each tree must be accessed individually. Once the flowers are reached, collecting the pollen is a challenge, particularly using mechanized methods. And, because it is difficult to identify the exact time when trees are in flower, spraying pollen on a receptor tree to manipulate crosses produces relatively low success rates.

25. W. T. Adams et al., "Pollen Contamination Trends in a Maturing Douglas-Fir Seed Orchard," *Canadian Journal of Forest Research* 27 (1997): 131–34; Silen and Copes, "Douglas-Fir Seed Orchard Problems."

26. See Kloppenburg, *First the Seed,* on science and industry, particularly chapter 2. There are some strong resonances here with Engels's ideas about this shifting interface of capital and science, particularly his notion that biology would be the last of the natural sciences to be drawn into the circuits of capitalist innovation. Kloppenburg made similar observations in his discussion of agriculture, and I am indebted to his groundbreaking work.

27. Ferrell and Woodward, "Effects of Seed Origin on Drought Resistance of Douglas-Fir"; and Silen and Copes, "Douglas-Fir Seed Orchard Problems."

28. C. S. Larsen, *Genetics in Silviculture* (London: Oliver and Boyd, 1956); R. Toda, "An Outline of the History of Forest Genetics," in *Advances in Forest Genetics,* ed. P. K. Khosla (New Delhi, India: Ambika, 1981), 4–12.

29. Boyd and Prudham, "Manufacturing Green Gold"; and F. I. Righter, "New Perspectives in Forest Tree Breeding," *Science* 104, no. 2688 (1946): 1–3.

30. See F. W. G. Baker, *Rapid Propagation of Fast-Growing Woody Species* (Wallingford, UK: C.A.B. International for CASAFA, 1992); T.- Y. Cheng and T. H. Voqui, "Regeneration of Douglas-Fir Plantlets through Tissue Culture," *Science* 198, no. 4315 (1977): 306–307; and R. L. M. Pierik and J. Prakash, *Plant Biotechnology, Commercial Prospects and Problems* (New Delhi, India: Oxford and IBH, 1993). The significance of the hybridity issue is not of merely biological interest. Rather, as Kloppenburg emphasized, hybridization offers a mechanism for de facto protection of improved varieties as proprietary inventions, because hybrid

crosses do not "breed true"; thus saved seed will not result in the regeneration of viable genotypes beyond the first generation. Although this strategy became a productive route for agroindustry to take hold of plants as exclusive commodities well before contemporary genetic engineering and attendant new possibilities in the realm of intellectual property rights, the same avenue was effectively blocked in commercial forest tree improvement.

31. See Ferrell and Woodward, "Effects of Seed Origin on Drought Resistance of Douglas-Fir."
32. See, for example, F. C. Sorensen, "Geographic Variation in Seedling Douglas-Fir (Pseudotsuga Menziesii) from the Western Siskiyou Mountains of Oregon," *Ecology* 64, no. 4 (1983): 696–702. This is also based on a key informant interview with a forest geneticist conducted in Centralia Washington on December 17, 1997.
33. Bordelon, "Genetic Improvement Opportunities"; and Silen and Wheat, "Progressive Tree Improvement Program in Coastal Douglas-Fir."
34. Cheng and Voqui, "Regeneration of Douglas-Fir Plantlets through Tissue Culture," 307.
35. Silen and Wheat, "Progressive Tree Improvement Program in Coastal Douglas-Fir," 81.
36. Zobel, "Forest Tree Improvement."
37. Righter, "New Perspectives in Forest Tree Breeding."
38. Hagenstein, *Growing 40,000 Homes a Year*; Silen and Copes, "Douglas-Fir Seed Orchard Problems."
39. Jayawickrama, *The Northwest Tree Improvement Cooperative*.
40. R. R. Silen, *A Simple Progressive Tree Improvement Program for Douglas-Fir* (Portland, OR: U.S. Department of Agriculture Forest Service, Pacific Northwest Range and Experiment Station, 1966); and Silen and Wheat, "Progressive Tree Improvement Program in Coastal Douglas-Fir." Silen went on to work with the forest genetics team of the U.S. Forest Service's Pacific Northwest Research Station, located on the campus of Oregon State University.
41. Silen and Copes, "Douglas-Fir Seed Orchard Problems."
42. Bordelon, "Genetic Improvement Opportunities"; and Silen and Wheat, "Progressive Tree Improvement Program in Coastal Douglas-Fir."
43. Jayawickrama, *The Northwest Tree Improvement Cooperative*.
44. Ibid. Also, information was obtained from a key informant interview conducted with a forest geneticist conducted on December 19, 1997, in Centralia, Washington.
45. Silen and Wheat, "Progressive Tree Improvement Program in Coastal Douglas-Fir." Also, information was obtained from a key informant interview conducted with a forest geneticist conducted on December 19, 1997, in Centralia, Washington.
46. Bordelon, "Genetic Improvement Opportunities"; St. Clair, "Evaluating Realized Genetic Gains from Tree Improvement."
47. Key informant interviews conducted with knowledgeable observers in Corvallis, Oregon, on September 3, 1997, September 17, 1997, and November 14, 1997.
48. Key informant interview conducted on November 14, 1997, in Corvallis, Oregon.
49. Jayawickrama, *The Northwest Tree Improvement Cooperative*.
50. As of December 2001, the membership included Longview Fibre Company, Menasha Forest Products Company, Oregon Department of Forestry, Oregon State University, Plum Creek Timber Company, Roseburg Resources, Simpson Timber Company, Stimson Timber Company, the Timber Company, the Bureau of Land Management, the Washington State Department of Natural Resources, Weyerhaeuser, and Willamette Industries. Associate members included the South Coast Lumber Company and Starker Forests; see Pacific Northwest Tree Improvement Research Co-operative, *Annual Report, 2000-2001* (Corvallis: Oregon State University, 2002).
51. Pacific Northwest Tree Improvement Research Co-operative, *Annual Report 1995–1996*
52. Key informant interview, conducted in Corvallis, Oregon, on November 17, 1997. See also Pacific Northwest Tree Improvement Research Co-operative, *Annual Report, 2000–2001*. On the issue of pollen drift in Douglas fir tree improvement, see Adams et al., "Pollen Contamination Trends in a Maturing Douglas-Fir Seed Orchard"; Silen and Copes, "Douglas-Fir Seed Orchard Problems."
53. Key informant interview conducted with a forest geneticist in Centralia, Washington, on December 17, 1997.
54. Key informant interview conducted with a forest geneticist in Centralia, Washington, on December 17, 1997.
55. B. Virgin, "Weyerhaeuser's Forestry Gamble Pays Off," *Oregonian*, 1997, E2.

56. Weyerhaeuser Company, *Form 10-K* (Washington, DC: Securities and Exchange Commission, 1997).

57. C. E. Twining, *George S. Long, Timber Statesman* (Seattle: University of Washington Press, 1994). Weyerhaeuser's dominant position in Northwest timberlands was consolidated only by its purchase of Willamette Industries after the research for this chapter was completed.

58. On these themes, see Demeritt, "Scientific Forest Conservation"; Greeley, "The Relation of Geography to Timber Supply"; and Robbins, *Lumberjacks and Legislators*.

59. National Science Foundation, *National Patterns of R&D Expenditures: 1996* (Washington, DC: Author, 1996). These industries are what used to be standard industrial classifications 24 and 26 prior to changes in the classification system used by the U.S. census.

60. P. V. Ellefson, "Forestry Research Undertaken by Private Organizations in Canada and the United States: A Review and Assessment," in *The Role of the Private Sector in Forestry Research: Recent Developments in Industrialized Countries*, ed. F. S. P. Ng, F. M. Schlegel, and R. Romeo (Rome: Food and Agriculture Organization of the United Nations, 1995), 73–144.

61. Demeritt, "Scientific Forest Conservation"; R. Rajala, *Clearcutting the Pacific Rain Forest: Production, Science, and Regulation* (Vancouver, Canada: UBC Press, 1998); W. G. Robbins, "Lumber Production and Community Stability: A View from the Pacific Northwest," *Journal of Forestry* 31, no. 4 (1987): 187–96; Robbins, *Lumberjacks and Legislators*.

62. Kloppenburg, *First the Seed*.

63. American Forest Council, *Forest Industry-Sponsored Research Cooperatives at U.S. Forestry Schools* (Washington, DC: Author, 1987).

64. Ellefson, "Forestry Research Undertaken by Private Organizations in Canada and the United States," 134

65. N. Rosenberg, *Inside the Black Box: Technology and Economics* (Cambridge Cambridgeshire and New York: Cambridge University Press, 1982), 141–42.

66. This view was propounded, for example, by Gordon Rausser, former dean of the College of Natural Resources at the University of California at Berkeley in his address at the conference "Knowledge Generation and Transfer: Implications for Agriculture in the 21st Century," held at the University of California, Berkeley, June 18, 1998 (author's notes). Rausser went on to negotiate and sign one of the most controversial deals in the history of the private appropriation of American public science, granting broad powers over the research agenda in the entire College of Natural Resources at Berkeley to biotech firm Novartis in exchange for $5 million in funding (see note 90 below).

67. Kloppenburg, *First the Seed*; Rajala, *Clearcutting the Pacific Rain Forest*; W. G. Robbins, *American Forestry: A History of National, State, and Private Cooperation* (Lincoln: University of Nebraska Press, 1985); Robbins, *Lumberjacks and Legislators*.

68. Again, for discussion, see Kloppenburg, *First the Seed*.

69. This terminology comes from Boyd, Prudham, and Schurman, "Industrial Dynamics and the Problem of Nature."

70. Forestry Research Task Force, U.S. Department of Agriculture and Association of State Universities and Land Grant Colleges, *A National Program of Research for Forestry* (Washington, DC: Research Program Development and Evaluation Staff, U.S. Department of Agriculture, 1967), 34

71. On the wider implications of biotechnology in terms of the social control of life and science, see L. Kay, "Problematizing Basic Research in Molecular Biology," in *Private Science: Biotechnology and the Rise of the Molecular Sciences*, ed. A. Thackray (Philadelphia: University of Pennsylvania Press, 1998), 20–38; L. E. Kay, *Who Wrote the Book of Life? A History of the Genetic Code* (Stanford, CA: Stanford University Press, 2000); D. Kevles, "Diamond V. Chakrabarty and Beyond: The Political Economy of Patenting Life," in *Private Science*, 65–79; S. Slaughter and L. L. Leslie, *Academic Capitalism: Politics, Policies, and the Entrepreneurial University* (Baltimore, MD: Johns Hopkins University Press, 1997). More specifically, on the role of biotechnology in opening "new economic spaces," see M. Kenney, "Biotechnology and the Creation of a New Economic Space," in *Private Science,* 131–43. On plant commodification in industrial agriculture, particularly in the United States, see W. Boyd, "Wonderful Potencies: Deep Structure and the Problem of Monopoly in Agricultural Biotechnology," in *Engineering Trouble: Biotechnology and Its Discontents*, ed. R. Schurman and D. Takahashi (Berkeley: University of California Press, 2003), 24–62; L. Busch, *Plants,*

Power, and Profit: Social, Economic, and Ethical Consequences of the New Biotechnologies (Cambridge, MA: Blackwell, 1991); and S. Krimsky and R. P. Wrubel, *Agricultural Biotechnology and the Environment: Science, Policy, and Social Issues* (Urbana: University of Illinois Press, 1996).

72. U.S. Congress Office of Technology Assessment, *Biotechnology in a Global Economy* (Washington, DC: Author, 1991).

73. S. H. Strauss, S. P. DiFazio, and R. Meilan, "Genetically Modified Poplars in Context," *Forestry Chronicle* 77, no. 2 (2001): 271–79. For an early take on the social implications of forest tree biotechnology, see also M. Sagoff, "On Making Nature Safe for Biotechnology," in *Assessing Ecological Risks of Biotechnology*, ed. L. Ginsburg (Stoneham, MA: Butterworth-Heineman, 1991), 341–65.

74. A. Séguin, G. Lapointe, and P. J. Charest, "Transgenic Trees," in *Forest Products Biotechnology*, ed. A. Bruce and J. W. Palfreyman (Bristol, PA: Taylor and Francis, 1998), 287–304.

75. Key informant interview conducted on September 3, 1997, in Corvallis, Oregon.

76. See S. Hee, "Intensive Silviculture Stewardship: The Weyerhaeuser Experience" (Silviculture Conference: Stewardship in the New Forest, Vancouver, Forestry Canada, 1991). Weyerhaeuser, along with Westvaco, is a leading firm in research on somatic embryogenesis in conifers. Together, these firms accounted for thirteen of the twenty-one patents for somatic embryogenesis issued between 1989 and March 1998 (data obtained on CD-ROM from the US Patent Office, Patent and Trademark Center, San Francisco Public Library, July 24, 1998). Weyerhaeuser also holds a total of nine patents related to the production of manufactured seed, a related technology for encasing the cultured embryos.

77. On poplars, see R. A. Donahue et al., "Growth, Photosynthesis, and Herbicide Tolerance of Genetically Modified Hybrid Poplar," *Canadian Journal of Forest Research* 24, no. 12 (1994): 2377–83; Séguin, Lapointe, and Charest, "Transgenic Trees"; and Strauss, DiFazio, and Meilan, "Genetically Modified Poplars in Context."

78. Tree Genetic Engineering Research Co-operative, *TGERC Profile, History, and Structure* (Corvallis: Oregon State University, 2002).

79. Information on these plantations was gleaned from site visits during the summer of 1997 and from key informant interviews, including one conducted on August 8, 1997, in Wallula, Washington; one conducted on August 11, 1997, in Corvallis, Oregon; and one conducted on September 5, 1997, in Clatskanie, Washington.

80. Strauss, DiFazio, and Meilan "Genetically Modified Poplars in Context"; S. H. Strauss, S. A. Knowe, and J. Jenkins, "Benefits and Risks of Transgenic Roundup Ready Cottonwoods," *Journal of Forestry* 95, no. 5 (1997): 12–19.

81. Key informant interview, September 3, 1997.

82. A. M. Brunner et al., "Genetic Engineering of Sexual Sterility in Shade Trees," *Journal of Arboriculture* 24, no. 5 (1998): 263–73; and S. H. Strauss, A. M. Rottmann, and L. A. Sheppard, "Genetic Engineering of Reproductive Sterility in Forest Trees," *Molecular Breeding* 1 (1995): 5–26.

83. Strauss, DiFazio, and Meilan, "Genetically Modified Poplars in Context"; Tree Genetic Engineering Research Co-operative, *TGERC Annual Report 1996–97* (Corvallis: Forest Research Laboratory, Oregon State University, 1997), 37.

84. R. R. James et al., "Environmental Effects of Genetically Engineered Woody Biomass Crops," *Biomass and Bioenergy* 14, no. 4 (1998): 403–14; Krimsky and Wrubel, *Agricultural Biotechnology and the Environment;* Strauss, DiFazio, and Meilan, "Genetically Modified Poplars in Context"; and S. H. Strauss, G. T. Howe, and B. Goldfarb, "Prospects for Genetic Engineering of Insect Resistance in Forest Trees," *Forest Ecology and Management* 43 (1991): 181–209.

85. Strauss, Knowe, and Jenkins, "Benefits and Risks of Transgenic Roundup Ready Cottonwoods."

86. Strauss, DiFazio, and Meilan, "Genetically Modified Poplars in Context."

87. Key informant interview, September 3, 1997.

88. Tree Genetic Engineering Research Co-operative, *TGERC Profile, History, and Structure.*

89. Sagoff, "On Making Nature Safe for Biotechnology."

90. Examples of the former are a hallmark of agricultural and health biotechnology research, including, for example, the highly controversial exclusive partnership signed in 1998 between the University of California–Berkeley and Novartis covering agricultural biotechnology.

The Swiss pharmaceutical, agrochemical, and biotechnology company Novartis signed an agreement with the Department of Plant and Microbial Biology at the University of California–Berkeley. The deal provided the department with $25 million in funding from Novartis and access to the company's gene sequencing and DNA resources. In return, Novartis received the right of first refusal over licensing patent rights from all research conducted using the funds and also gained two seats on a five-member departmental panel established to allocate funds to research projects. The deal was and remains highly controversial at Berkeley, and in wider circles, because of the degree to which it transfers to a single firm governance over research at one of America's leading, publicly supported land grant universities. This is part of a more widespread turn toward exclusive partnerships between universities (public and private) and private companies in research funding and control over and dissemination of research results. Although the TGERC is less dramatic and far reaching, it does blur the distinction between public and private science in disturbing ways. Alongside such developments, capital has also shown some forms of enthusiasm for the GE tree project less centered on the appropriation of public science. This is exemplified by the formation of Aborgen in January 2000; ArborGen is a wholly private joint venture by industry giants Westvaco, International Paper, and Fletcher Challenge together with a New Zealand genomics company by the name of Genesis Research and Development; K. Brown, "Industry Hugs Biotech Trees," *Technology Review* 104, no. 2 (2001): 34. The goal of ArborGen, according to its partners, is to be the first to commercialize GE trees.

91. Boyd, Prudham, and Schurman, "Industrial Dynamics and the Problem of Nature."
92. L. C. Duchesne, "Impact of Biotechnology on Forest Ecosystems," *Forestry Chronicle* 69, no. 3 (1993): 307–13; S. R. Radosevich, C. M. Ghersa, and G. Comstock, "Concerns a Weed Scientist Might Have about Herbicide-Tolerant Crops," *Weed Technology* 6 (1992): 635–39; Strauss, Knowe, and Jenkins, "Benefits and Risks of Transgenic Roundup Ready Cottonwoods."
93. S. T. Freidman and G. S. Foster, "Forest Genetics on Federal Lands in the United States: Public Concerns and Policy Responses," *Canadian Journal of Forest Resources* 27 (1997): 401–408.
94. Strauss, DiFazio, and Meilan, "Genetically Modified Poplars in Context."
95. Letter to Dr. Steve Strauss, Oregon State University forestry professor and leader of the TGERC, from a group identified only as "concerned OSU students and alumni," March 23, 2001. Quoted from Genetix Alert News Release, "GE Trees Destroyed at Oregon State University," posted to transgenictrees@iatp.org on March 25, 2001, by mritchie@iatp.org.
96. For example, in the United Kingdom in July 1999, protestors destroyed GE poplars being evaluated by AstraZeneca; in July 2000, activists destroyed what they described as GE trees planted by the Mead Corporation near Milo, Maine.
97. O'Connor, *Natural Causes*. See also G. Bridge, "The Social Regulation of Resource Access and Environmental Impact: Nature and Contradiction in the US Copper Industry," *Geoforum* 31, no. 2 (2000): 237–56; G. Bridge and P. McManus, "Sticks and Stones: Environmental Narratives and Discursive Regulation in the Forestry and Mining Sectors," *Antipode* 32, no. 1 (2000): 10–47.
98. A. N. H. Creager, "Biotechnology and Blood: Edwin Cohn's Plasma Fractionation Project, 1940–1953," *Private Science,* 39–62; Slaughter and Leslie, *Academic Capitalism;* and S. Wright, *Molecular Politics: Developing American and British Regulatory Policy for Genetic Engineering, 1972–1982* (Chicago: University of Chicago Press, 1994).
99. D. J. Haraway, *Modest-Witness, Second-Millennium: Femaleman Meets Oncomouse: Feminism and Technoscience* (New York: Routledge, 1997); Harvey, *Justice, Nature, and the Geography of Difference.*
100. See L. Levidow, "Democratizing Technology—Or Technologizing Democracy? Regulating Agricultural Biotechnology in Europe," *Technology in Society* 20 (1998): 211–26, for this argument made forcefully in relation to the regulation of biotechnology.

Chapter 6

1. For more on Earth First! and the politics of logging in the Pacific Slope, specifically in the context of the California redwoods, see J. London, "Common Roots and Entangled Limbs: Earth First! and the Growth of Post-Wilderness Environmentalism on California's North Coast," *Antipode* 30, no. 2 (1998): 155–76.

2. Protests of this sort became increasingly common in the ancient forest campaign but were augmented by litigation over Forest Service and the BLM compliance with NEPA, the Endangered Species Act, and the various laws governing sustained-yield management of federal forests. For more information on the campaign in the courts, see the epilogue, and also B. G. Marcot and J. W. Thomas, *Of Spotted Owls, Old Growth, and New Policies: A History since the Interagency Scientific Committee Report* (Portland, OR: U.S. Department of Agriculture, Forest Service, Pacific Northwest Forest Research Station, 1997); and V. M. Sher, "Travels with Strix: The Spotted Owl's Journey through the Federal Courts," *Public Land Law Review* 14 (1993): 41–79.

3. R. Rajala, *Clearcutting the Pacific Rain Forest: Production, Science, and Regulation* (Vancouver, Canada: UBC Press, 1998); W. G. Robbins, *Hard Times in Paradise: Coos Bay, Oregon, 1850–1986* (Seattle: University of Washington Press, 1988); W. G. Robbins, *Lumberjacks and Legislators: Political Economy of the U.S. Lumber Industry, 1890–1941* (College Station: Texas A&M University Press, 1982); and M. Williams, *Americans and Their Forests: A Historical Geography* (Cambridge Cambridgeshire, and New York: Cambridge University Press, 1989).

4. D. Demeritt, "Scientific Forest Conservation and the Statistical Picturing of Nature's Limits in the Progressive-Era United States," *Environment and Planning D-Society and Space* 19, no. 4 (2001): 431–59; and J. C. Scott, *Seeing Like a State: How Certain Schemes to Improve the Human Condition Have Failed* (New Haven, CT: Yale University Press, 1998).

5. W. S. Prudham, "Timber and Town: Post-War Federal Forest Policy, Industrial Organization, and Rural Change in Oregon's Illinois Valley," *Antipode* 30, no. 2 (1998): 177–96; W. G. Robbins, "Lumber Production and Community Stability: A View from the Pacific Northwest," *Journal of Forestry* 31, no. 4 (1987): 187–96; and C. H. Schallau and R. M. Alston, "The Commitment to Community Stability: A Policy or Shibboleth," *Environmental Law* 17 (1987): 429–81.

6. The notion of a structured coherence applied to sustained-yield forestry draws on Barnes, Hayter, and Hay in reference to quite parallel political economic arrangements presiding over the exploitation of forest resources in Port Alberni, British Columbia, during the postwar; see T. J. Barnes, R. Hayter, and E. Hay, "Stormy Weather: Cyclones, Harold Innis, and Port Alberni, BC," *Environment and Planning A* 33 (2001): 2127–47. However, their use of the concept ultimately draws on D. Harvey, *The Urbanization of Capital: Studies in the History and Theory of Capitalist Urbanization* (Baltimore, MD: Johns Hopkins University Press, 1985).

7. I lived in the Illinois Valley for most of the summer of 1996, with return trips in 1997 and 1998. I conducted numerous semistructured interviews and also engaged in participation observation research and informal conversations and interactions with locals to learn about the area and its history. I am indebted to many people who helped me with the research, including, in particular, Erik Jules and Marcus Kaufman. Go TORCH!

8. Census data from 1990 for the Illinois Valley were compiled from the census county divisions of Cave Junction and Wilderville.

9. The Illinois Valley is home to a community called Takilma, founded by a group of back-to-the-land hippies who established residence in the early 1970s. This community is probably one of the reasons for the discrepancy between the census and the actual population of the area, because more people live in the various nontraditional circumstances of Takilma than the census is likely to capture. Moreover, many residents also have various political or economic reasons for not reporting information to the census. More generally, there are people living in relative isolation in the valley who are unlikely to be counted by the census or who have various reasons to avoid being contacted by anyone from the government. It also bears mentioning that my fieldwork in the area was conducted in 1996 and 1997, and some growth, particularly in retirees, likely occurred between 1990 and when I was there. Josephine County grew by 21 percent in the 1990s, and its population of people age sixty-five years and older swelled to 20 percent of the county's population, compared with 13 percent in Oregon overall. The area has become something of a magnet for rural retirements. Data from the U.S. Census Bureau, "State and County Quick Facts," http://quickfacts.census.gov/qfd/states/41/41033.html (accessed January 24, 2004).

10. This is according to a local biologist who has conducted extensive fieldwork in the area on the effects of clear-cut logging on local plant communities. I interviewed him several times

during the summer of 1998. For reference material on local biota, see J. F. Franklin and C. T. Dyrness, *Natural Vegetation of Oregon and Washington* (Portland, OR: U.S. Department of Agriculture, Forest Service, 1973); R. H. Whittaker, "Vegetation of the Siskiyou Mountains, Oregon and California," *Ecological Monographs* 30 (1960): 279–338; and J. P. Smith and J. O. Sawyer, "Endemic Vascular Plants of Northwestern California and Southwestern Oregon," *Madrono* 35 (1988): 54–69.

11. Information on the Applegate Trail obtained from the Douglas County Historical Society.
12. R. White, *"It's Your Misfortune and None of My Own": A History of the American West* (Norman: University of Oklahoma Press, 1991).
13. J. B. Dunn, ed., *Land in Common: An Illustrated History of Jackson County* (Medford: Southern Oregon Historical Society, 1993).
14. I indicate the whiteness of settlers here to highlight the fact that by no means were these lands empty upon settler arrival. Rather, preexisting Indian communities were very much present but were removed by the twin ravages of forced exodus and disease. Moreover, the racial dynamics of land and labor continued to revolve around privilege and power favoring whites to the exclusion of nonwhites, including Chinese workers brought in to work mines and railroads and, later, black migrants from the South seeking work and land. The 1857 state constitution contained language specifically restricting African Americans: "No free negro or mulatto, not residing in this state at the time of this constitution, shall come, reside or be within this state, or hold any real estate, or make any contracts, or maintain any suit therein"; Dunn, *Land in Common,* 28. Institutionalized discrimination against African Americans was maintained until quite recently in southern Oregon, with "sundown laws" that restricted the movements of blacks after dark still on the books in the Rogue Valley into the 1960s; see B. A. Brown, *In Timber Country: Working People's Stories of Environmental Conflict and Urban Flight* (Philadelphia, PA: Temple University Press, 1995).
15. E. S. Pomeroy, *The Pacific Slope: A History of California, Oregon, Washington, Idaho, Utah, and Nevada* (Lincoln: University of Nebraska Press, 1991).
16. Information about Waldo was obtained from the Illinois Valley Historical Society, from archival materials and from exhibits in the museum.
17. Pomeroy, *The Pacific Slope,* 172.
18. Water cannons were large guns capable of directing highly pressurized water streams at banks of ore material, washing it into sluice boxes. Information on gold mining techniques used in the area were obtained from the Illinois Valley Historical Society's collection and from an interview conducted on July 16, 1996, with a local historian. Also, see L. McLane, *First There Was Twogood: A Pictorial History of Northern Josephine County* (Grants Pass, OR: Josephine County Historical Society, 1996).
19. From time to time, there are proposals to resurrect industrial mining in the area. When I was in the area, for example, there was a proposal to establish a nickel-laterite strip mine on Forest Service land along Rough and Ready Creek.
20. Dunn, *Land in Common;* E. Richardson, U.S. Bureau of Land Management, Oregon State Office and Forest History Society, *BLM's Billion-Dollar Checkerboard: Managing the O and C Lands* (Santa Cruz, CA, and Washington, DC: Forest History Society, 1980); U.S. Bureau of Corporations, *The Lumber Industry* (Washington, DC: Government Printing Office, 1913).
21. The federal commitment to small-scale production was further advanced by passage of the Timber and Stone Act of 1878, allowing for the sale of public lands in Oregon and other western states in 160-acre parcels of land valuable chiefly for timber (and stone) production; that is, not suitable for farming; S. T. Dana and S. K. Fairfax, *Forest and Range Policy, Its Development in the United States* (New York: McGraw-Hill, 1980).
22. This problem also applied to the practical application of the Timber and Stone Act.
23. Richardson et al., *BLM's Billion-Dollar Checkerboard;* U.S. Bureau of Corporations, *The Lumber Industry.*
24. White, *"It's Your Misfortune and None of My Own."*
25. Robbins, *Hard Times in Paradise.*
26. Dicken and Dicken, *The Making of Oregon.*
27. As a result of this bypass, Waldo lost the county seat to Grants Pass in 1886, and by 1900, with the local decline in gold mining, Waldo was largely deserted.

28. Aspects of this characterization are based on the recollections of a local resident who once operated a small mill near Cave Junction; interview conducted on July 12, 1996, in Cave Junction.

29. It is telling that this man remembers more mills in the valley in 1949 than the West Coast Lumbermen's Association reported for the entire county. It would be easy to dismiss his recollections as inaccurate, except this individual had a remarkable ability to remember details, including names and dates. Moreover, the size and mobility of mills and a distinct tendency among owners of small mills to avoid taxation means that it is highly unlikely that the census of even industry associations could obtain an accurate count of operating mills. This introduces an error in official statistics, but if anything, it results in an underestimate of mills and therefore an overestimate of average mill size. This is significant given that the lumber sector, by official statistics, still stands out for remarkably low-scale economies (see chapter 3).

30. Interview conducted on July 18, 1996, in Cave Junction.

31. This is based on archival information in the Illinois Valley Historical Society Collection, an interview conducted July 18, 1996, in Cave Junction, and an interview conducted with a longtime resident and industry figure on July 17, 1996, in Selma, Oregon.

32. Robbins, "Lumber Production and Community Stability"; Robbins, *Lumberjacks and Legislators.*

33. This is from the preamble to draft legislation as recorded in *Hearings on H.R. 4830 before the Committee on Agriculture, House of Representatives,* 68th Cong., 1st sess. (Washington, DC: U.S. Government Printing Office).

34. See Dana and Fairfax, *Forest and Range Policy.*

35. W. B. Greeley, "The Relation of Geography to Timber Supply," *Economic Geography* 1, no. 1 (1925): 1-14.

36. See, for example, Williams, *Americans and Their Forests.*

37. Greeley, "The Relation of Geography to Timber Supply."

38. Rajala, *Clearcutting the Pacific Rain Forest.*

39. V. H. Jensen, *Lumber and Labor* (New York: Farrar and Rinehart, 1945).

40. W. H. Gibbons, *Logging in the Douglas Fir Region,* government print off (Washington, DC, 1918), 9.

41. See, for example, Jensen, *Lumber and Labor;* A. M. Prouty, *More Deadly Than War! Pacific Coast Logging, 1827-1981* (New York: Garland, 1988); Robbins, *Hard Times in Paradise;* White, *"It's Your Misfortune and None of My Own."*

42. M. Hibbard and J. Elias, "The Failure of Sustained-Yield Forestry and the Decline of the Flannel-Shirt Frontier," in *Forgotten Places: Uneven Development in Rural America,* ed. T. A. Lyson and W. W. Falk (Lawrence: University Press of Kansas, 1993), 195–215; B. Warf, "Regional Transformation, Everyday Life, and Pacific Northwest Lumber Production," *Annals of the Association of American Geographers* 78, no. 2 (1988): 326–46.

43. See also W. Boyd and S. Prudham, "Manufacturing Green Gold: Industrial Tree Improvement and the Power of Heredity in the Post-War United States," in *Industrializing Organisms: Introducing Evolutionary History,* ed. S. Schrepfer and P. Scranton (New York: Routledge, 2003), 107–42.

44. Dana and Fairfax, *Forest and Range Policy.*

45. Ibid.

46. Rajala, *Clearcutting the Pacific Rain Forest;* Robbins, *Lumberjacks and Legislators.*

47. During the early New Deal, private capital collaborated in the development of forest practice rules as part of the Lumber Industry Code (part of the National Industrial Recovery Act) but insisted on rules that did not mandate reforestation. When the National Industrial Recovery Act was ultimately struck down by the Supreme Court, persistent fears in the industry that forest practices would come under federal oversight then led to a push for legislation by local states. As a result, it was in fact Oregon that passed the nation's first state forest practices legislation—the Oregon Forest Conservation Act—in 1941, followed by Washington in 1945. Although Oregon's forest practices law did not initially require artificial reforestation (i.e., tree planting or seeding), it did establish restocking standards for cutover lands based on natural regeneration from nearby seed trees; W. D. Hagenstein, *Growing 40,000 Homes a Year* (Berkeley: School of Forestry and Conservation, University of California, Berkeley, 1973).

48. D. A. Adams, *Renewable Resource Policy: The Legal-Institutional Foundations* (Washington, DC: Island Press, 1993); and Dana and Fairfax, *Forest and Range Policy.*

49. G. P. Marsh and G. Marshall, *Man and Nature, or, Physical Geography as Modified by Human Action* (New York: Scribner, 1864). On its influence in this respect, see F. W. Cubbage, J. O'Laughlin, and C. S. Bullock, *Forest Resource Policy* (New York: John Wiley, 1993), 290.

50. As Steen pointed out, the section actually derived from an 1888 bill that sought to secure land to actual settlers as a counter to fraudulence and land consolidation; see H. K. Steen, "The Origins and Significance of the National Forest System," in *Origins of the National Forests: A Centennial Symposium,* ed. H. K. Steen (Durham, NC: Forest History Society, 1992), 3–9, especially p. 6. The political momentum also came from a desire to retain and conserve forest resources on the public domain, for public use; see B. De Voto, "The Easy Chair—The Sturdy Corporate Homesteader; Western Land Grabs," *Harper's* 206 (1953): 57–60. Henry S. Graves, Forest Division Chief from 1910 to 1920, characterized the reversal of disposal policies thus: "When the policy of deeding away the public timberlands was at last found an unsafe one for the Nation, it was changed and the bulk of the remaining public timberlands were withdrawn from public appropriation and segregated as national forests. In this way about 155 million acres, nearly all in the western mountains, were reserved. . . . The public forests are being protected from fire, the timber is used as it is called for by economic conditions, and the cutting is conducted by such methods as leave the land in favorable condition for the next crop of timber." Forest History Society, "Henry S. Graves," http://www.lib.duke.edu/forest/usfscoll/people/Graves/Graves.html (accessed January 5, 2004).

51. Dana and Fairfax, *Forest and Range Policy.*

52. G. C. Coggins, C. F. Wilkinson, and J. D. Leshy, *Federal Public Land and Resources Law* (Westbury, NY: Foundation Press, 1993).

53. This excludes private land within the National Forests. Total National Forest acreage in Oregon is 17.5 million acres, also fifth among states; U.S. Bureau of the Census, *Statistical Abstract of the United States, 1993* (Washington, DC: Author, 1994).

54. W. B. Smith et al., *Forest Resources of the United States, 1997* (St. Paul, MN: U.S. Department of Agriculture, Forest Service, 2001).

55. It is important to keep in mind that much of the forested land of the coastal Northwest makes very poor farmland. Valley bottoms have rich soils and are excellent for cultivation, for example, the Willamette Valley, but upland slopes are often too steep and too rocky and have very shallow acidic soils.

56. Richardson et al., *BLM's Billion-Dollar Checkerboard.*

57. In 1913 a federal district court decision found in the government's favor and ordered the lands forfeited. This was reversed by the Supreme Court, which found that the grant was made not on the basis of a "condition subsequent" but instead on the basis of a covenant. The Supreme Court decision therefore argued that the only possible way to revest the lands based on a violation of congressional direction would be for Congress to specifically order it; see Dana and Fairfax, *Forest and Range Policy.*

58. The BLM was created in 1946 and solidified as a permanent federal agency with a presidentially appointed secretary in 1976; Cubbage, O'Laughlin, and Bullock, *Forest Resource Policy.*

59. Smith et al., *Forest Resources of the United States, 1997.*

60. The Organic Act of 1897 had established a basis for federal timber programs, but actual harvests remained low for decades after. This was in part because of industry opposition to federal timber flooding the market and thereby depressing prices in an industry already plagued by overproduction and frequent price collapses.

61. Robbins, *Hard Times in Paradise;* Robbins, "Lumber Production and Community Stability"; Robbins, *Lumberjacks and Legislators.*

62. D. T. Mason, "Sustained Yield and American Forest Problems," *Journal of Forestry* 25, no. 10 (1927): 625–28.

63. N. V. Brasnett, *Planned Management of Forests* (London: Allen and Unwin, 1953).

64. B. Fernow, *Economics of Forestry* (New York: Thomas Y. Crawell, 1902); Society of American Foresters, *Forest Terminology* (Washington, DC: Author, 1944).

65. See Demeritt, "Scientific Forest Conservation"; Scott, *Seeing Like a State.*

66. On the Progressive period and resource management more broadly, see S. P. Hays, *Conservation and the Gospel of Efficiency: The Progressive Conservation Movement, 1890–1920* (New York: Atheneum, 1980).

67. Public No. 405, 75th Cong., Chapter 876, 1st sess., H.R. 7618.

68. The Sustained Yield Management Act went further than the O&C Act in its endorsement of sustained yield, instituting public-private working circles that would give exclusive access to National Forest timber to individual firms in exchange for the pooling of private and public timber into joint working circles to be managed according to sustained-yield principles. Yet only one such cooperative working circle was ever created. See Hibbard and Elias, "The Failure of Sustained-Yield Forestry."

69. W. J. Mead, *Competition and Oligopsony in the Douglas Fir Lumber Industry* (Berkeley: University of California Press, 1966).

70. There are intriguing and strong parallels between this model of social regulation in the Pacific Slope region of the United States and that under provincial jurisdiction in British Columbia during the postwar. In British Columbia, about 95 percent of forestland is publicly owned and controlled by the province, and the timber on these lands has been offered up to capital in exchange for the jobs that local processing would entail, together with conditions on forest management according to sustained-yield maxim and weakly enforced provisions guaranteeing timber volumes to specific communities. This political economic arrangement has been referred to as the "iron triangle" and "exploitation axis" of timber extraction in the province, creating a codependence between capital, the state, and workers and communities forged in the exploitation of the province's rich forest resources. See variously Barnes, Hayter, and Hay, "Stormy Weather"; R. Hayter, *Flexible Crossroads: The Restructuring of British Columbia's Forest Economy* (Vancouver, Canada: UBC Press, 2000); Rajala, *Clearcutting the Pacific Rain Forest*; D. J. Salazar and D. K. Alper, "Perceptions of Power and the Management of Environmental Conflict: Forest Politics in British Columbia," *Social Science Journal* 33, no. 4 (1996): 381–99.

71. K. Hackworth and B. Greber, "Timber Derived Revenues: Importance to Local Governments in Oregon," in *Assessment of Oregon's Forests 1988,* ed. G. J. Lettman (Salem: Oregon State Department of Forestry, 1988), 205–25.

72. Ibid.

73. Because firms are allowed some flexibility in when they actually harvest timber bought in Forest Service auctions, the total amount sold and cut in any year may not be the same. Moreover, planned sale and harvest levels may diverge, because the Forest Service may offer timber for sale that is not actually purchased.

74. Various annual statistical reports of the West Coast Lumbermen's Association and the Western Wood Products Association were used to compile these data. Representative samples are *West Coast Lumbermen's Association Industrial Facts* and *Western Wood Products Association Statistical Yearbook of the Western Lumber Industry.*

75. Interview with a longtime resident of the Illinois Valley and former mill worker and operator, conducted in Cave Junction, on July 18, 1996.

76. See chapter 4, and also Jensen, *Lumber and Labor;* and A. J. Van Tassel, *Mechanization in the Lumber Industry: A Study of Technology in Relation to Resources and Employment Opportunity* (Philadelphia, PA: Work Projects Administration, National Research Project on Employment Opportunities and Recent Changes in Industrial Techniques, 1940).

77. Data taken from the County Business Patterns for various years, available from the U.S. Department of the Census; U.S. Department of Commerce Bureau of the Census, *County Business Patterns* (Washington, DC: U.S. Bureau of the Census, 2000).

78. I do not wish to overstate the case here. There are other sources of income, particularly in the growing professional sector serving an aging local population. Also, there is considerable unrecorded income in the area from legal and illicit activities, including everything from mushroom picking, to firewood cutting and gathering, to cultivation of narcotic plants, including mushrooms and marijuana. However, no one familiar with the area and its history could fail to acknowledge its gradual slide into economically and socially depressed conditions during the 1970s and 1980s, as my fieldwork in the summer of 1996 made very plain.

79. Data taken from the 1990 census for census tract 3614, the Cave Junction County Census Division.

80. Personal communication, Josephine County Commission on Children and Families, 1996.

81. Interview conducted on June 14, 1996, with a local resident and economic development officer; interview conducted on July 20, 1996, with a member of the local town council.

82. Conditions were bad enough to motivate out-migration by locals in search of work elsewhere, particularly by men, who comprised 60 percent of the county's workforce in 1980.

Some found work in Grants Pass and were able to commute or relocate. Others found work outside the area, and they pulled their families with them. Still others found work elsewhere, promising to return later or arrange for a subsequent family relocation. But many simply left, and left their families behind, partly explaining why single-parent households rose from 10.6 percent of Josephine County households in 1980 to 13.0 percent of households in 1990.

83. Demeritt, "Scientific Forest Conservation."

84. See, for example, K. E. Franzsreb, "Perspectives on the Landmark Decision Designating the Northern Spotted Owl (*Strix Occidentalis Caurina*) as a Threatened Subspecies," *Environmental Management* 17, no. 4 (1993): 445–52; Marcot and Thomas, *Of Spotted Owls, Old Growth, and New Policies;* D. Simberloff, "The Spotted Owl Fracas: Mixing Academic, Applied, and Political Ecology," *Ecology* 68, no. 4 (1987): 766–72.

85. On these points, see P. W. Hirt, *A Conspiracy of Optimism: Management of the National Forests since World War Two* (Lincoln: University of Nebraska Press, 1994); and R. O'Toole, *Reforming the Forest Service* (Washington, DC: Island Press, 1988).

86. On fall down, see J. Beuter, K. Johnson, and H. L. Scheurman, *Timber for Oregon's Tomorrow: An Analysis of Reasonably Possible Occurrences* (Corvallis: Oregon State University, Forest Research Laboratory, 1977); R. Clapp, "The Resource Cycle in Forestry and Fishing," *Canadian Geographer* 42, no. 2 (1998): 129–44; Demeritt, "Scientific Forest Conservation"; M. P. Marchak, *Green Gold: The Forest Industry in British Columbia* (Vancouver, Canada: University of British Columbia Press, 1983); M. P. Marchak, S. L. Aycock, and D. M. Herbert, *Falldown: Forest Policy in British Columbia* (Vancouver: David Suzuki Foundation Ecotrust Canada, 1999).

87. The density of softwood timber growing stock on Oregon National Forest lands dropped from more than four thousand cubic feet per acre in 1952 to just more than thirty-two hundred cubic feet per acre in 1992 as the result of old-growth liquidation. Thus more area is required to generate the same volume of wood. This is compounded by the decreasing proportion of merchantable wood in smaller diameter logs. Data from D. S. Powell et al., *Forest Resources of the United States, 1992* (Fort Collins, CO: U.S. Department of Agriculture, Forest Service, 1993).

88. On community resource management institutions in general, see C. C. Gibson, E. Ostrom, and M. A. McKean, *People and Forests: Communities, Institutions, and Governance* (Cambridge, MA: MIT Press, 2000).

89. D. A. Clary, "What Price Sustained Yield? The Forest Service, Community Stability, and Timber Monopoly under the 1944 Sustained Yield Act," in *American Forests: Nature, Culture, and Politics,* ed. C. Miller (Lawrence: University Press of Kansas, 1997), 209–28; Robbins, "Lumber Production and Community Stability"; Schallau and Alston, "The Commitment to Community Stability"; and T. R. Waggener, "Community Stability as Forest Management Objective," *Journal of Forestry* 75 (1977): 710–14.

90. W. R. Freudenburg and R. Gramling, "Natural Resources and Rural Poverty: A Closer Look," *Society and Natural Resources* 7 (1994): 5–22.

91. K. Marx, *Capital: A Critique of Political Economy, Volume 1* (New York: Vintage Books, 1977), 783.

92. For general discussion of the flexibility turn in capitalism, see B. Harrison, *Lean and Mean: The Changing Landscape of Corporate Power in the Age of Flexibility* (New York: Basic Books, 1994); D. Harvey, *The Condition of Postmodernity : An Enquiry into the Origins of Cultural Change* (Oxford and Cambridge, MA: Blackwell, 1989); M. J. Piore and C. F. Sabel, *The Second Industrial Divide: Possibilities for Prosperity* (New York: Basic Books, 1984); A. Saxenian, *Regional Advantage: Culture and Competition in Silicon Valley and Route 128* (Cambridge, MA: Harvard University Press, 1994); and A. Sayer and R. Walker, *The New Social Economy: Reworking the Division of Labor* (Cambridge, MA: Blackwell, 1992). In the forest sector, see Barnes, Hayter, and Hay, "Stormy Weather"; Hayter, *Flexible Crossroads;* R. Hayter, "High Performance Organizations and Employment Flexibility: A Case Study of In Situ Change at the Powell River Paper Mill, 1980–1994," *Canadian Geographer* 41, no. 1 (1997): 26–40; R. Hayter and T. Barnes, "The Restructuring of British Columbia's Coastal Forest Sector: Flexibility Perspectives," in *Troubles in the Rainforest: British Columbia's Forest Economy in Transition,* vol. 33, ed. T. J. Barnes and R. Hayter (Victoria, Canada: Western Geographical Press, 1997), 181-202; and G. Norcliffe and J. Bates, "Imple-

menting Lean Production in an Old Industrial Space: Restructuring at Corner Brook, New-foundland, 1984-94," *Canadian Geographer* 41, no. 1 (1997): 41–60. Although there is some evidence of a shift in markets, and some implications for firm structures in the forest indus-try, the flexibility turn is hardly a return to cottage production, and the laws of competition seemingly still apply.

93. K. Marx and F. Engels, *The Communist Manifesto* (London: Verso, 1988).

94. See, for example, D. Harvey, *The Limits to Capital* (Oxford: Blackwell, 1982); D. Harvey, *Spaces of Hope* (Edinburgh: Edinburgh University Press, 2000); and Neil Smith, *Uneven Development: Nature, Capital, and the Production of Space* (Oxford: Blackwell, 1984).

95. See, for example, G. Arrighi and H. S. Fernand Braudel Center for the Study of Economies and Civilizations, *Capitalism and the Modern World-System: Rethinking the Non-Debates of the 1970s* (Binghamton: Fernand Braudel Center for the Study of Economies Historical Systems and Civilizations, State University of New York at Binghamton, 1997); Harvey, *The Condition of Postmodernity*; Harvey, *The Limits to Capital*; and J. A. Schumpeter, *Business Cycles: A Theoretical, Historical, and Statistical Analysis of the Capitalist Process* (New York and London: McGraw-Hill, 1939), as well as, of course, K. Polanyi, *The Great Transforma-tion* (Boston: Beacon Press, 1944).

96. G. Bridge, "The Social Regulation of Resource Access and Environmental Impact: Nature and Contradiction in the US Copper Industry," *Geoforum* 31, no. 2 (2000): 237–56. See also Boyd, Prudham, and Schurman, "Industrial Dynamics and the Problem of Nature."

97. O. Hirschman, "A Generalized Linkage Approach to Development with Special Reference to Staples," *Economic Development and Cultural Change* 25, Suppl. (1977): 67–98; and H. A. Innis, *Essays in Canadian Economic History* (Toronto: University of Toronto Press, 1956).

98. On this and related points, see variously, B. Barham, "Strategic Capacity Investments and the Alcoa-Alcan Monopoly, 1888–1945," in *States, Firms, and Raw Materials: The World Economy and Ecology of Aluminum*, ed. B. Barham, S. G. Bunker, and D. O'Hearn (Madi-son: University of Wisconsin Press, 1994), 69–110; B. Barham, S. G. Bunker, and D. O'Hearn, *States, Firms, and Raw Materials: The World Economy and Ecology of Aluminum* (Madi-son: University of Wisconsin Press, 1994); B. Bacham and O. Coomes, "Reinterpreting the Amazon Rubber Boom: Investment, the State, and Dutch Disease," *Latin Amercian Research Review* 29, no. 2 (1994): 73–109; T. Barnes et al., "Focus: A Geographical Appreciation of Harold A. Innis," *Canadian Geographer* 37, no. 4 (1993): 352–64; Barnes, Hayter, and Hay, "Stormy Weather"; S. G. Bunker, "Staples, Links, and Poles in the Construction of Regional Development Theory," *Sociological Forum* 4, no. 4 (1989): 589–610; W. Freudenberg, "Ad-dictive Economies: Extractive Industries and Vulnerable Localities in a Changing World Economy," *Rural Sociology* 57 (1992): 305–32; Freudenburg and Gramling, "Natural Re-sources and Rural Poverty"; G. Halseth, "Resource Town Employment: Perceptions in Small Town British Columbia," *Tijdschrift Voor Economische En Sociale Geografie* 90, no. 2 (1999): 196–210; R. Hayter and T. Barnes, "Innis' Staples Theory, Exports, and Recession: British Columbia, 1981–86," *Economic Geography* 66 (1990): 156–73; and M. Nord, "Natural Re-sources and Persistent Rural Poverty: In Search of the Nexus," *Society and Natural Resources* 7 (1994): 205–20.

99. Production data taken from Pacific Northwest Research Station, Forest Service, U.S. De-partment of Agriculture, *Production, Prices, Employment, and Trade*, RB-PNW-106 (Port-land, OR: U.S. Department of Agriculture, Forest Service, various years). Employment data taken from U.S. Bureau of the Census (1958–1997), *County Business Patterns*, "Oregon," http://quickfacts.census.gov/qfd/index.html (accessed May 20, 2003).

100. On this recession and its political economic circumstances, see G. Arrighi, *The Long Twen-tieth Century: Money, Power, and the Origins of Our Times* (London and New York: Verso, 1994).

101. See W. Turner, "Lumber Industry Woes Dim Good Life in Oregon," *New York Times*, March 3, 1982, A16.

102. Data from U.S. Bureau of the Census (1958–1997), *County Business Patterns*, "Oregon," http://quickfacts.census.gov/qfd/index.html; and D. Warren, *Harvest, Employment, Exports, and Prices in Pacific Northwest Forests, 1965–2000* (Portland, OR: U.S. Department of Ag-riculture, Forest Service, Pacific Northwest Research Station, 2002); D. Warren, *Production, Prices, Employment and Trade in Northwest Forest Industries, All Quarters 2000*, including various years of the annual report.

103. On technical change in the industry, see B. J. Greber, "Impacts of Technological Change on Employment in the Timber Industries of the Pacific Northwest," *Western Journal of Applied Forestry* 8, no. 1 (1993): 34–37; and B. J. Greber and D. E. White, "Technical Change and Productivity Growth in the Lumber and Wood Products Industry," *Forest Science* 28, no. 1 (1982): 135–47. More generally, see Freudenburg and Gramling, "Natural Resources and Rural Poverty: A Closer Look." For similar perspectives in the British Columbia forest industry, see Hayter, *Flexible Crossroads;* Marchak, *Green Gold;* M. P. Marchak, *Logging the Globe* (Montreal and Buffalo: McGill-Queen's University Press, 1995); and Marchak, Aycock, and Herbert, *Falldown.*

104. T. M. Power, *Lost Landscapes and Failed Economies: The Search for a Value of Place* (Washington, DC: Island Press, 1996).

105. E. P. Thompson, *The Making of the English Working Class* (Harmondsworth, UK: Penguin, 1968), 198.

106. On the economic impacts of the spotted owl decision, see V. Alaric and D. C. Le Master, "Economic Effects of Northern Spotted Owl Protection," *Journal of Forestry* 90 (1992): 31–35; K. R. Dixon and T. C. Juelson, "The Political Economy of the Spotted Owl," *Ecology* 68, no. 4 (1987): 772–76; C. A. Montgomery, G. M. Brown Jr., and D. M. Adams, "The Marginal Cost of Species Preservation: The Northern Spotted Owl," *Journal of Environmental Economics and Management* 26, no. (2) (1994): 111–28; and W. C. Stewart, California Department of Forestry and Fire Protection Strategic and Resources Planning Program and Forest and Rangeland Resources Assessment Program, *Predicting Employment Impacts of Changing Forest Management in California* (Sacramento: Forest and Rangeland Resources Assessment Program, California Department of Forestry and Fire Protection, 1993).

107. I take this term from Maarten Hajer's influential analysis of ecological modernization and the regulation of acid rain in the 1980s, even though I disagree with what I see as an overly idealist epistemology that confers great causal power to discourses relatively untethered from material interests, institutions, and political groups and movements. See M. Hajer, *The Politics of Environmental Discourse* (New York: Oxford University Press, 1995).

108. R. White, "Are You an Environmentalist or Do You Work for a Living? Work and Nature," in *Uncommon Ground: Toward Reinventing Nature,* ed. W. Cronon (New York: W. W. Norton, 1995), 171–85.

109. Interviews with representatives of the Northwest Forestry Association and the Oregon Forest Industries Council were conducted in Portland, in March 1995, and in Salem, on July 16, 1997, respectively.

Epilogue

1. G. Pinchot, *The Training of a Forester* (Philadephia, PA: Lippincott, 1917), 14; cited in D. Demeritt, "Scientific Forest Conservation and the Statistical Picturing of Nature's Limits in the Progressive-Era United States," *Environment and Planning D-Society and Space* 19, no. 4 (2001): 431–59, especially p. 445.

2. "Adaptive management" is a term associated most with ecologist Carl Walters of the University of British Columbia, circulated in scientific and resource management circles largely by a widely cited article by Walters and Holling in 1990; see C. J. Walters and C. S. Holling, "Large-Scale Management Experiments and Learning by Doing," *Ecology* 71, no. 6 (1990): 2060–68. It has become a key concept in ecosystem management parlance in part to reflect the inherent complexity and uncertainty of trying to manage complex, unpredictable systems—thus the idea of adaptation and flexibility in management. Also, the term is meant to capture the spirit of a more precautionary approach to management (a "do no harm" or at least "do no irreversible harm" ethic) combined with a commitment to intentional environmental transformation and management, not preservation. Adaptive management suggests caution and an intentional admission of incomplete knowledge prior to the commencement of management activities from which information may be gleaned for improved practices. This closes the gap between science and practises, and turns management into a sort of large-scale experiment in learning by doing. See C. Walters et al., "Ecosystem Modeling for Evaluation of Adaptive Management Policies in the Grand Canyon," *Conservation Ecology* 4, no. 2 (2000); C. J. Walters and R. Hilborn, "Ecological Optimization and Adaptive Management," *Annual Review of Ecology and Systematics* 9 (1978): 157–88.

3. B. G. Marcot and J. W. Thomas, *Of Spotted Owls, Old Growth, and New Policies: A History since the Interagency Scientific Committee Report* (Portland, OR: U.S. Department of Agriculture, Forest Service, Pacific Northwest Forest Research Station, 1997); and U.S. Forest Service and US. Bureau of Land Management, *Final Supplemental Environmental Impact Statement on Management of Habitat for Late-Successional and Old-Growth Forest Related Species within the Range of the Northern Spotted Owl.*

4. In their overview of the New Forestry, Swanson and Franklin argued that the new approach actually moves past so-called featured species altogether (including the examples they provided of the northern spotted owl and Douglas fir). They stated that the New Forestry moves instead "to ecosystems, and from the scale of forest stands to landscapes and the entire region"; F. J. Swanson and J. F. Franklin, "New Forestry Principles from Ecosystem Analysis of Pacific-Northwest Forests," *Ecological Applications* 2, no. 3 (1992): 262–74, especially p. 262. This is an important view, given the authors' position at the centre of the New Forestry revolution, but I am not sure that the story of the spotted owl as a catalyst for ecosystem management demonstrates that the age of indicator species is over, even if the science has purportedly moved on.

5. J. F. Franklin, "The New Forestry," *Journal of Soil and Water Conservation* 44, no. 6 (1989): 549.

6. Clearly, for instance, the changing view of disturbance and succession within New Forestry, and the role of clear-cutting as a form of disturbance, is tied to a move away from the more deterministic models of stability and succession under the influence of Clementsian ecology; see D. Demeritt, "Ecology, Objectivity and Critique in Writings on Nature and Human Societies," *Journal of Historical Geography* 20, no. 1 (1994): 22–37; and D. Worster, *Nature's Economy: A History of Ecological Ideas* Cambridge and New York: Cambridge University Press, 1994). Gurus of the New Forestry and ecosystem management would be among the first to point to these wider influences; see, for example, Franklin, "The New Forestry"; J. F. Franklin, "Preserving Biodiversity: Species, Ecosystems or Landscapes," *Ecological Applications* 3, no. 2 (1993): 202–05; J. F. Franklin, "Toward a New Forestry," *American Forests* 95, nos. 11, 12 (1989): 37–44; and Swanson and Franklin, "New Forestry Principles."

7. On these broader influences, see, for example, D. Bengston, "Changing Forest Values and Ecosystem Management," *Society and Natural Resources* 7, no. 6 (1994): 515–33; S. Hays, *Beauty, Health, and Permanence: Environmental Politics in the United States 1955–1985* (New York: Cambridge University Press, 1987); and R. Nash, *Wilderness and the American Mind* (New Haven, CT: Yale University Press, 1982).

8. Franklin, "Toward a New Forestry."

9. J. F. Franklin, *Forest Stewardship in an Ecological Age* (Syracuse: SUNY College of Environmental Science and Forestry, 1992); and J. F. Franklin, C. S. Bledsoe, and J. T. Callahan, "Contributions to the Long-Term Ecological Research Program," *BioScience* 40, no. 7 (1990): 509–23.

10. The IBP dates to 1964, with congressional funding through the National Science Foundation in the United States since 1970. The IBP was established in response the lack of basic knowledge about the earth's basic biological and ecological systems, yet students of the sociology of science might be quick to point out that the program was also an attempt to civilize and governmentalize global biology under the auspices of big science, an example of what Foucault described as biopower exercised on the earth as a "body." On the IBP and its importance in elevating ecology to Big Science, see C. Kwa, "Representations of Nature Mediating between Ecology and Science Policy: The Case of the International Biological Programme," *Social Studies of Science* 17, no. 3 (1987): 413–42. On the notion of biopower, see M. Foucault, *The History of Sexuality* (New York: Vintage Books, 1980).

11. In fact, it was in the Andrews forest during the 1970s that the first scientific fieldwork on the northern spotted owl was conducted by Oregon State University graduate student Eric Forsman; E. D. Forsman, "Habitat Utilization by Spotted Owls in the West-Central Cascades of Oregon" (master's thesis, Dept. of Biology, Oregon State University, Corvallis, 1980), 95; E. Forsman, "A Preliminary Investigation of the Northern Spotted Owl in Oregon," in *Biology* (Corvallis: Oregon State University, 1976), 127.

12. Franklin, Bledsoe, and Callahan, "Contributions to the Long-Term Ecological Research Program"; and F. Franklin et al., *Ecological Characteristics of Old-Growth Douglas-Fir Forests* (U.S. Department of Agriculture Forest Service, 1981).

13. Franklin, *Forest Stewardship in an Ecological Age;* Franklin, "Toward a New Forestry"; Swanson and Franklin, "New Forestry Principles from Ecosystem Analysis of Pacific-Northwest Forests."
14. This contrast is based on points offered by Swanson and Franklin, "New Forestry Principles from Ecosystem Analysis of Pacific-Northwest Forests," although in the article, the comparison with sustained-yield forestry is not always quite so explicit.
15. Franklin, "Toward a New Forestry."
16. Franklin, *Forest Stewardship in an Ecological Age;* Franklin, "Toward a New Forestry"; P. M. Frenzen, A. M. Delano, and C. M. Crisafulli, *Mount St. Helens, Biological Research Following the 1980 Eruption: An Indexed Bibliography and Research Abstracts (1980-1993)* (Portland, OR: U.S. Department of Agriculture, Forest Service, Pacific Northwest Forest Research Station, 1994); and P. M. Frenzen and J. F. Franklin, "Establishment of Conifers From Seed on Tephra Deposited by the 1980 Eruption of Mt. St. Helens, Washington," *American Midland Naturalist* 114, no. 1 (1985): 84–97.
17. Swanson and Franklin, "New Forestry Principles from Ecosystem Analysis of Pacific-Northwest Forests."
18. These comparisons and observations based primarily on Franklin, *Forest Stewardship in an Ecological Age;* Franklin, "Preserving Biodiversity"; and Swanson and Franklin "New Forestry Principles from Ecosystem Analysis of Pacific-Northwest Forests."
19. Franklin, "The New Forestry," 549.
20. The Gang of Four got its nickname from a derisive comment from an industry spokesman, but it was actually called the Scientific Panel on Late Successional Forest Ecosystems; see Marcot and Thomas, *Of Spotted Owls, Old Growth, and New Policies.*
21. On the notion of ecosocialization, see N. Low, "Ecosocialization and Environmental Planning: A Polanyian Approach," *Environment and Planning A* 34 (2002): 43–60.
22. Reading even the 1994 Final Supplemental Environmental Impact Statement, key to a transition from industrial, sustained-yield forestry to the new, ecosystem-based management, it would be easy to mistake the goal as one of narrowly ensuring viable populations of spotted owls remain in the Pacific Northwest for the foreseeable future U.S. Forest Service and U.S. Bureau of Land Management, *Final Supplemental Environmental Impact Statement.*
23. J. D. Proctor, "Whose Nature? The Contested Moral Terrain of Ancient Forests," in *Uncommon Ground: Toward Reinventing Nature,* ed. W. Cronon (New York: W. W. Norton, 1995), 269–97.
24. In fact, as Simberloff noted in his excellent paper on the science and the politics of the spotted owl controversy, other species cited as likely to parallel the spotted owl in key respects vis-à-vis dependence on old growth include the northern goshawk, vaux's swift, the silver-haired bat, the red tree vole, and the northern flying squirrel; D. Simberloff, "The Spotted Owl Fracas: Mixing Academic, Applied, and Political Ecology," *Ecology* 68, no. 4 (1987): 766–72. The marbled murrelet was also to emerge as a concern later, as were numerous species and populations of Pacific salmon whose spawning and rearing habitat was seriously degraded by the effects of logging on streams.
25. Ibid.
26. A. L. Hungerford, "Changing the Management of Public Land Forests: The Role of the Spotted Owl Injunctions (1993 Ninth Circuit Environmental Review)," *Environmental Law* 24, no. 3 (1994): 1395–43.
27. R. J. Gutierrez and A. Carey, *Ecology and Management of the Spotted Owl in the Pacific Northwest* (Arcata, CA: U.S. Forest Service, Southwest Research Station, 1984).
28. A. B. Carey, "A Summary of the Scientific Basis for Spotted Owl Management," in *Ecology and Management of the Spotted Owl in the Pacific Northwest,* ed. R. J. Gutierrez and A. B. Carey (Arcata, CA: U.S. Department of Agriculture Forest Service, 1985), PNW-185, 100–14. See also Simberloff, "The Spotted Owl Fracas." The 1994 Final Supplemental Environmental Impact Statement for managing old growth on Forest Service and BLM land in the Douglas fir region estimated that old-growth or late successional forests covered perhaps 20 percent of the regional landscape, as compared with 75 to 80 percent prior to 1800. The combined Forest Service and BLM statement also estimated that a total of 7.4 million acres of suitable owl habitat remained, roughly double the estimate in the Carey paper; see U.S. Forest Service and U.S. Bureau of Land Management, *Final Supplemental Environmental Impact Statement.*

29. U.S. Forest Service and U.S. Bureau of Land Management, *Final Supplemental Environmental Impact Statement.*
30. Ibid. With rates of annual habitat loss ranging from 1 to 3 percent, the annual survivorship for adult northern spotted owls was estimated to fall between 91 percent and 99 percent, with a median value of 95 percent at the 95 percent confidence interval and based on an aggregation of studies. In short, the best guess in the mid-1990s was that spotted owl populations were declining at 5 percent per year. If accurate, this would result in a halving of the existing population in fifteen years. Even at a rate of decline of 1 percent per year, the population would be reduced to half its early 1990s level by the middle of the twenty-first century.
31. The ISC helped bring the controversy to a head by prescribing instead the establishment of large Habitat Conservation Areas to protect nesting sites, entailing dramatic reductions or the outright elimination of logging within these areas; J. W. Thomas et al., *A Conservation Strategy for the Northern Spotted Owl: A Report to the Interagency Scientific Committee to Address the Conservation of the Northern Spotted Owl* (Washington, DC: U.S. Forest Service, U.S. Fish and Wildlife Service, and the National Park Service, 1990), 427.
32. B. W. Cashore, *Governing Forestry: Environmental Group Influence in British Columbia and the US Pacific Northwest* (Ph.D. dissertation, Toronto, Canada: University of Toronto, Graduate Department of Political Science, 1997), 329.
33. It does bear noting here that the Supreme Court overturned some key Ninth Circuit Court decisions, reinforcing the power of the Forest Service and the BLM; see V. M. Sher, "Travels with Strix: The Spotted Owl's Journey through the Federal Courts," *Public Land Law Review* 14 (1993): 41–79. Moreover, the discretion afforded these agencies to interpret prevailing scientific evidence, and the fact that the environmental impact statement mechanism under the NEPA remains consultative vis-à-vis the public, means that agency power over the administration of enabling laws and regulations is quite intact. Moreover, Congress also changed the ESA subsequent to the spotted owl case to strengthen protections of private-property owners against so-called federal takings and to give the government more flexibility in dealing with the problem of endangered species; see, for example, S. C. Peterson, *The Modern Ark: A History of the Endangered Species Act* (Ph.D. dissertation, Dept. of History, University of Wisconsin–Madison, 2000), 293. Nevertheless, the owl crisis certainly pointed to the potential created by federal regulations that enabled direct challenges by citizen groups.
34. See, for example, Cashore, *Governing Forestry;* Hungerford, "Changing the Management of Public Land Forests"; Marcot and Thomas, *Of Spotted Owls, Old Growth, and New Policies;* Peterson, *The Modern Ark;* and Sher, "Travels with Strix."
35. W. Dietrich, *The Final Forest: The Battle for the Last Great Trees of the Pacific Northwest* (New York: Simon & Schuster, 1992); Proctor, "Whose Nature?"; and T. Satterfield, *Anatomy of a Conflict: Identity, Knowledge, and Emotion in Old-Growth Forests* (Vancouver and Toronto: University of British Columbia Press, 2002).
36. R. Freeman, "The Ecofactory: The United States Forest Service and the Political Construction of Ecosystem Management," *Environmental History* 7, no. 4 (2002): 632-58; R. Freeman, *The U.S. Forest Service and the Political Construction of Ecosystem Management* (Ph.D. dissertation, University of Montana School of Forestry, Missoula, 1998), 288; Hungerford, "Changing the Management of Public Land Forests"; and Marcot and Thomas, *Of Spotted Owls, Old Growth, and New Policies.*
37. Cashore, *Governing Forestry;* and Hungerford, "Changing the Management of Public Land Forests."
38. V. M. Sher, "Ancient Forests, Spotted Owls, and the Demise of Federal Environmental Law," *Environmental Law Reporter* 20, no. 11 (1990): 10469-70; and Simberloff, "The Spotted Owl Fracas."
39. Simberloff, "The Spotted Owl Fracas."
40. BLM plans were even less conservation oriented, with lower numbers of set asides and smaller areas in each.
41. Sher, "Travels with Strix."
42. In the case of the BLM, the agency's timber management plans for the Pacific Northwest had been passed in 1983 with a ten-year horizon. In 1987 the agency elected not to review the plans in light of new evidence that spotted owls were on the decline, and therefore it decided not to complete a supplemental environmental impact statement. This was the basis

of an initial lawsuit filed by the Portland Audubon Society, claiming a violation of NEPA. The spotted owl's listing under the ESA strengthened the case against the agency's timber management plans; see ibid.

43. Freeman, "The Ecofactory"; and Freeman, *The U.S. Forest Service and the Political Construction of Ecosystem Management.*

44. See Hungerford, "Changing the Management of Public Land Forests"; and Sher, "Travels with Strix." Congress specifically passed Section 314 of the 1988 Department of the Interior Appropriations Act that restricted judicial review (lawsuits) against the BLM to individual agency activities, ostensibly precluding the challenge of the BLM's timber management plans. The Ninth Circuit ruled the timber management plans were individual actions and therefore not blocked from challenge by the rider, and the Supreme Court then overturned the Ninth Circuit, but by this time, the rider had expired. Congress also passed the infamous Section 318 of the Department of the Interior and Related Agencies Appropriations Act for Fiscal Year 1990, six months after a sweeping injunction had been issued by Judge William Dwyer of the federal district court for western Washington blocking 140 Forest Service timber sales in spotted owl habitat (*Seattle Audubon Society v. Robertson*). Section 318 blocked further judicial review of either Forest Service or BLM sales, removed the temporary injunction issued by Dwyer, and set the federal lands timber sale level for 1990 at 7.7 billion board feet, including 5.8 billion board feet from Oregon and Washington public lands.

45. In April 1993 the BLM actually backed down from its attempt to follow through with these exemptions as the Ninth Circuit Court held hearings on political interference in the agency's functioning by White House staff; Sher "Travels with Strix."

46. *Seattle Audubon Society v. Robertson,* No. C89-160WD, 1991 WL 180099 (W.D. Wash. Mar. 7, 1991), quoted in ibid., 66. Dwyer also noted the politics of the case, and the ways in which this had undermined the process of administering National Forest and federal environmental policy. Specifically, he wrote, "More is involved here than a simple failure by an agency to comply with its governing statute. The most recent violation of NFMA exemplifies a deliberate and systematic refusal by the Forest Service and the FWS [Fish and Wildlife Service] to comply with the laws protecting wildlife. This is not the doing of scientists, foresters, rangers, and others at the working levels of these agencies. It reflects decisions made by higher authorities in the executive branch of government" (*Seattle Audubon Society v. Evans,* 771 F.Supp. 1081 (W.D. Wash.) at 1090; cited in Sher, "Travels with Strix," 68.

47. *Seattle Audubon Society v. Mosely,* 798 F.Supp. 1484 (W.D. Wash. 1992) (Memorandum Decision and Injunction).

48. *Portland Audubon Society v. Lujan.*

49. For good summaries of this complex lineage of litigation, see Hungerford, "Changing the Management of Public Land Forests"; and Sher "Travels with Strix." The Ninth Circuit cases are *Seattle Audubon Society v. Babbitt* and *Portland Audubon Society v. Espy.*

50. Marcot and Thomas, *Of Spotted Owls, Old Growth, and New Policies.*

51. Freeman, "The Ecofactory."

52. U.S. Forest Service and United States, Bureau of Land Management, *Final Supplemental Environmental Impact Statement.*

53. Marcot and Thomas, *Of Spotted Owls, Old Growth, and New Policies;* U.S. Forest Service and U.S. Bureau of Land Management, *Final Supplemental Environmental Impact Statement.*

54. Neil Smith, *Uneven Development: Nature, Capital, and the Production of Space* (Oxford: Blackwell, 1984); and R. Williams, *The Country and the City* (London: Verso, 1973).

55. What I mean here is that, as Smith and others pointed out, wilderness would not be wilderness if it were not scarce. And it would not be scarce without capitalism. Wilderness is in this sense a social reaction to capitalism, one product of a Polanyian dual movement courtesy of the various wilderness advocacy movements of the past 150 years. Yet wilderness and wilderness advocacy do not challenge commodity production per se. Instead they create spaces that are refuges from it, without any need to acknowledge the underlying reasons why such a refuge is necessary. On these and related critiques of wilderness, see W. Cronon, "The Trouble with Wilderness or, Getting Back to the Wrong Nature," *Environmental History* 1, no. 1 (1996): 7-28; and Smith, *Uneven Development.*

56. R. White, "Are You an Environmentalist or Do You Work for a Living? Work and Nature," in *Uncommon Ground: Toward Reinventing Nature,* ed. W. Cronon (New York: W. W. Norton, 1995), 171–85.

57. I refer here, for example, to the embrace of community forestry concepts by the Forest Service and the BLM, although the political implications of this "Small Is Beautiful" turn in forest policy are somewhat ambiguous (i.e., not all communities are progressive or necessarily proconservation). Among the new breed of ENGOs embracing different ideas of social nature is the impressive Rogue Institute for Ecology and Economy (http:// www.forestsandcommunities.org/display.php3?id=201).

58. I speak here of the fundamental political questions concerning the division of labor under American capitalism during the second half of the past century, including issues such as conception versus execution in the workplace, the scope of collective bargaining concerns, and the embrace of a wider social justice agenda by the labor movement; see, for example, H. Braverman, *Labor and Monopoly Capital: The Degradation of Work in the Twentieth Century* (New York: Monthly Review Press, 1975); and M. Davis, *Prisoners of the American Dream: Politics and Economy in the History of the US Working Class* (London: Verso, 1986). The shape of compromise and struggle over these questions are among the primary drivers of Gramscii's idea of Fordism as a peculiarly American variant of capitalist society—on Grasmci and Fordism, see M. Burawoy, "For a Sociological Marxism: The Complementary Convergence of Antonio Gramsci and Karl Polanyi," *Politics and Society* 31, no. 2 (2003): 193–261. I am trying to suggest that these questions cannot be divorced from the emergence of a very particular ideological landscape in the politics of nature, one largely devoid of a contending alternative to capitalist nature as a set of use values and therefore vulnerable to the dichotomy between wilderness and commodity production.

59. Related points are made in S. Bernstein, "Ideas, Social Structure, and the Compromise of Liberal Environmentalism," *European Journal of International Relations* 6, no. 4 (2000): 464–512; S. Bernstein, "Liberal Environmentalism and Global Environmental Governance," *Global Environmental Politics* 2, no. 3 (2002): 1–16; and J. McCarthy and S. Prudham, "Neoliberal Nature and the Nature of Neoliberalism," *Geoforum* (2004).

Bibliography

Adam, B. *Timescapes of Modernity: The Environment and Invisible Hazards.* London: Routledge, 1998.
———. "The Temporal Gaze: The Challenge for Social Theory in the Context of GM Food." *British Journal of Sociology* 51, no. 1 (2000): 125–42.
Adams, D. A. *Renewable Resource Policy: The Legal-Institutional Foundations.* Washington, DC: Island Press, 1993.
Adams, W. T., V. D. Hipkins, J. Burczyk, and W. K. Randall. "Pollen Contamination Trends in a Maturing Douglas-Fir Seed Orchard." *Canadian Journal of Forest Research* 27 (1997): 131–34.
Aglietta, M. *A Theory of Capitalist Regulation: The US Experience.* London: NLB, 1979.
Alaric, V., and D. C. Le Master. "Economic Effects of Northern Spotted Owl Protection." *Journal of Forestry* 90 (1992): 31–35.
American Forest Council. *Forest Industry-Sponsored Research Cooperatives at U.S. Forestry Schools.* Washington, DC: Author, 1987.
Amin, A. "Post-Fordism: Models, Fantasies and Phantoms of Transition." In *Post-Fordism: A Reader,* edited by A. Amin, 1–39. Oxford and Cambridge, MA: Blackwell, 1994.
Arrighi, G. *The Long Twentieth Century: Money, Power, and the Origins of Our Times.* London and New York: Verso, 1994.
Arrighi, G., and H. S. Fernand Braudel Center for the Study of Economies, and Civilizations. *Capitalism and the Modern World-System: Rethinking the Non-Debates of the 1970s.* Binghamton: Fernand Braudel Center for the Study of Economies Historical Systems and Civilizations, State University of New York at Binghamton, 1997.
Atterbury Consultants. *Major Forest Landownership: Western Oregon and Southwest Washington.* Map. Portland, OR: Author, 1996.
Ayres, R. U., and L. Ayres. *Industrial Ecology: Towards Closing the Materials Cycle.* Cheltenham, UK, and Brookfield, VT: Elgar, 1996.
Baker, F. W. G. *Rapid Propagation of Fast-Growing Woody Species.* Wallingford, UK: C.A.B. International for CASAFA, 1992.
Bakker, K. J. "Privatising Water, Producing Scarcity: The Yorkshire Drought of 1995." *Economic Geography* 76, no. (2000): 4–27.
Baldwin, R. F. *Plywood Manufacturing Practices.* San Francisco, CA: Miller Freeman, 1981.
Barham, B. "Strategic Capacity Investments and the Alcoa-Alcan Monopoly, 1888-1945." In *States, Firms, and Raw Materials: The World Economy and Ecology of Aluminum,* edited by B. Barham, S. G. Bunker, and D. O'Hearn, 69-110. Madison: University of Wisconsin Press, 1994.
Barham, B., S. G. Bunker, and D. O'Hearn. *States, Firms, and Raw Materials: The World Economy and Ecology of Aluminum.* Madison: University of Wisconsin Press, 1994.

Barham, B., and O. Coomes. "Reinterpreting the Amazon Rubber Boom: Investment, the State, and Dutch Disease." *Latin American Research Review* 29, no. 2 (1994): 73–109.

Barnes, T., and R. Hayter. "The Little Town That Did: Flexible Accumulation and Community Response in Chemainus, British Columbia." *Regional Studies* 26 (1992): 647–63.

Barnes, T., I. Parker, C. Harris, and M. Gertler. "Focus: A Geographical Appreciation of Harold A. Innis." *Canadian Geographer* 37, no. 4 (1993): 352–64.

Barnes, T. J., and R. Hayter. "Economic Restructuring, Local Development and Resource Towns: Forest Communities in Coastal British Columbia." *Canadian Journal of Regional Science* 17, no. 3 (1994): 289–310.

Barnes, T. J., R. Hayter, and E. Hay. "Stormy Weather: Cyclones, Harold Innis, and Port Alberni, BC." *Environment and Planning A* 33 (2001): 2127–47.

Barnet, R. J., and J. Cavanagh. *Global Dreams: Imperial Corporations and the New World Order.* New York: Simon & Schuster, 1994.

Barnett, H. J., and C. Morse. *Scarcity and Growth; the Economics of Natural Resource Availability.* Washington, DC, and Baltimore, MD: Resources for the Future by Johns Hopkins Press, 1963.

Baumann, M. G. D. "Air Quality and Composite Wood Products." *Women in Natural Resources* 20, no. 4 (1999): 4–6.

Beck, U. *World Risk Society.* Malden, MA: Polity Press, 1999.

Beck, U., and M. Ritter. *Risk Society: Towards a New Modernity.* London and Newbury Park, CA, and New Delhi: Sage Publications, 1992.

Bengston, D. "Changing Forest Values and Ecosystem Management." *Society and Natural Resources* 7, no. 6 (1994): 515–33.

Bengston, D. N., and H. M. Gregerson. "Technical Change in the Forest-Based Sector." In *Emerging Issues in Forest Policy,* edited by P. N. Nemetz, 187–211. Vancouver, Canada: UBC Press, 1992.

Benton, T. "Marxism and Natural Limits: An Ecological Critique and Reconstruction." *New Left Review* 178 (1989): 51–86.

Bernstein, S. "Ideas, Social Structure, and the Compromise of Liberal Environmentalism." *European Journal of International Relations* 6, no. 4 (2000): 464–512.

———. "Liberal Environmentalism and Global Environmental Governance." *Global Environmental Politics* 2, no. 3 (2002): 1–16.

Berry, S. "Social Institutions and Access to Resources." *Africa* 59, no. 1 (1989): 41–55.

Beuter, J., K. Johnson, and H. L. Scheurman. *Timber for Oregon's Tomorrow: An Analysis of Reasonably Possible Occurrences.* Corvallis: Oregon State University, Forest Research Laboratory, 1977.

Bigbee, K. "Emissions Control: Background and Status of EPA's Maximum Achievable Control Technology Rule." *Engineered Wood Journal* 3, no. 1 (2000).

Blaikie, P. M., and H. C. Brookfield. *Land Degradation and Society.* London and New York: Methuen, 1987.

Boise Cascade Corporation. *Form 10-K.* Washington, DC: Securities and Exchange Commission, 1997.

Bordelon, M. A. "Genetic Improvement Opportunities." In *Assessment of Oregon's Forests 1988,* edited by G. J. Lettman, 231–39. Salem: Oregon State Department of Forestry, 1988.

Bourhill, B., and Oregon Department of Forestry. *History of Oregon's Timber Harvests and or Lumber Production: State Data, 1849 to 1992, County Data, 1925 to 1992.* Salem: Oregon State Department of Forestry, 1994.

Bowles, S., and H. Gintis. "The Revenge of Homo-Economicus: Contested Exchange and the Revival of Political Economy." *Journal of Economic Perspectives* 7, no. 1 (1993): 83–102.

Boyce, S. G. "The Good Old New Forestry." *Bioscience* 41, no. 2 (1991): 67.

Boyd, W. "Making Meat: Science, Technology, and American Poultry Production." *Technology and Culture* 42, no. 4 (2001): 631–64.

———. "Wonderful Potencies: Deep Structure and the Problem of Monopoly in Agricultural Biotechnology." In *Engineering Trouble: Biotechnology and Its Discontents,* edited by R. Schurman and D. Takahashi, 24–62. Berkeley: University of California Press, 2003.

Boyd, W., and S. Prudham. "Manufacturing Green Gold: Industrial Tree Improvement and the Power of Heredity in the Post-War United States." In *Industrializing Organisms: Introducing Evolutionary History,* edited by S. Schrepfer and P. Scranton, 107–42. New York: Routledge, 2003.

Boyd, W., S. Prudham, and R. Schurman. "Industrial Dynamics and the Problem of Nature." *Society and Natural Resources* 14, no. 7 (2001): 555–70.

Boyd, W., and M. Watts. "Agro-Industrial Just-in-Time." In *Globalising Food: Agrarian Questions and Global Restructuring,* edited by D. Goodman and M. Watts, 192–225. London and New York: Routledge, 1997.

Boyer, R. *The Regulation School: A Critical Introduction.* New York: Columbia University Press, 1990.

Brasnett, N. V. *Planned Management of Forests.* London: Allen and Unwin, 1953.

Braun, B. *The Intemperate Rainforest: Nature, Culture, and Power on Canada's West Coast.* Minneapolis: University of Minnesota Press, 2002.

Braverman, H. *Labor and Monopoly Capital: The Degradation of Work in the Twentieth Century.* New York: Monthly Review Press, 1975.

Brenner, R., and M. Glick. "The Regulation Approach: Theory and History." *New Left Review* 188 (1991): 45–120.

Bridge, G. "The Social Regulation of Resource Access and Environmental Impact: Nature and Contradiction in the US Copper Industry." *Geoforum* 31, no. 2 (2000): 237–56.

———. "Resource Triumphalism: Postindustrial Narratives of Primary Commodity Production." *Environment and Planning A* 33, no. 12 (2001): 2149–73.

Bridge, G., and P. McManus. "Sticks and Stones: Environmental Narratives and Discursive Regulation in the Forestry and Mining Sectors." *Antipode* 32, no. 1 (2000): 10–47.

Brown, B. A. *In Timber Country: Working People's Stories of Environmental Conflict and Urban Flight.* Philadelphia, PA: Temple University Press, 1995.

Brown, K. "Industry Hugs Biotech Trees." *Technology Review* 104, no. 2 (2001): 34.

Brunner, A. M., R. Mohamed, R. Meilan, L. A. Sheppard, W. H. Rottman, and S. H. Strauss. "Genetic Engineering of Sexual Sterility in Shade Trees." *Journal of Arboriculture* 24, no. 5 (1998): 263–73.

Bryant, R. L. "Political Ecology: An Emerging Research Agenda in Third-World Studies." *Political Geography* 11, no. 1 (1992): 12–36.

Bunker, S. G. "Staples, Links, and Poles in the Construction of Regional Development Theory." *Sociological Forum* 4, no. 4 (1989): 589–610.

Burawoy, M. "For a Sociological Marxism: The Complementary Convergence of Antonio Gramsci and Karl Polanyi." *Politics and Society* 31, no. 2 (2003): 193–261.

Busch, L. *Plants, Power, and Profit: Social, Economic, and Ethical Consequences of the New Biotechnologies.* Cambridge, MA: Blackwell, 1991.

Carey, A. B. "A Summary of the Scientific Basis for Spotted Owl Management." In *Ecology and Management of the Spotted Owl in the Pacific Northwest,* edited by R. J. Gutierrez and A. B. Carey, PNW-185: 100–14. Arcata, CA: U.S. Department of Agriculture Forest Service, 1985.

Cashore, B. W. "Governing Forestry: Environmental Group Influence in British Columbia and the U.S. Pacific Northwest." PhD diss., Graduate Department of Political Science, University of Toronto, Ontario, 1997.

Castells, M. *The Rise of the Network Society.* Cambridge, MA: Blackwell, 1996.

Castree, N. "The Nature of Produced Nature: Materiality and Knowledge Construction in Marxism." *Antipode* 27, no. 1 (1995): 12–48.

———. "Commodifying What Nature?" *Progress in Human Geography* 27, no. 3 (2003): 273–97.

Castree, N., and B. Braun. "The Construction of Nature and the Nature of Construction: Analytical and Political Tools for Building Survivable Futures." In *Remaking Reality: Nature at the Millennium,* edited by B. Braun and N. Castree, 3–42. London and New York: Routledge, 1998.

Chandler, A. D. *The Visible Hand: The Managerial Revolution in American Business.* Cambridge, MA: Belknap Press, 1977.

Chandler, A. D., and T. Hikino. *Scale and Scope: The Dynamics of Industrial Capitalism.* Cambridge, MA: Belknap Press, 1990.

Chase, A. *In a Dark Wood: The Fight over Forests and the Rising Tyranny of Ecology.* Boston: Houghton Mifflin, 1995.

Cheng, T. -Y., and T. H. Voqui. "Regeneration of Douglas-Fir Plantlets through Tissue Culture." *Science* 198, no. 4315 (1977): 306–307.

Ching, K. K. *Controlled Pollination of Douglas-Fir.* Corvallis: Oregon State University, Forest Research Center, 1960.

Ching, K. K., T. M. Ching, and D. P. Lavender. *Flower Induction in Douglas-Fir (Pseudotsuga Menziezii (Mirb.) Franco) by Fertilizer and Water Stress. Part I: Changes in Free Amino Acid Composition in Needles.* Siberian Branch, Nauka, USSR: Sexual Reproduction of Conifers, Academy of Science, 1973.

Clapp, R. "The Resource Cycle in Forestry and Fishing." *Canadian Geographer* 42, no. 2 (1998): 129–44.

Clary, D. A. "What Price Sustained Yield? The Forest Service, Community Stability, and Timber Monopoly under the 1944 Sustained Yield Act." In *American Forests: Nature, Culture, and Politics,* edited by C. Miller, 209–28. Lawrence: University Press of Kansas, 1997.

Cleveland, C. J., and D. I. Stern. "Productive and Exchange Scarcity: An Empirical Analysis of the U.S. Forest Products Industry." *Canadian Journal of Forest Research* 23 (1993): 1537–49.

Coase, R. "The Nature of the Firm." *Economica* 4 (1937): 386–405.

Coggins, G. C., C. F. Wilkinson, and J. D. Leshy. *Federal Public Land and Resources Law.* Westbury, NY: Foundation Press, 1993.

Copes, D., F. Sorensen, and R. Silen. "Douglas-Fir Seedling Grows 8 Feet Tall in Two Seasons." *Journal of Forestry* 67, no. 3 (1969): 174–75.

Corbridge, S. "Marxism, Post-Marxism, and the Geography of Development." In *New Models in Geography: The Political-Economy Perspective,* edited by R. Peet and N. J. Thrift, 224–54. London and Winchester, MA: Unwin-Hyman, 1989.

Cour, R. M. *The Plywood Age: A History of the Fir Plywood Industry's First Fifty Years.* Portland, OR: Douglas Fir Plywood Association by Binfords and Mort, 1955.

Cox, T. R. *Mills and Markets: A History of the Pacific Coast Lumber Industry to 1900.* Seattle: University of Washington Press, 1974.

Creager, A. N. H. "Biotechnology and Blood: Edwin Cohn's Plasma Fractionation Project, 1940–1953." In *Private Science: Biotechnology and the Rise of the Molecular Sciences,* edited by A. Thackray, 39–62. Philadelphia: University of Pennsylvania Press, 1998.

Cronon, W. *Nature's Metropolis: Chicago and the Great West.* New York: W. W. Norton, 1991.

———. "The Trouble with Wilderness." In *Uncommon Ground: Toward Reinventing Nature,* edited by W. Cronon, 69–90. New York: W.W. Norton, 1995.

———. "The Trouble with Wilderness or, Getting Back to the Wrong Nature." *Environmental History* 1, no. 1 (1996): 7–28.

Cubbage, F. W., J. O'Laughlin, and C. S. Bullock. *Forest Resource Policy.* New York: John Wiley, 1993.

Dana, S. T., and S. K. Fairfax. *Forest and Range Policy: Its Development in the United States.* New York: McGraw-Hill, 1980.

Davis, M. *Prisoners of the American Dream: Politics and Economy in the History of the US Working Class.* London: Verso, 1986.

Debell, D. S., and R. O. Curtis. "Silviculture and New Forestry in the Pacific-Northwest." *Journal of Forestry* 91, no. 12 (1993): 25–30.

Demeritt, D. "Ecology, Objectivity and Critique in Writings on Nature and Human Societies." *Journal of Historical Geography* 20, no. 1 (1994): 22–37.

———. "Science, Social Constructivism and Nature." In *Remaking Reality: Nature at the Millennium,* edited by B. Braun and N. Castree, 173–93. London and New York: Routledge, 1998.

———. "The Construction of Global Warming and the Politics of Science." *Annals of the Association of American Geographers* 91, no. 2 (2001): 307–37.

———. "Scientific Forest Conservation and the Statistical Picturing of Nature's Limits in the Progressive-Era United States." *Environment and Planning D-Society & Space* 19, no. 4 (2001): 431–59.

De Voto, B. "The Easy Chair—The Sturdy Corporate Homesteader; Western Land Grabs." *Harper's* 206 (1953): 57–60.

Dicken, P. *Global Shift: Transforming the World Economy.* New York: Guilford Press, 1998.

Dicken, S. N., and E. F. Dicken. *The Making of Oregon: A Study in Historical Geography.* Portland: Oregon Historical Society, 1979.

Dietrich, W. *The Final Forest: The Battle for the Last Great Trees of the Pacific Northwest.* New York: Simon & Schuster, 1992.

Dixon, K. R., and T. C. Juelson. "The Political Economy of the Spotted Owl." *Ecology* 68, no. 4 (1987): 772–76.

Doak, D. "Spotted Owls and Old Growth Logging in the Pacific Northwest." *Conservation Biology* 3, no. 4 (1989): 389–96.

Donahue, R. A., T. D. Davis, C. H. Michler, D. E. Riemenschneider, D. R. Carter, P. E. Marquardt, D. Sankhla, B. E. Haissig, and J. G. Isebrands. "Growth, Photosynthesis, and Herbicide Tolerance of Genetically Modified Hybrid Poplar." *Canadian Journal of Forest Research* 24, no. 12 (1994): 2377–83.

Dryzek, J. S. *The Politics of the Earth: Environmental Discourses.* Oxford and New York: Oxford University Press, 1997.

Duchesne, L. C. "Impact of Biotechnology on Forest Ecosystems." *Forestry Chronicle* 69, no. 3 (1993): 307–13.

Dunn, J. B., ed. *Land in Common: An Illustrated History of Jackson County.* Medford: Southern Oregon Historical Society, 1993.

Egan, A., and C. Alerich " 'Danger Trees' in Central Appalachian Forests in the United States: An Assessment of Their Frequency of Occurrence." *Journal of Safety Research* 29, no. 2 (1998): 77–85.

Ellefson, P. V. "Forestry Research Undertaken by Private Organizations in Canada and the United States: A Review and Assessment." In *The Role of the Private Sector in Forestry Research: Recent Developments in Industrialized Countries,* edited by F. S. P. Ng, F. M. Schlegel, and R. Romeo, 73–144. Rome: Food and Agriculture Organization of the United Nations, 1995.

Ellefson, P. V., and R. N. Stone. *U.S. Wood-Based Industry: Industrial Organization and Performance.* New York: Praeger, 1984.

Engels, F. *The Condition of the Working Class in England.* Oxford and New York: Oxford University Press, 1993.

Escobar, A. *Encountering Development: The Making and Unmaking of the Third World.* Princeton, NJ: Princeton University Press, 1995.

———. "Constructing Nature: Elements for a Poststructural Political Ecology." In *Liberation Ecologies: Environment, Development, Social Movements,* edited by R. Peet and M. Watts, 46–68. New York: Routledge, 1996.

Falk, R. H., and F. Colling. "Laminating Effects in Glued-Laminated Timber Beams." *Journal of Structural Engineering* 121, no. 12 (1995): 1857–63.

Fernow, B. *Economics of Forestry.* New York: Thomas Y. Crawell, 1902.

Ferrell, W. K., and S. E. Woodward. "Effects of Seed Origin on Drought Resistance of Douglas-Fir (Pseudotsuga Menziesii) (Mirb.) Franco." *Ecology* 47, no. 3 (1966): 499–503.

Fishlow, A. *American Railroads and the Transformation of the Ante-Bellum Economy.* Cambridge: Harvard University Press, 1965.

Forestry Research Task Force, U.S. Department of Agriculture, and Association of State Universities and Land Grant Colleges. *A National Program of Research for Forestry.* Washington, DC: Research Program Development and Evaluation Staff, U.S. Department of Agriculture, 1967.

Forsman, E. D. "A Preliminary Investigation of the Northern Spotted Owl in Oregon." Master's thesis, Department of Biology, Oregon State University, Corvallis, 1976.

———. "Habitat Utilization by Spotted Owls in the West-Central Cascades of Oregon." PhD diss., Department of Biology, Oregon State University, Corvallis, 1980.

Foucault, M. *The History of Sexuality.* New York: Vintage Books, 1980.

Franklin, J. F. "The New Forestry." *Journal of Soil and Water Conservation* 44, no. 6 (1989): 549.

———. "Toward a New Forestry." *American Forests* 95, nos. 11, 12 (1989): 37–44.

———. *Forest Stewardship in an Ecological Age.* Syracuse: SUNY College of Environmental Science and Forestry, 1992.

———. "Preserving Biodiversity: Species, Ecosystems or Landscapes." *Ecological Applications* 3, no. 2 (1993): 202–205.

———., C. S. Bledsoe, and J. T. Callahan. "Contributions to the Long-Term Ecological Research Program." *BioScience* 40, no. 7 (1990): 509–23.

———, K. Cromack Jr., W. Denison, A. McKee, C. Maser, J. Sedell, F. Swanson, and G. Juday. *Ecological Characteristics of Old-Growth Douglas-Fir Forests.* U.S. Department of Agriculture, Forest Service, USA General Technical Report PNW-118, 1981.

——— and C. T. Dyrness. *Natural Vegetation of Oregon and Washington.* Portland, OR: U.S. Department of Agriculture Forest Service, 1973.

Franzsreb, K. E. "Perspectives on the Landmark Decision Designating the Northern Spotted Owl (*Strix Occidentalis Caurina*) as a Threatened Subspecies." *Environmental Management* 17, no. 4 (1993): 445–52.

Freeman, H. G., and W. C. Grendon. "Formaldehyde Detection and Control in the Wood Industry." *Forest Products Journal* 21, no. 9 (1971): 54–57.

Freeman, R. "The U.S. Forest Service and the Political Construction of Ecosystem Management." PhD diss., School of Forestry, University of Montana, Missoula, 1998.

———. "The Ecofactory: The United States Forest Service and the Political Construction of Ecosystem Management." *Environmental History* 7, no. 4 (2002): 632–58.

Freidman, S. T., and G. S. Foster "Forest Genetics on Federal Lands in the United States: Public Concerns and Policy Responses." *Canadian Journal of Forest Resources* 27 (1997): 401–408.

Frenzen, P. M., A. M. Delano, and C. M. Crisafulli, *Mount St. Helens, Biological Research Following the 1980 Eruption: And Indexed Bibliography and Research Abstracts (1980–1993)*. U.S. Department of Agriculture, Forest Service, Pacific Northwest Forest Research Station, Portland, OR. GTR-342 94-157, 1994.

Frenzen, P. M., and J. F. Franklin. "Establishment of Conifers Form Seed on Tephra Deposited by the 1980 Eruption of Mt. St. Helens, Washington." *American Midland Naturalist* 114, no. 1 (1985): 84–97.

Freudenberg, W. "Addictive Economies: Extractive Industries and Vulnerable Localities in a Changing World Economy." *Rural Sociology* 57 (1992): 305–32.

Freudenburg, W. R., and R. Gramling. "Natural Resources and Rural Poverty: A Closer Look." *Society and Natural Resources* 7 (1994): 5–22.

Friedland, W. H., A. E. Barton, and R. J. Thomas. *Manufacturing Green Gold: Capital, Labor, and Technology in the Lettuce Industry*. Cambridge Cambridgeshire and New York: Cambridge University Press, 1981.

Galarza, E. *Merchants of Labor: The Mexican Bracero Story: An Account of the Managed Migration of Mexican Farm Workers in California 1942 1960*. San Jose, CA: Rosicrician Press, 1964.

Gedney, D. R. "The Private Timber Resource." In *Assessment of Oregon's Forests 1988*, edited by G. J. Lettman, 53–57. Salem: Oregon State Department of Forestry, 1988.

Gedney, D. R., and S. E. Corder. "Residue Use Is Basis for Pulp Industry Expansion." *The Timberman* 57 (1956).

Georgia-Pacific Corporation. *Form 10-K405*. Washington, DC: Securities and Exchange Commission, 1999.

Gertler, M. "The Limits to Flexibility: Comments on the Post-Fordist Vision of Production and Its Geography." *Transactions, Institute of British Geographers* 13 (1988): 419 32.

Gibbons, W. H. *Logging in the Douglas Fir Region*. Washington, DC: U.S. Government Printing Office, 1918.

Gibbs, D. "Integrating Sustainable Development and Economic Restructuring: A Role for Regulation Theory?" *Geoforum* 27, no. 1 (1996): 1 10.

———. "Ecological Modernisation, Regional Economic Development, and Regional Development Agencies." *Geoforum* 31 (2000): 9 19.

Gibbs, D., and A. E. G., Jonas. "Governance and Regulation in Local Environmental Policy: The Utility of a Regime Approach." *Geoforum* 31, no. 3 (2000): 299 313.

Gibson, C. C., E. Ostrom, and M. A. McKean. *People and Forests: Communities, Institutions, and Governance*. Cambridge: MIT Press, 2000.

Giddens, A. *Central Problems in Social Theory: Action, Structure, and Contradiction in Social Analysis*. Berkeley: University of California Press, 1979.

———. *A Contemporary Critique of Historical Materialism*. Houndmills, UK: Macmillan, 1995.

Gillis, A. M. "The New Forestry—An Ecosystem Approach to Land Management." *Bioscience* 40, no. 8 (1990): 558–562.

Goldfarb, B., and J. B. Zaerr. "Douglas-Fir." In *Biotechnology in Agriculture and Forestry*, edited by Y. P. S. Bajaj, 526 48. Berlin: Springer-Verlag, 1989, vol. 5, 562–548.

Goodman, D., and M. R. Redclift. *Refashioning Nature: Food, Ecology, and Culture*. London and New York: Routledge, 1991.

Goodman, D., B. Sorj, and J. Wilkinson. *From Farming to Biotechnology: A Theory of Agro-Industrial Development*. Oxford and New York: Blackwell, 1987.

Goodman, D., and M. Watts. "Reconfiguring the Rural or Fording the Divide? Capitalist Restructuring and the Global Agro-Food System." *Journal of Peasant Studies* 22, no. 1 (1994): 1–49.

Graham, J., and K. St. Martin. "Resources and Restructuring in the International Solid Wood Products Industry." *Geoforum* 20, no. 24 (1989): 11–24.

Greber, B. J. "Impacts of Technological Change on Employment in the Timber Industries of the Pacific Northwest." *Western Journal of Applied Forestry* 8, no. 1 (1993): 34–37.

Greber, B. J., and D. E. White. "Technical Change and Productivity Growth in the Lumber and Wood Products Industry." *Forest Science* 28, no. 1 (1982): 135–47.

Greeley, W. B. "The Relation of Geography to Timber Supply." *Economic Geography* 1, no. 1 (1925): 1–14.

Gutierrez, R. J., and A. Carey. *Ecology and Management of the Spotted Owl in the Pacific Northwest.* U.S. Forest Service, Southwest Research Station, Arcata, CA, PNW-185, 1984.

Hackworth, K., and B. Greber. "Timber Derived Revenues: Importance to Local Governments in Oregon." In *Assessment of Oregon's Forests 1988,* edited by G. J. Lettman, 205–25. Salem: Oregon State Department of Forestry, 1988.

Hagenstein, W. D. *Growing 40,000 Homes a Year.* Berkeley: School of Forestry and Conservation, University of California, Berkeley, 1973.

Hajer, M. *The Politics of Environmental Discourse.* New York: Oxford University Press, 1995.

Halseth, G. "Resource Town Employment: Perceptions in Small Town British Columbia." *Tijdschrift Voor Economische En Sociale Geografie* 90, no. 2 (1999): 196–210.

Hannum, J. *Oregon Lumber and Wood Products Employment as a Percentage of Total Manufacturing Employment,* personal communication, 1995.

Haraway, D. J. *Modest-Witness, Second-Millennium: Femaleman Meets Oncomouse: Feminism and Technoscience.* New York: Routledge, 1997.

Harrison, B. *Lean and Mean: The Changing Landscape of Corporate Power in the Age of Flexibility.* New York: Basic Books, 1994.

Hartwick, E. "Geographies of Consumption: A Commodity-Chain Approach." *Environment and Planning D-Society & Space* 16, no. 4 (1998): 423–37.

Hartwick, E. R. (2000). "Towards a Geographical Politics of Consumption." *Environment and Planning A* 32, no. 7: 1177–92.

Harvey, D. "Population, Resources and the Ideology of Science." *Economic Geography* 50 (1974): 256–77.

———. *The Limits to Capital.* Oxford: Blackwell, 1982.

———. *The Urbanization of Capital: Studies in the History and Theory of Capitalist Urbanization.* Baltimore, MD: Johns Hopkins University Press, 1985.

———. *The Condition of Postmodernity: An Enquiry into the Origins of Cultural Change.* Oxford and Cambridge, MA: Blackwell, 1989.

———. *The Urban Experience.* Baltimore, MD: Johns Hopkins University Press, 1989.

———. *Justice, Nature, and the Geography of Difference.* Cambridge, MA: Blackwell, 1996.

———. *Spaces of Hope.* Edinburgh, UK: Edinburgh University Press, 2000.

Hays, S. *Beauty, Health, and Permanence: Environmental Politics in the United States 1955–1985.* New York: Cambridge University Press, 1987.

Hays, S. P. *Conservation and the Gospel of Efficiency: The Progressive Conservation Movement, 1890–1920.* New York: Atheneum, 1980.

Hayter, R. "High Performance Organizations and Employment Flexibility: A Case Study of In Situ Change at the Powell River Paper Mill, 1980–1994." *Canadian Geographer* 41, no. 1 (1997): 26–40.

———. *Flexible Crossroads: The Restructuring of British Columbia's Forest Economy.* Vancouver, Canada: UBC Press, 2000.

Hayter, R., and T. Barnes. "Innis' Staples Theory, Exports, and Recession: British Columbia, 1981-86." *Economic Geography* 66 (1990): 156–73.

———. "The Restructuring of British Columbia's Coastal Forest Sector: Flexibility Perspectives." In *Troubles in the Rainforest: British Columbia's Forest Economy in Transition,* vol. 33, edited by T. J. Barnes and R. Hayter, 181–202. Victoria, Canada: Western Geographical Press, 1997.

———. "Troubles in the Rainforest: British Columbia's Forest Economy in Transition." In *Troubles in the Rainforest: British Columbia's Forest Economy in Transition*, edited by T. Barnes and R. Hayter, 1–11. Victoria, Canada: Western Geographical Press, 1997.

Hecht, S. B., and A. Cockburn. *The Fate of the Forest: Developers, Destroyers, and Defenders of the Amazon.* London and New York: Verso, 1989.

Hee, S. "Intensive Silviculture Stewardship: The Weyerhaeuser Experience. " Presented at the Silviculture Conference: Stewardship in the New Forest, Forestry Canada, Vancouver, 1991.

Heilman, R. L. *Overstory— Zero: Real Life in Timber Country.* Seattle, WA: Sasquatch Books, 1995.

Henderson, G. "Nature and Fictitious Capital: The Historical Geography of an Agrarian Question." *Antipode* 30, no. 2 (1998): 73–118.

Henderson, G. L. *California & the Fictions of Capital.* New York: Oxford University Press, 1999.

Hibbard, M., and J. Elias. "The Failure of Sustained-Yield Forestry and the Decline of the Flannel-Shirt Frontier." In *Forgotten Places: Uneven Development in Rural America,* edited by T. A. Lyson and W. W. Falk, 195–215. Lawrence: University Press of Kansas, 1993.

Hirschman, O. "A Generalized Linkage Approach to Development with Special Reference to Staples." *Economic Development and Cultural Change* 25, Suppl. (1977): 67–98.

Hirt, P. W. *A Conspiracy of Optimism: Management of the National Forests since World War Two.* Lincoln: University of Nebraska Press, 1994.

Holmes, J. "The Organization and Locational Structure of Production Subcontracting." In *Production, Work, Territory: The Geographical Anatomy of Industrial Capitalism*, edited by A. J. Scott and M. Storper, 80–106. Boston: Allen & Unwin, 1986.

———. "In Search of Competitive Efficiency: Labour Process Flexibility in Canadian Newsprint Mills." *Canadian Geographer* 41, no. 1 (1997): 7–25.

Howd, C. R., and U.S. Bureau of Labor Statistics. *Industrial Relations in the West Coast Lumber Industry : December, 1923.* Washington, DC: U.S. Government Printing Office, 1924.

Hungerford, A. L. "Changing the Management of Public Land Forests: The Role of the Spotted Owl Injunctions. (1993 Ninth Circuit Environmental Review)." *Environmental Law* 24, no. 3 (1994): 1395–434.

Imam, S. H., S. H. Gordon, L. J. Mao, and L. Chen. "Environmentally Friendly Wood Adhesive from a Renewable Plant Polymer: Characteristics and Optimization." *Polymer Degradation and Stability* 73, no. 3 (2001): 529–33.

Innis, H. A. *Essays in Canadian Economic History.* Toronto, Canada: University of Toronto Press, 1956.

Innis, M. Q. *An Economic History of Canada.* Toronto, Canada: Ryerson Press, 1954.

James, R. R., S. P. DiFazio, A. M. Brunner, and S. H. Strauss. "Environmental Effects of Genetically Engineered Woody Biomass Crops." *Biomass and Bioenergy* 14, no. 4 (1998): 403–14.

Jayawickrama, K. *The Northwest Tree Improvement Cooperative.* Audiovisual presentation. Wilsonville, OR: 2001.

Jensen, D., and G. Draffan. *Railroads and Clearcuts: Legacy of Congress's 1864 Pacific Railroad Land Grant.* Spokane, WA: Inland Empire Public Lands Council, 1995.

Jensen, V. H. *Lumber and Labor.* New York: Farrar and Rinehart, 1945.

Kauppinen, T. "Occupational Exposure to Chemical Agents in the Plywood Industry." *Annals of Occupational Hygiene* 30, no. 1 (1986): 19–29.

Kauppinen, T., T. J. Partanen, M. M. Nurminen, J. I. Nickels, S. G. Hernberg, T. R. Hakulinen, E. I. Pukkala, and E. T. Savonen. "Respiratory Cancer and Chemical Exposures in the Wood Industry: A Nested Case-Control Study." *British Journal of Industrial Medicine* 43, no. 2 (1986): 84–90.

Kautsky, K. *The Agrarian Question.* London and Winchester, MA: Zwan Publications, 1988.

Kay, L. "Problematizing Basic Research in Molecular Biology." In *Private Science: Biotechnology and the Rise of the Molecular Sciences,* edited by A. Thackray, 20-38. Philadelphia: University of Pennsylvania Press, 1998.

Kay, L. E. *Who Wrote the Book of Life? A History of the Genetic Code.* Stanford, CA: Stanford University Press, 2000.

Kendrick, J. W. *Productivity Trends in the United States.* Princeton, NJ: Princeton University Press, 1961.

Kenney, M. "Biotechnology and the Creation of a New Economic Space." In *Private Science: Biotechnology and the Rise of the Molecular Sciences,* edited by A. Thackray, 131-43. Philadelphia: University of Pennsylvania Press, 1998.

Kevles, D. Diamond V. Chakrabarty and Beyond: The Political Economy of Patenting Life." In *Private Science: Biotechnology and the Rise of the Molecular Sciences,* edited by A. Thackray, 65–79. Philadelphia: University of Pennsylvania Press, 1998.

Kloppenburg, J. R. *First the Seed: The Political Economy of Plant Biotechnology, 1492–2000.* Cambridge Cambridgeshire and New York: Cambridge University Press, 1988.

Krimsky, S., and R. P. Wrubel. *Agricultural Biotechnology and the Environment: Science, Policy, and Social Issues.* Urbana: University of Illinois Press, 1996.

Kunesh, R. H., and J. W. Johnson. "Effect of Single Knots on Tensile Strength of 2- by 8-Inch Douglas-Fir Dimension Lumber." *Forest Products Journal* 22, no. 1 (1972): 32–36.

Kwa, C. "Representations of Nature Mediating between Ecology and Science Policy: The Case of the International Biological Programme." *Social Studies of Science* 17, no. 3 (1987): 413–42.

Larsen, C. S. *Genetics in Silviculture.* London: Oliver and Boyd, 1956.

Leff, E. *Green Production: Toward an Environmental Rationality.* New York and London: Guilford Press, 1995.

Le Master, D. C. *Mergers among the Largest Forest Products Firms 1950–1970.* Pullman: Washington State University, College of Agriculture Research Center, 1977.

LeVan, S. L. "Life Cycle Assessment: Measuring Environmental Impact." Presented at Forest Products Society 49th Annual Meeting, 7–16, Portland, Oregon, 1995.

Levidow, L. "Democratizing Technology—Or Technologizing Democracy? Regulating Agricultural Biotechnology in Europe." *Technology in Society* 20 (1998): 211–26.

Lewontin, R. C. *Biology as Ideology: The Doctrine of DNA.* Concord, Canada: Anansi, 1991.

London, J. "Common Roots and Entangled Limbs: Earth First! And the Growth of Post-Wilderness Environmentalism on California's North Coast." *Antipode* 30, no. 2 (1998): 155–76.

Low, N. "Ecosocialisation and Environmental Planning: A Polanyian Approach." *Environment and Planning A* 34 (2002): 43–60.

Lucia, E. *The Big Woods: Logging and Lumbering, from Bull Teams to Helicopters, in the Pacific Northwest.* Garden City, NY: Doubleday, 1975.

MacDonald, P., and M. Clow. " 'Just One Damn Machine after Another'? Technological Innovation and the Industrialization of Tree Harvesting Systems." *Technology in Society* 21 (1999): 323–44.

Mackie, G. *The Rise and Fall of the Forest Workers' Cooperatives of the Pacific Northwest.* Master's thesis, Department of Political Science, University of Oregon, Eugene, 1990.

Makinen, M., P. Kalliokoski, and J. Kangas. "Assessment of Total Exposure to Phenol-Formaldehyde Resin Glue in Plywood Manufacturing." *International Archives of Occupational and Environmental Health* 72, no. 5 (1999): 309–14.

Malaka, T., and A. M. Kodama. "Respiratory Health of Plywood Workers Exposed to Formaldehyde." *Archives of Environmental Health* 45 (1993): 288–94.

Mann, S. A. *Agrarian Capitalism in Theory and Practice.* Chapel Hill: University of North Carolina Press, 1990.

Mann, S. A., and J. M. Dickinson. "Obstacles to the Development of a Capitalist Agriculture." *Journal of Peasant Studies* 5, no. 4 (1978): 466–81.

Manock, E. R., G. A. Choate, and D. R. Gedney. *Oregon Timber Industries: Wood Consumption and Mill Characteristics.* Salem: Oregon State Department of Forestry, 1968.

Marchak, M. P. *Green Gold: The Forest Industry in British Columbia.* Vancouver, Canada: University of British Columbia Press, 1983.

———. *Logging the Globe.* Montreal, Canada, and Buffalo, NY: McGill-Queen's University Press, 1995.

Marchak, M. P., S. L. Aycock, and D. M. Herbert. *Falldown: Forest Policy in British Columbia.* Vancouver: David Suzuki Foundation, Ecotrust Canada, 1999.

Marcot, B. G., and J. W. Thomas. *Of Spotted Owls, Old Growth, and New Policies: A History since the Interagency Scientific Committee Report.* U.S. Department of Agriculture, Forest Service, Pacific Northwest Forest Research Station, Portland, OR, GTR-408 97-068, 1997.

Margosian, R. "Initial Formaldehyde Emission Levels for Particleboard Manufactured in the United States." *Forest Products Journal* 40, no. 6 (1990): 19–20.

Marsh, G. P., and G. Marshall. *Man and Nature, or, Physical Geography as Modified by Human Action.* New York: Scribner, 1864.

Marx, K. *Capital: A Critique of Political Economy. Volume 2. The Process of Circulation of Capital.* New York: International Publishers, 1967.

———. *Capital: A Critique of Political Economy, Volume 1*. New York: Vintage Books, 1977.

Marx, K., and F. Engels. *The Communist Manifesto*. London: Verso, 1988.

Mason, D. T. "Sustained Yield and American Forest Problems." *Journal of Forestry* 25, no. 10 (1927): 625–28.

McCarthy, J. "First World Political Ecology: Lessons from the Wise Use Movement." *Environment and Planning A* 34 (2002): 1281–302.

McCarthy, J., and S. Prudham. "Neoliberal Nature and the Nature of Neoliberalism." *Geoforum* 35, no. 3 (2004): 275–283.

McCarthy, J. P. "The Political and Moral Economy of Wise Use." PhD diss., Department of Geography, University of California–Berkeley, 1999.

McLane, L. *First There Was Twogood: A Pictorial History of Northern Josephine County*. Grants Pass, OR: Josephine County Historical Society, 1996.

Mead, W. J. *Competition and Oligopsony in the Douglas Fir Lumber Industry*. Berkeley: University of California Press, 1966.

Mead, W. J., M. Schniepp, R. B. Watson, and U.S. Forest Service, Division of Timber Management. *Competitive Bidding for U.S. Forest Service Timber in the Pacific Northwest, 1963-83*. Washington, DC: U.S. Department of Agriculture, Forest Service Timber Management, 1983.

Meadows, D. H., and Club of Rome. *The Limits to Growth: A Report for the Club of Rome's Project on the Predicament of Mankind*. New York: Universe Books, 1972.

Miller Freeman. *1995 Lockwood-Post 'S Directory of the Pulp, Paper and Allied Trades*. San Francisco, CA: Author, 1994.

———. *Directory of the Wood Products Industry 1999*. San Francisco, CA: Author, 1999.

Mitchell, M. C. "Unionism and Productivity in the Western Sawmill Industry." PhD diss., Department of Economics, University of Oregon, Eugene, 1988.

Montgomery, C. A., G. M. Brown Jr., and D. M. Adams. "The Marginal Cost of Species Preservation: The Northern Spotted Owl." *Journal of Environmental Economics and Management* 26, no. 2 (1994): 111–28.

Montrey, H., and J. M. Utterback. "Current Status and Future of Structural Panels in the Wood Products Industry." *Technological Forecasting and Social Change* 38 (1990): 15–35.

Moody, R. C., R. Hernandez, and J. Y. Liu. "Glued Structural Members." In *Wood Handbook—Wood as an Engineering Material*. U.S. Department of Agriculture Forest Service, Forest Products Laboratory, Madison, WI, FPL-GTR-113, 1999.

Morgenstern, E. K. *Geographic Variation in Forest Trees: Genetic Basis and Application of Knowledge in Silviculture*. Vancouver, Canada: UBC Press, 1996.

Munger, T. T., and W. G. Morris. *Growth of Douglas-Fir Trees of Known Seed Source*. Portland, OR: U.S. Department of Agriculture, Pacific Northwest Forest Experiment Station, 1936.

Namkoong, G., R. A. Usanis, and R. R. Silen. "Age-Related Variation in Genetic Control of Height Growth in Douglas-Fir." *Theoretical and Applied Genetics* 42 (1972): 151–59.

Nash, R. *Wilderness and the American Mind*. New Haven, CT: Yale University Press, 1982.

National Science Foundation. *National Patterns of R&D Expenditures: 1996*. Washington, DC: Author, 96-333, 1996.

Neurath, P. *From Malthus to the Club of Rome and Back: Problems of Limits to Growth, Population Control, and Migrations*. Armonk, NY: M. E. Sharpe, 1994.

Norcliffe, G., and J. Bates. "Implementing Lean Production in an Old Industrial Space: Restructuring at Corner Brook, Newfoundland, 1984–94." *Canadian Geographer* 41, no. 1 (1997): 41–60.

Nord, M. "Natural Resources and Persistent Rural Poverty: In Search of the Nexus." *Society and Natural Resources* 7 (1994): 205–20.

Norgaard, R. "Economic Indicators of Resource Scarcity: A Critical Essay." *Journal of Environmental Economics and Management* 18 (1990): 19–25.

O'Connor, J. "Capitalism, Nature, Socialism: A Theoretical Introduction." *Capitalism, Nature, Socialism* 1, no. 1 (1988): 11–38.

———. *Natural Causes: Essays in Ecological Marxism*. New York: Guilford Press, 1998.

Ò hUallachàin, B., and R. A. Matthews. "Restructuring of Primary Industries: Technology, Labor, and Corporate Strategy and Control in the Arizona Copper Industry." *Economic Geography* 72, no. 2 (1996): 196–215.

Ó hUallachàin, B., and D. Wasserman. "Vertical Integration in a Lean Supply Chain: Brazilian Automobile Component Parts." *Economic Geography* 75, no. 1 (1999): 21–42.

Oregon Bureau of Labor. *"Vamonos Pal Norte" (Let's Go North) a Social Profile of the Spanish Speaking Migratory Farm Laborer.* Salem: Author, 1958.
———._*And Migrant Problems Demand Attention; the Final Report of the 1958-59 Migrant Farm Labor Studies in Oregon Including Material from the Preliminary Report of the Bureau of Labor (July 1959) Entitled "We Talked to the Migrants . . ."* Salem: Author, 1959.
Oregon Department of Forestry. *Timber Harvest Report.* Salem: Author, 1996.
———. *1995 Annual Reports.* Salem: Author, 1997.
———. *Timber Harvest Report.* Salem: Oregon Department of Forestry, 2002.
Oregon Interagency Committee on Migratory Labor. *1965 Report of the Interagency Committee on Migratory Labor.* Salem: Oregon Department of Agriculture, 1966.
Oregon State University. *Seasonal Agricultural Labor in Oregon; Task Force Report.* Salem: Author, 1968.
O'Toole, R. *Reforming the Forest Service.* Washington, DC: Island Press, 1988.
Pacific Northwest Tree Improvement Research Co-operative. *Annual Report, 1995–1996.* Corvallis: Forest Research Laboratory, Oregon State University, 1996.
———. *Annual Report, 2000-2001.* Corvallis: Oregon State University, 2002.
Padjen, E. S. "Engineered Lumber's Strengths." *Architecture* 86, no. 2 (1997): 104–108.
Peet, R., and M. Watts. "Introduction: Development Theory and Environment in an Age of Market Triumphalism." *Economic Geography* 69, no. 3 (1993): 227–53.
———. *Liberation Ecologies: Environment, Development, Social Movements.* London and New York: Routledge, 1996.
Peluso, N. L. *Rich Forests, Poor People: Resource Control and Resistance in Java.* Berkeley: University of California Press, 1992.
Perkins, J. H. *Geopolitics and the Green Revolution: Wheat, Genes, and the Cold War.* New York and Oxford: Oxford University Press, 1997.
Perry, T. D. *Modern Plywood.* New York: Pitman, 1948.
Peterson, S. C. "The Modern Ark: A History of the Endangered Species Act." PhD diss., Department of History, University of Wisconsin, Madison, 2000.
Pierik, R. L. M., and J. Prakash. *Plant Biotechnology, Commercial Prospects and Problems.* New Delhi: Oxford & IBH, 1993.
Pinchot, G. *The Training of a Forester.* Philadelphia, PA: Lippincott, 1917.
Piore, M. J., and C. F. Sabel. *The Second Industrial Divide: Possibilities for Prosperity.* New York: Basic Books, 1984.
Polanyi, K. *The Great Transformation.* Boston: Beacon Press, 1944.
Pomeroy, E. S. *The Pacific Slope: A History of California, Oregon, Washington, Idaho, Utah, and Nevada.* Lincoln: University of Nebraska Press, 1991.
Powell, D. S., J. L. Faulkner, D. R. Darr, Z. Zhu, and D. W. MacCleery. *Forest Resources of the United States, 1992.* U.S. Department of Agriculture, Forest Service, Fort Collins, CO, RM-234, 1993.
Power, T. M. *Lost Landscapes and Failed Economies: The Search for a Value of Place.* Washington, DC: Island Press, 1996.
Preston, D., and B. Alverts. "Oregon's BLM Timber Resources." In *Assessment of Oregon's Forests 1988,* edited by G. J. Lettman, 33–41. Salem: Oregon State Department of Forestry, 1998.
Proctor, J. D. "Whose Nature? The Contested Moral Terrain of Ancient Forests." In *Uncommon Ground: Toward Reinventing Nature,* edited by W. Cronon, 269–97. New York: W. W. Norton, 1995.
———. "The Social Construction of Nature: Relativist Accusations, Pragmatist and Critical Realist Responses." *Annals of the Association of American Geographers* 88, no. 3 (1998): 352–76.
Prouty, A. M. *More Deadly Than War! Pacific Coast Logging, 1827–1981.* New York: Garland, 1988.
Prudham, W. S. "Timber and Town: Post-War Federal Forest Policy, Industrial Organization, and Rural Change in Oregon's Illinois Valley." *Antipode* 30, no. 2 (1998): 177–96.
———. "Regional Science, Political Economy, and the Environment." *Canadian Journal of Regional Science* 25, no. 2 (2003): 171–206.
Radosevich, S. R., C. M. Ghersa, and G. Comstock. "Concerns a Weed Scientist Might Have About Herbicide-Tolerant Crops." *Weed Technology* 6 (1992): 635–39.
Rajala, R. *Clearcutting the Pacific Rain Forest: Production, Science, and Regulation.* Vancouver, Canada: UBC Press, 1998.

Richardson, E., U.S. Bureau of Land Management, Oregon State Office and Forest History Society. *BLM's Billion-Dollar Checkerboard: Managing the O and C Lands.* Santa Cruz, CA, and Washington, DC: Forest History Society, 1980.

Righter, F. I. "New Perspectives in Forest Tree Breeding." *Science* 104, no. 2688 (1946): 1–3.

Ritters, K. H. "Early Genetic Selection in Douglas-Fir: Interactions with Shade, Drought, and Stand Density." PhD diss., College of Forestry, Oregon State University, Corvallis, 1986.

Robbins, P. "The Practical Politics of Knowing: State Environmental Knowledge and Local Political Economy." *Economic Geography* 76, no. 2 (2000): 126–44.

Robbins, W. G. *Lumberjacks and Legislators: Political Economy of the U.S. Lumber Industry, 1890-1941.* College Station: Texas A&M University Press, 1982.

———. *American Forestry: A History of National, State, and Private Cooperation.* Lincoln: University of Nebraska Press, 1985.

———. "Lumber Production and Community Stability: A View from the Pacific Northwest." *Journal of Forestry* 31, no. 4 (1987): 187–96.

———. *Hard Times in Paradise: Coos Bay, Oregon, 1850–1986.* Seattle: University of Washington Press, 1988.

———. *Colony and Empire: The Capitalist Transformation of the American West.* Lawrence: University Press of Kansas, 1994.

Roe, E. "Why Ecosystem Management Can't Work without Social Science: An Example from the California Northern Spotted Owl Controversy." *Environmental Management* 20, no. 5 (1996): 667–74.

Rosenberg, N. *Inside the Black Box: Technology and Economics.* Cambridge Cambridgeshire and New York: Cambridge University Press, 1982.

———. *Exploring the Black Box: Technology, Economics, and History.* Cambridge and New York: Cambridge University Press, 1994.

RTI International. "Plywood and Composite Wood Products: Final Background Report." In *Emission Factor Documentation for AP-42.* Research Triangle Park, NC: U.S. Environmental Protection Agency Office of Air Quality Planning and Standards, Emission Factors and Inventory Group, 2003.

Ruderman, F. *Production, Prices, Employment, and Trade in Northwest Forest Industries.* U.S. Department of Agriculture, Forest Service, Pacific Northwest Research Station, Portland, OR, 1974–85.

Sagoff, M. "On Making Nature Safe for Biotechnology." In *Assessing Ecological Risks of Biotechnology,* edited by L. Ginsburg, 341-65. Stoneham, MA: Butterworth-Heinemann, 1991.

Salazar, D. J., and D. K. Alper. "Perceptions of Power and the Management of Environmental Conflict: Forest Politics in British Columbia." *Social Science Journal* 33, no. 4 (1996): 381–99.

Samuels, R. M. "Expanding the Use of Wood Residues for Pulp Production." *Forest Products Journal* 7, no. 8 (1957): 253–55.

Satterfield, T. *Anatomy of a Conflict: Identity, Knowledge, and Emotion in Old-Growth Forests.* Vancouver and Toronto, Canada: University of British Columbia Press, 2002.

Saxenian, A. *Regional Advantage: Culture and Competition in Silicon Valley and Route 128.* Cambridge, MA: Harvard University Press, 1994.

Sayer, A. "Postfordism in Question." *International Journal of Urban and Regional Research* 13 (1989): 666–95.

Sayer, A., and R. Walker. *The New Social Economy: Reworking the Division of Labor.* Cambridge, MA: Blackwell, 1992.

Sayer, R. A. *Radical Political Economy: A Critique.* Oxford and Cambridge, MA: Blackwell, 1995.

Schallau, C. H., and R. M. Alston. "The Commitment to Community Stability: A Policy or Shibboleth." *Environmental Law* 17 (1987): 429–81.

Schmidt, A. *The Concept of Nature in Marx.* London: NLB, 1971.

Schumpeter, J. A. *Business Cycles: A Theoretical, Historical, and Statistical Analysis of the Capitalist Process.* New York and London: McGraw-Hill, 1939.

Scott, A. J. "Industrial Organization and Location: Division of Labor, the Firm and Spatial Process." *Economic Geography* 63 (1987): 215–31.

———. "Flexible Production Systems and Regional Development: The Rise of New Industrial Spaces in North America and Western Europe." *International Journal of Urban and Regional Research* 12 (1988): 171–85.

———. *New Industrial Spaces: Flexible Production Organization and Regional Development in North America and Western Europe.* London: Pion, 1988.

Scott, J. C. *Seeing Like a State: How Certain Schemes to Improve the Human Condition Have Failed.* New Haven, CT: Yale University Press, 1998.

Sedjo, R. *The Forest Sector: Important Innovations.* Washington, DC: Resources for the Future, 1997.

Sedjo, R. A., and K. S. Lyon. *The Long-Term Adequacy of World Timber Supply.* Washington, DC, and Baltimore, MD: Resources for the Future, distributed worldwide by Johns Hopkins University Press, 1990.

Séguin, A., G. Lapointe, and P. J. Charest. "Transgenic Trees." In *Forest Products Biotechnology,* edited by A. Bruce and J. W. Palfreyman, 287–304. Bristol, PA: Taylor and Francis, 1998.

Sher, V. M. "Ancient Forests, Spotted Owls, and the Demise of Federal Environmental Law." *Environmental Law Reporter* 20, no. 11 (1990): 10469–70.

———. "Travels with Strix: The Spotted Owl's Journey through the Federal Courts." *Public Land Law Review* 14 (1993): 41–79.

Silen, R. R. *A Simple Progressive Tree Improvement Program for Douglas-Fir.* Portland, OR: U.S. Department of Agriculture, Forest Service, Pacific Northwest Range and Experiment Station, 1966.

Silen, R. R., and D. L. Copes. "Douglas-Fir Seed Orchard Problems—A Progress Report." *Journal of Forestry* 70, no. 3 (1972): 145–47.

Silen, R. R., and J. G. Wheat. "Progressive Tree Improvement Program in Coastal Douglas-Fir." *Journal of Forestry* 77 (1979): 78–83.

Simberloff, D. "The Spotted Owl Fracas: Mixing Academic, Applied, and Political Ecology." *Ecology* 68, no. 4 (1987): 766–72.

Simon, J. L. *The Ultimate Resource.* Princeton, NJ: Princeton University Press, 1981.

———. *The Ultimate Resource 2.* Princeton, NJ: Princeton University Press, 1996.

Simon, S. "Update on Formaldehyde Emission as It Relates to the Production of Panel Products." Proceedings of the 1982 Canadian Waferboard Symposium, Special Publication SP508E, Forintek Canada Corp., 1984.

Skog, K. E., P. J. Ince, D. J. S. Dietzman, and C. D. Ingram. "Wood Products Technology Trends: Changing the Face of Forestry." *Journal of Forestry* 93, no. 12 (1995): 30-33.

Slappendel, C., I. Laird, I. Kawachi, S. Marshall, and C. Cryer. "Factors Affecting Work-Related Injury among Forestry Workers: A Review." *Journal of Safety Research* 24, no. 1 (1993): 19–32.

Slaughter, S., and L. L. Leslie. *Academic Capitalism: Politics, Policies, and the Entrepreneurial University.* Baltimore, MD: Johns Hopkins University Press, 1997.

Smith, D. M., B. C. Larson, M. J. Kelty, and P. M. S. Ashton. *The Practice of Silviculture: Applied Forest Ecology.* New York: Wiley, 1997.

Smith, J. P., and J. O. Sawyer. "Endemic Vascular Plants of Northwestern California and Southwestern Oregon." *Madrono* 35 (1988): 54–69.

Smith, M. *The U.S. Paper Industry and Sustainable Production: An Argument for Restructuring.* Cambridge: MIT Press, 1997.

Smith, N. *Uneven Development: Nature, Capital, and the Production of Space.* Oxford: Blackwell, 1984.

Smith, W. B., J. S. Vissage, D. R. Barr, and R. M. Sheffield. *Forest Resources of the United States, 1997.* St. Paul, MN: U.S. Department of Agriculture, Forest Service, 2001.

Society of American Foresters. *Forest Terminology.* Washington, DC: Author, 1944.

Socolow, R. H. *Industrial Ecology and Global Change.* Cambridge and New York: Cambridge University Press, 1994.

Somers, H. M., and A. Somers. *Workmen's Compensation: Prevention, Insurance, and Rehabilitation of Occupational Disability.* New York: Wiley, 1954.

Sorensen, F. C. "Geographic Variation in Seedling Douglas-Fir (Pseudotsuga Menziesii) from the Western Siskiyou Mountains of Oregon." *Ecology* 64, no. 4 (1983): 696–702.

Spelter. *Capacity, Production, and Manufacture of Wood-Based Panels in the United States and Canada.* U.S. Department of Agriculture, Forest Service, Forest Products Laboratory, Madison, WI, FPL-GTR-90, 1996.

St. Clair, J. B. "Evaluating Realized Genetic Gains from Tree Improvement." Presented at IUFRO S4.01 Conference, College of Forestry, Virginia Polytechnic Institute, Blacksburg, Virginia, 1993.

Steen, H. K. "The Origins and Significance of the National Forest System." In *Origins of the National Forests: A Centennial Symposium,* edited by H. K. Steen, 3–9. Durham, NC: Forest History Society, 1992.

Stewart, W. C., California Department of Forestry and Fire Protection, Strategic and Resources Planning Program and Forest and Rangeland Resources Assessment Program. *Predicting Employment Impacts of Changing Forest Management in California.* Sacramento: Forest and Rangeland Resources Assessment Program, California Department of Forestry and Fire Protection, 1993.

Storper, M. *The Regional World: Territorial Development in a Global Economy.* New York: Guilford Press, 1997.

Strauss, S. H., S. P. DiFazio, and R. Meilan. "Genetically Modified Poplars in Context." *Forestry Chronicle* 77, no. 2 (2001): 271–79.

Strauss, S. H., G. T. Howe, and B. Goldfarb. "Prospects for Genetic Engineering of Insect Resistance in Forest Trees." *Forest Ecology and Management* 43 (1991): 181–209.

Strauss, S. H., S. A. Knowe, and J. Jenkins. "Benefits and Risks of Transgenic Roundup Ready Cottonwoods." *Journal of Forestry* 95, no. 5 (1997): 12–19.

Strauss, S. H., A. M. Rottmann, and L. A. Sheppard. "Genetic Engineering of Reproductive Sterility in Forest Trees." *Molecular Breeding* 1 (1995): 5–26.

Suchsland, O., and G. E. Woodson. *Fibreboard Manufacturing Practices in the United States.* Washington, DC: U.S. Department of Agriculture, 1986.

Swanson, F. J., and J. F. Franklin. "New Forestry Principles from Ecosystem Analysis of Pacific-Northwest Forests." *Ecological Applications* 2, no. 3 (1992): 262–74.

Sygnatur, E. F. "Logging Is Perilous Work." *Compensation and Working Conditions* Winter (1998): 3–9.

Thomas, J. W., E. D. Forsman, J. B. Lint, E. C. Meslow, B. R. Noon, and J. Verner. *A Conservation Strategy for the Northern Spotted Owl: A Report to the Interagency Scientific Committee to Address the Conservation of the Northern Spotted Owl.* Washington, DC: U.S. Forest Service, U.S. Fish and Wildlife Service, and the National Park Service, 1990.

Thompson, E. P. *The Making of the English Working Class.* Harmondsworth, UK: Penguin, 1968.

Tickell, A., and J. Peck. "Accumulation, Regulation, and the Geographies of Post-Fordism: Missing Links in Regulation Research." *Progress in Human Geography* 16 (1992): 190–218.

Toda, R. "An Outline of the History of Forest Genetics." In *Advances in Forest Genetics,* edited by P. K. Khosla, 4–12. New Delhi, India: Ambika, 1981.

Tree Genetic Engineering Research Co-operative. *TGERC Annual Report 1996-97.* Corvallis: Forest Research Laboratory, Oregon State University, 1997.

———. *TGERC Profile, History, and Structure.* Corvallis: Oregon State University, 1999.

Twining, C. E. *George S. Long, Timber Statesman.* Seattle: University of Washington Press, 1994.

U.S. Bureau of the Census. *Facts for Industry: Softwood Plywood and Veneer 1956.* Washingtion, DC: Author, M24H-06, 1957.

———. *Statistical Abstract of the United States, 1993.* Washington, DC: Author, 1994.

U.S. Bureau of Corporations. *The Lumber Industry.* Washington, DC: U.S. Government Printing Office, 1913.

U.S. Bureau of Labor Statistics. *Selected Industries: Employment and Annual Rates of Change in Output Per Hour, Selected Periods.* Washington, DC: Author, June 27, 1997. http://www.stats.bls.gov/news.release/prin.t01.htm.

U.S. Census Bureau. *1997 Economic Census, Manufacturing, Industry Series: Engineered Wood Member (except Trusses) Manufacturing.* Washington, DC: U.S. Department of Commerce, 1999.

———. *1997 Economic Census, Manufacturing, Industry Series: Reconstituted Wood Product Manufacturing.* Washington, DC: U.S. Department of Commerce, 1999.

U.S. Congress House Committee on Agriculture, S. o. F. *Reforestation Efforts in Western Oregon: Hearing before the Subcommittee on Forests of the Committee on Agriculture, House of Representatives.* 95th Cong., 1st sess., July 8, 1977, Roseburg, OR. Washington, DC: U.S. Government Printing Office, iv, 257.

———. *Use of Illegal Aliens in Government Reforestation Contracts: Hearing before the Subcommittee on Forests of the Committee on Agriculture, House of Representatives.* 96th Cong., 2nd sess., May 15, 1980, Eugene, OR. Washington, DC: U.S. Government Printing Office, 1981.

U.S. Congress House Committee on Government Operations, I., Justice, Transportation, and Agriculture Subcommittee. *Allegations of Contract Abuse in the U.S. Forest Service Refor- estation Program: Hearing before the Information, Justice, Transportation, and Agriculture Subcommittee of the Committee on Government Operations, House of Representatives.* 103rd Cong., 1st sess., June 30, 1993. Washington, DC: U.S. Government Printing Office, 1994.

U.S. Congress Office of Technology Assessment. *Biotechnology in a Global Economy.* Washing- ton, DC: Author, 1991.

U.S. Consumer Products Safety Commission. *An Update on Formaldehyde: 1997 Revision.* Wash- ington, DC: Author, 2003. http://www.cpsc.gov/CPSCPUB/PUBS/725.html

U.S. Department of Agriculture Forest Service. *Forest Resources of the United States 2002* (Draft Tables). May 22, 2003. http://www.ncrs.fs.fed.us/4801/fiadb/rpa–tabler/Draft_RPA_2002_ Forest_Resource_Tables.pdf

U.S. Department of Commerce, U.S. Bureau of the Census. *1992 Census of Manufactures.* Wash- ington, DC: U.S. Department of Commerce Economics and Statistics Administration Bu- reau of the Census: For sale by Supt. of Docs. U.S. G.P.O., 1995.

———. *1992 Census of Manufactures. Industry Series. Millwork, Plywood, and Structural Wood Members, Not Elsewhere Classified.* Washington, DC: Author, 1995.

———. *1992 Census of Manufacturers. Industry Series. Logging Camps, Sawmills, and Planing Mills.* Washington, DC: Author, MC92-I-24A, 1995a.

———. *1992 Census of Manufacturers. Industry Series. Millwork, Plywood, and Structural Wood Members, Not Elsewhere Classified.* Washington, DC: Author, MC92-I-24b, 1995b.

———. *1992 Census of Manufactures, Subject Series.* Washington, DC: Author, 1996.

———. *1997 Economic Census Manufacturing Industry Series: Logging.* Washington, DC: Au- thor, 2000.

———. *County Business Patterns.* Washington, DC: Author, 2000.

———. *Concentration Ratios in Manufacturing, 1997 Census.* Washington, DC: Author, 2001.

U.S. Department of Labor, Bureau of Labor Statistics. *Census of Fatal Occupational Injuries, 1995.* Washington, DC: Author, 1997.

U.S. Environmental Protection Agency. *Profile of the Lumber and Wood Products Industry.* Wash- ington, DC: Office of Compliance, Office of Enforcement and Compliance Assurance, U.S. EPA, EPA/310-R-95-006, 1995.

———. "Wood Products Industry." In *Compilation of Air Pollutant Emission Factors, AP-42, 5th Edition, Volume 1: Stationary Point and Area Sources.* Washington, DC: Author, August 15, 2002. http://www.epa.gov/ttn/chief/ap42/ch10/htm.

———. *Technology Transfer Network, Clearinghouse for Inventories & Emission Factors.* Wash- ington, DC: Author, 2003.

U.S. Forest Service. *Forest Resources of the United States 2002 (Draft).* http://ncrs2.fs.fed.us/ 4801/fiadb/rpa_tabler/2002_rpa_draft_tables.htm.

——— and U.S. Bureau of Land Management. *Final Supplemental Environmental Impact State- ment on Management of Habitat for Late-Successional and Old-Growth Forest Related Spe- cies within the Range of the Northern Spotted Owl.* Washington, DC: U.S. Department of Agriculture, Forest Service; U.S. Department of the Interior Bureau of Land Management, 1994.

Van Tassel, A. J. *Mechanization in the Lumber Industry: A Study of Technology in Relation to Resources and Employment Opportunity.* Philadelphia, PA: Work Projects Administration, National Research Project on Employment Opportunities and Recent Changes in Industrial Techniques, 1940.

Van Vliet, A. "Strength of Second-Growth Douglas-Fir in Tension Parallel to the Grain." *Forest Products Journal* 9, no. 4 (1959): 143–48.

Vick, C. B. "Adhesive Bonding of Wood Materials." In *Wood Handbook—Wood as an Engineer- ing Material,* 463. U.S. Department of Agriculture, Forest Service, Forest Products Labora- tory, Madison, WI, FPL-GTR-13, 1999.

Virgin, B. "Weyerhaeuser's Forestry Gamble Pays Off." *The Oregonian,* 1997, E2.

Waggener, T. R. "Community Stability as Forest Management Objective." *Journal of Forestry* 75 (1977): 710–14.

Walker, P. "Reconsidering 'Regional' Political Ecologies: Toward a Political Ecology of the Rural American West." *Progress in Human Geography* 27, no. 1 (2003): 7–24.

Walker, R. A. "Regulation and Flexible Specialization as Theories of Capitalist Development." In *Spatial Practices: Markets, Politics, and Community Life,* edited by H. Liggett and D. Perry, 167–208. Thousand Oaks, CA: Sage Publications, 1995.

———. "California's Golden Road to Riches: Natural Resources and Regional Capitalism, 1848-1940." *Annals of the Association of American Geographers* 91, no. 1 (2001): 167–99.

Walters, C., J. Korman, L. E. Stevens, and B. Gold. "Ecosystem Modeling for Evaluation of Adaptive Management Policies in the Grand Canyon." *Conservation Ecology* 4, no. 2 (2000).

Walters, C. J., and R. Hilborn. "Ecological Optimization and Adaptive Management." *Annual Review of Ecology and Systematics* 9 (1978): 157–88.

Walters, C. J., and C. S. Holling. "Large-Scale Management Experiments and Learning by Doing." *Ecology* 71, no. 6 (1990): 2060–68.

Ward, F. R. *Oregon's Forest Products Industry, 1992.* Portland, OR: U.S. Department of Agriculture, Forest Service Pacific Northwest Research Station, 1995.

———. *Oregon's Forest Products Industry, 1994. Resource bulletin PNW; RB-216.* Portland, OR: U.S. Department of Agriculture, Forest Service Pacific Northwest Research Station, 1997.

Warf, B. "Regional Transformation, Everyday Life, and Pacific Northwest Lumber Production." *Annals of the Association of American Geographers* 78, no. 2 (1988): 326–46.

Warren, D. *Harvest, Employment, Exports, and Prices in Pacific Northwest Forests, 1965-2000.* U.S. Department of Agriculture, Forest Service, Pacific Northwest Research Station, Portland, OR, PNW-GTR-547, 2002.

———. *Production, Prices, Employment and Trade in Northwest Forest Industries, All Quarters 2000.* Portland, OR: U.S. Department of Agriculture, Forest Service, Pacific Northwest Research Station, 2002.

Watts, M. *Silent Violence: Food, Famine, & Peasantry in Northern Nigeria.* Berkeley: University of California Press, 1983.

Watts, M., and R. Peet. "Conclusion: Toward a Theory of Liberation Ecology." In *Liberation Ecologies: Environment, Development, Social Movements,* edited by R. Peet and M. Watts. New York: Routledge, 1996.

Watts, M. J. "Life under Contract: Contract Farming, Agrarian Restructuring, and Flexible Accumulation." In *Living under Contract: Contract Farming and Agrarian Transformation in Sub-Saharan Africa,* edited by P. D. Little and M. J. Watts, 21–77. Madison: University of Wisconsin Press, 1994.

Webber, M. J., and D. L. Rigby. *The Golden Age Illusion: Rethinking Postwar Capitalism.* New York: Guilford Press, 1996.

Weigand, J. F. *Composition, Volume, and Prices for Major Softwood Lumber Types in Western Oregon and Washington 1971-2020.* U.S. Department of Agriculture, Forest Service, Pacific Northwest Research Station, Portland, OR, PNW-RP-509, 1998.

Wells, M. J. *Strawberry Fields: Politics, Class, and Work in California Agriculture.* Ithaca, NY: Cornell University Press, 1996.

West Coast Lumbermen's Association. *Industrial Facts.* Portland, OR: Author, 1964.

Western Wood Products Association. *Statistical Yearbook of the Western Lumber Industry.* Portland, OR: Author, 1998.

Weyerhaeuser Company. *Form 10-K.* Washington, DC: Securities and Exchange Commission, 1997.

———. *Form 10-K.* Washington, DC: Securities and Exchange Commission, 1999.

White, R. *"It's Your Misfortune and None of My Own": A History of the American West.* Norman: University of Oklahoma Press, 1991.

———. "'Are You an Environmentalist or Do You Work for a Living?': Work and Nature." In *Uncommon Ground: Toward Reinventing Nature,* edited by W. Cronon, 171–85. New York: W. W. Norton, 1995.

Whittaker, R. H. "Vegetation of the Siskiyou Mountains, Oregon and California." *Ecological Monographs* 30 (1960): 279–338.

Widenor, M. "Pattern Bargaining in the Pacific Northwest Lumber and Sawmill Industry, 1980-1990." In *Labor in a Global Economy: Perspectives from the U.S. and Canada,* edited by S. Hecker and M. Hallock, 252-62. Eugene: Labor Education Research Center, University of Oregon, 1991.

———. "Diverging Patterns: Labor in the Pacific Northwest Wood Products Industry." *Industrial Relations* 34, no. 3 (1995): 441–63.

Willamette Industries Inc. *Form 10-K*. Washington, DC: Securities and Exchange Commission, 1997.

———. *Form 10-K*. Washington, DC: Securities and Exchange Commission, 1999.

Willems-Braun, B. "Buried Epistemologies: The Politics of Nature in (Post)Colonial British Columbia." *Annals of the Association of American Geographers* 87, no. 1 (1997): 3–31.

Williams, M. *Americans and Their Forests: A Historical Geography*. Cambridge Cambridgeshire and New York: Cambridge University Press, 1989.

Williams, R. *The Country and the City*. London: Verso, 1973.

Williston, E. M. *Lumber Manufacturing: The Design and Operation of Sawmills and Planer Mills*. San Francisco, CA: Miller Freeman, 1988.

Wood-Products Sub-council. "Principal Pollution Problems Facing the Solid Wood Product Industry." *Forest Products Journal* 21, no. 9 (1971): 33–36.

Worster, D. *Nature's Economy: A History of Ecological Ideas*. Cambridge and New York: Cambridge University Press, 1994.

Wright, S. *Molecular Politics: Developing American and British Regulatory Policy for Genetic Engineering, 1972–1982*. Chicago: University of Chicago Press, 1994.

WTD Industries. *Form 10-K405*. Washington, DC: Securities and Exchange Commission, 1998.

Yaffee, S. L. *The Wisdom of the Spotted Owl: Policy Lessons for a New Century*. Covelo, CA: Island Press, 1994.

Yergin, D. *The Prize: The Epic Quest for Oil, Money, and Power*. New York: Simon & Schuster, 1991.

Young, J. A., and J. M. Newton. *Capitalism and Human Obsolescence: Corporate Control vs. Individual Survival in Rural America*. Montclair, NJ: LandMark Studies, 1980.

Youngquist, J. A. "Wood-Based Composites and Panel Products." In *Wood Handbook—Wood as an Engineering Material*, 463. Madison, WI: U.S. Department of Agriculture, Forest Service, Forest Products Laboratory, 1999.

Youngs, R. L. "Reconstituted Wood Materials—New Opportunities and New Responsibilities." Proceedings of the 1982 Waferboard Symposium, Special Publication Sp508e, Forintek Canada Corp., 1984.

Zaremba, J. *Economics of the American Lumber Industry*. New York: R. Speller, 1963.

Zinkhan, F. C. *Timberland Investments: A Portfolio Perspective*. Portland, OR: Timber Press, 1992.

Zinn, T. W., D. Cline, and W. F. Lehmann. "Long-Term Study of Formaldehyde Emission Decay from Particleboard." *Forest Products Journal* 40, no. 6 (1990): 15–18.

Zobel, B., and J. B. Jett. *Genetics of Wood Production*. Berlin and New York: Springer-Verlag, 1995.

Zobel, B. J. "Forest Tree Improvement—Past and Present." In *Advances in Forest Genetics*, edited by P. K. Khosla, 11–22. New Delhi, India: Ambika, 1981.

Index